X-ray microanalysis is a widely used technique in studies of cell and tissue elemental composition. It has applications in many different areas of biology, including animal and plant cell physiology, medicine, pathology and studies of environmental pollution.

The editors present an up-to-date look at the use of X-ray microanalysis in biology. Experts from around the world give state of the art accounts of research in the field and discuss the potential of the technique for future research. Four main themes are tackled – detection and quantification of X-rays, associated techniques, specimen preparation, and applications.

The book is clearly written and highly illustrated and is essential reading for researchers, postgraduates and advanced undergraduates in the field.

X-ray microanalysis in biology:

experimental techniques and applications

X-ray microanalysis in biology: *experimental techniques and applications*

EDITED BY

David C. Sigee
Department of Cell and Structural Biology, University of Manchester

A. John Morgan
School of Pure and Applied Biology, University of Wales

Adrian T. Sumner
MRC Human Genetics Unit, Western General Hospital, Edinburgh

Alice Warley
Department of Physiology, St Thomas's Hospital Campus, London

CAMBRIDGE
UNIVERSITY PRESS

Published by the Press Syndicate of the University of Cambridge
The Pitt Building, Trumpington Street, Cambridge CB2 1RP
40 West 20th Street, New York, NY 10011-4211, USA
10 Stamford Road, Oakleigh, Victoria 3166, Australia

© Cambridge University Press 1993

First published 1993

Printed in Great Britain at the University Press, Cambridge

A catalogue record for this book is available from the British Library

Library of Congress cataloguing in publication data

X-ray microanalysis in biology: experimental techniques and
applications / edited by David C. Sigee ... [et al.].
     p.   cm.
Includes index.
ISBN 0 521 41530 6 hardback
1. X-ray microanalysis. 2. Biology – Technique. I. Sigee, D. C.
QH324.9.X2X18   1993
574.19'285—dc20 92–23129 CIP

ISBN 0 521 41530 6 hardback

Contents

Contributors

A.L. Beckers
Department of Medical Informatics, Faculty of Medicine and Health
Sciences, Erasmus University, P.O. Box 1738, 3000 DR Rotterdam, The
Netherlands

C.J.G. Blok-van Hoek
Tracor Europa B.V., Jan van Eycklaan 10, P.O. Box 333, 3720 AH
Bilthoven, The Netherlands

R.A. Dalmeyer
Laboratory for electron microscopy, University of Leiden, Rijnsburgerweg
10, 2333 AA Leiden, The Netherlands

W.C. de Bruijn
A.E.M. Unit, Pathologic Institut I, Faculty of Medicine and Health
Sciences, Erasmus University, P.O. Box 1738, 3000 DR Rotterdam, The
Netherlands

C. de Rouffignac
Département de biologie cellulaire et moléculaire, CEA Saclay, F91191 Gif
sur Yvette Cedex, France

T.J. Flowers
School of Biological Sciences, University of Sussex, Falmer, Brighton BN1
9QG, UK

E.S. Gelsema
Department of Medical Informatics, Faculty of Medicine and Health
Sciences, Erasmus University, P.O. Box 1738, 3000 DR Rotterdam, The
Netherlands

G.W. Grime
Scanning Proton Microprobe Unit, Nuclear Physics Laboratory,
Department of Physics, Keble Road, Oxford OX1 3RH, UK

B.L. Gupta
Department of Zoology, Cambridge University, Downing Street,
Cambridge CB2 3EJ, UK

M.A. Hajibagheri
Electron Microscopy Unit, Imperial Cancer Research Fund, London
WC2A 3PX, UK

N. Hodson
Department of Cell and Structural Biology, University of Manchester, Stopford Building, Oxford Road, Manchester M13 9PT, UK

J.F. Jongkind
Department of Cell Biology and Genetics, Faculty of Medicine and Health Sciences, Erasmus University, P.O. Box 1738, 3000 DR Rotterdam, The Netherlands

H.K. Koerten
Laboratory for electron microscopy, University of Leiden, Rijnsburgerweg 10, 2333 AA Leiden, The Netherlands

J.P. Landsberg
Scanning Proton Microprobe Unit, Nuclear Physics Laboratory, Department of Physics, Keble Road, Oxford OX1 3RH, UK

B. McDonald
Department of Neuropathology, Radcliffe Infirmary, Oxford, UK

A.J. Morgan
School of Pure and Applied Biology, University of Wales College of Cardiff, P.O. Box 915, Cardiff CF1 3TL, Wales, UK

J.A. Nott
Plymouth Marine Laboratory, Citadel Hill, Plymouth PL1 2PB, UK

M. Okumura
Laboratory for Otobiology and Biocompatibility, University of Leiden, Rijnsburgerweg 10, 2333 AA Leiden, The Netherlands

J.F.M. Pinxter
Tracor Europa B.V., Jan van Eycklaan 10, P.O. Box 333, 3720 AH Bilthoven, The Netherlands

N. Roinel
Département de biologie cellulaire et moléculaire, CEA Saclay, F91191 Gif sur Yvette Cedex, France

G.M. Roomans
Department of Human Anatomy, University of Uppsala, Box 571, S-75123 Uppsala, Sweden

R.B. Shen
Edax International, 150 West Center Ct., Schaumburg, Illinois 60195, USA

D.C. Sigee
Department of Cell and Structural Biology, University of Manchester, Stopford Building, Oxford Road, Manchester M13 9PT, UK

C.W.J. Sorber
A.E.M. Unit, Pathologic Institut I, Faculty of Medicine and Health
Sciences, Erasmus University, P.O. Box 1738, 3000 DR Rotterdam, The
Netherlands

A.T. Sumner
MRC Human Genetics Unit, Western General Hospital, Crewe Road,
Edinburgh EH4 2XU, UK

G.A.M. Trommelen-Ketelaars
A.E.M. Unit, Pathologic Institut I, Faculty of Medicine and Health
Sciences, Erasmus University, P.O. Box 1738, 3000 DR Rotterdam, The
Netherlands

C.A. van Blitterswijk
Laboratory for Otobiology and Biocompatibility, University of Leiden,
Rijnsburgerweg 10, 2333 AA Leiden, The Netherlands

I. van den Brink
Laboratory for Otobiology and Biocompatibility, University of Leiden,
Rijnsburgerweg 10, 2333 AA Leiden, The Netherlands

T. von Zglinicki
Institute of Pathology, Charite, Schumannstr. 20/21, 01040 Berlin,
Germany

A. Warley
Department of Physiology, Block 9, U.M.D.S., St Thomas's Hospital
Campus, Lambeth Palace Road, London SE1 7EH, UK

F. Watt
Scanning Proton Microprobe Unit, Nuclear Physics Laboratory,
Department of Physics, Keble Road, Oxford OX1 3RH, UK

C. Winters
School of Pure and Applied Biology, University of Wales College of
Cardiff, P.O. Box 915, Cardiff CF1 3TL, Wales, UK

J. Wood
Oxford Instruments, Analytical Systems Division, Halifax Road, High
Wycombe, Bucks. HP12 3SE, UK

J. Wroblewski
Department of Medical Cell Genetics, Karolinska Institutet, S-104
Stockholm, Sweden

R. Wroblewski
Department of Pathology, Karolinska Institutet, S-104 Stockholm, Sweden

K. Zierold
Max-Planck-Institut fur Systemphysiologie, Rheinlanddamm 201, 4600
Dortmund, Germany

Preface

The technique of electron probe X-ray microanalysis (XRMA), by which the elemental composition of specimens can be determined on a microscopical scale, has now been applied to biological materials for about 30 years, and in that time has made many valuable contributions to a variety of biological problems. To improve awareness of the technique, and to provide a discussion forum for those interested in biological XRMA, the Biological X-ray Microanalysis Group was formed in Britain some five years ago. Although its primary purpose was to hold regular meetings, the possibility of producing a book which reviewed both developments in the equipment and the wide variety of applications for which XRMA could be used was another aim, now realised in the present volume. This book arose out of a meeting held by the Biological X-ray Microanalysis Group in Manchester, England, in April 1991. What we have tried to produce is not simply another set of conference proceedings, but a well balanced and integrated series of chapters describing the hardware and software used for XRMA, the necessary procedures for specimen preparation and quantification, and the enormous range of applications of the technique in biology.

It might be supposed that after 30 years' application of biological X-ray microanalysis, the field might be settling down to a comfortable but perhaps rather routine maturity. The contributions in this book show, however, that this is far from being so, with many exciting developments in all aspects of the subject. Continual improvements in solid state silicon detectors and their associated electronics over the years have produced substantial improvements in spectral resolution, while at the same time permitting much higher X-ray count rates than formerly. The recent introduction of germanium detectors (Blok-van Hoek & Pinxter) foreshadows further dramatic improvements in this field. The development of software commercially means that quantitative results are available more easily (Shen), and sophisticated quantitative multi-element mapping procedures are now becoming available (Wood). It is nevertheless essential to realise that the most sophisticated equipment and software produced by the manufacturers is of limited use if the specimen is not prepared properly (Gupta, Zierold) and the pitfalls of quantification are not fully understood (Warley). Problems of radiation damage may also vitiate results if appropriate precautions are not taken (von Zglinicki). As well as electron probe X-ray microanalysis, there are now various complementary techniques of elemental analysis available, each with its own advantages and disadvantages for biological work: for example, the scanning proton microprobe, described by Watt *et al.*, and electron energy-loss spectroscopy, which is compared with XRMA by de Bruijn *et al.* It seems likely that these different techniques of analysis together will be able to provide much more information to the biologist (if he can gain access to such sophisticated equipment) than any single method.

Biological XRMA is associated particularly with studies of diffusible ions in cells, a field in which it is pre-eminent (von Zglinicki, Zierold, Gupta, Wroblewski & Wroblewski). As well as animal cells, it has been used to study ions in plant cells (Hajibagheri & Flowers) and in bacteria (Sigee & Hodson). This is only a part of the broad range of problems to which XRMA has brought its unique insights: others include biomaterials research, in which the tissue reactions to the degradation of prostheses can be analysed (Koerten *et al.*); the study of extraneous mineral particles in pathological conditions; uptake of pollutants by animal cells (Nott); histochemistry (Sumner); pathology (Roomans) and analysis of food. All these require both elemental and structural analysis at the microscopical level, for which purpose XRMA is still the method of choice, but there are nevertheless further biological applications of XRMA. The body fluids of many organisms are produced on such a minute scale that it requires a microscopical method to analyse them, even though no structural information is required; such methods are described by Roinel & de Rouffignac and by Morgan & Winters.

We hope that this book may induce readers to realise that electron probe XRMA can be used much more extensively in biology in the future than at present, not only as a technique which yields meaningful biological data that correlate structure with composition, but also as a tool that complements other microscopical, microprobe and biochemical approaches.

D.C. Sigee, *Manchester*
A.J. Morgan, *Cardiff*
A.T. Sumner, *Edinburgh*
Alice Warley, *London*

SECTION A
DETECTION AND QUANTIFICATION OF X-RAYS

The technique of X-ray microanalysis has existed for considerably longer than is, perhaps, generally realised, having been invented in 1951 and first applied to biological material at the beginning of the 1960s (Hall, 1986). Much very useful pioneering work was carried out using wavelength-dispersive detectors. However, there can be no doubt that the introduction of the energy-dispersive detector based on the Si(Li) detector crystal, with its capability of multi-element analysis, and ease of operation led to the technique becoming more widely accepted. For the detection of the majority of elements the ease of use of the EDS detector offsets its lower resolution, and both the continued improvements in the Si(Li) detectors and the introduction of germanium detectors, will help to improve the sensitivity of the technique.

Biological specimens are many and varied with their own special problems for quantification, and perhaps the second most important development in X-ray microanalysis for biologists has been the commercial availability of software for carrying out quantitative routines. The procedure most commonly used for quantification is the continuum normalisation method of Hall (Hall, 1971). With the advent of the microcomputer in the early 1980s the first software for applying the Hall technique to biological specimens became available, and now the importance of quantification of spectra from biological specimens is recognised with all major detector manufacturers offering this option with their software.

A major advantage of X-ray microanalysis is that it is a visual technique. This is fully exploited in the mapping systems that present information in a pictorial manner which is easily understood by the human eye. Although the early dot maps were visually appealing these could be easily misinterpreted, especially maps from biological specimens with their regional variations in mass thickness. The need for fully quantitative maps which would avoid these problems was recognised by workers in the United States, and the first fully quantitative maps were produced by groups who developed their own software (see e.g. Ingram *et al.*, 1989). Again this major advance has been taken up by a commercial supplier.

The living biological specimen typically consists largely of water, which needs to be removed or stabilised before examination and analysis in the electron microscope can take place. This removal of water and/or its replacement by resin can lead to some uncertainties in the interpretation of data. It is, therefore, essential that all results are viewed with a cautious eye.

1

This approach is exemplified in the work of Dr T.A. Hall. All of us who carry out quantitative analysis can be counted as students of Dr Hall, and we were indeed lucky to have him with us at this meeting to guide us in his own gentle and inimitable way, cautiously along the thorny paths of data interpretation.

References

Hall, T.A. (1971). The microprobe assay of chemical elements. In *Physical techniques in biochemical research*. Vol. 1A., ed. G. Oster, pp. 157–275. New York: Academic Press.

  (1986). The history and current status of biological electron-probe X-ray micro-analysis. *Micron and Microscopica Acta*, **17**, 91–100.

Ingram, P., Nassar, R., LeFurgey, A., Davilla, S. & Sommer, J.A. (1989). Quantitative X-ray elemental mapping of dynamic physiologic events in skeletal muscle. In *Electron probe microanalysis applications in biology and medicine*. Springer series in biophysics Vol. 4, pp. 251–64. Berlin: Springer-Verlag.

1 Recent developments in X-ray detectors and their relevance in biological X-ray microanalysis

C.J.G. Blok-van Hoek and J.F.M. Pinxter

A Introduction

In 1949, Castaing and Guinier combined for the first time the technique of electron optics and X-ray analysis for the production of the first electron probe analyser (Castaing & Guinier, 1949). Developments continued during the years from a static beam instrument into a scanning device. However, all the analysis was initially limited to thick specimens and it was only at a later date that the first prototype microanalyser was specifically developed for thin specimens (Duncumb, 1962). The detection and counting of X-rays was done by using conventional wavelength dispersive crystal spectrometers. A major breakthrough in X-ray analysis came in 1968 with the development of the solid state energy dispersive X-ray detector (Fitzgerald, Keil & Heinrich, 1968). This type of detector provided higher collection efficiency than the wavelength dispersive spectrometer as well as more reproducible results because mechanical movements were no longer necessary.

Nowadays X-ray microanalysis is a well used technique for analysis of biological material in combination with the transmission and scanning electron microscopes. In this chapter new developments on the energy dispersive detectors will be discussed. Special attention will be given to the advantage of using a high purity germanium crystal instead of Si(Li) crystal in the EDS detector.

B X-ray generation

When a focused electron beam hits a specimen several processes occur at the surface and in the specimen. In Fig. 1.1 a schematic presentation of the different interactions is presented. The X-rays released from the specimen can be categorised in two groups:

1. characteristic X-rays,
2. continuum X-rays.

Characteristic X-rays are generated as a result of a collision between an electron of the incident beam and an electron in one of the inner shells of an atom. The transition of an electron from one of the outer shells into a generated gap gives an X-ray photon release with an energy characteristic for

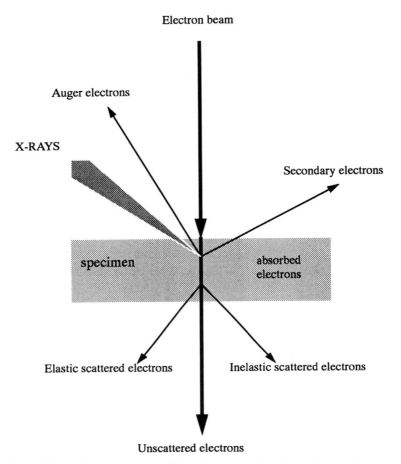

Fig. 1.1.   Schematic presentation of signals generated at the surface, and internally, during electron beam–specimen interaction.

that atom. This is in contrast to the continuum or 'bremsstrahlung' which is formed during the deceleration of the beam electrons in the coulombic field of the atoms. The continuum forms a background energy range extended from zero to the incident beam energy.

Characteristic X-rays do not always come from the specimen. Several extraneous sources can give spurious X-ray signals, especially in transmission electron microscopes (Fig. 1.2). The sources which can be minimised, or even eliminated are (Goodhew & Chescoe, 1980):

1. X-ray generation in the condensor aperture: these X-rays can cause fluorescence in the specimen, grid or somewhere else in the specimen chamber;

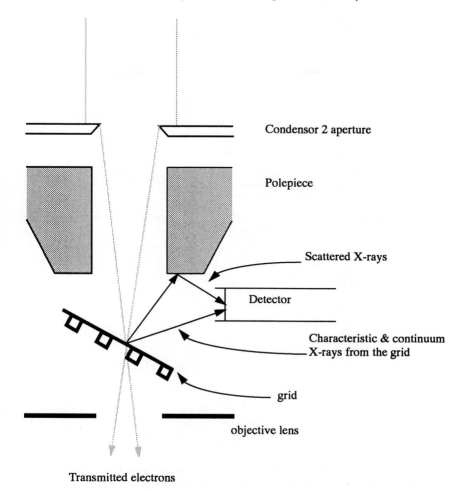

Fig. 1.2.   Spurious signals caused by grid and polepiece.

2. X-rays scattering from components below the specimen (objective aperture, anticontaminator, objective pole piece);
3. X-rays from the final pole piece.

Spurious X-rays generated by the specimen itself are more difficult to eliminate. Examples are: backscattering of electrons against the lower condensor pole piece which generate X-ray excitation; and fluorescence of distant regions of the specimen, grid and chamber by hard X-rays generated in the specimen. To minimise the detection of those spurious X-rays, the geometry between microscope and detector has to be optimised. The relative

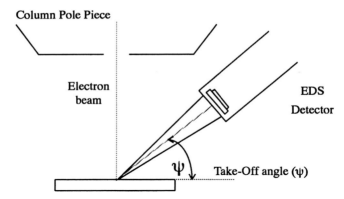

Fig. 1.3.   Detector take-off angle, (ψ), which is the angle between sample exication point and the centre of the crystal.

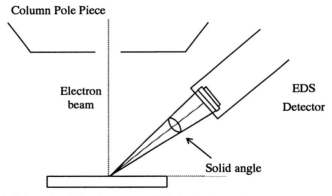

Fig. 1.4.   Solid angle of the detector, which is the three-dimensional cone-shaped angle between the X-ray point source and the surface of the crystal.

position of the detector crystal and the sample affects the number of X-ray counts, sensitivity, and absorption of the X-rays by the sample.

Two types of angles therefore have to be considered.

1. Take off angle; which is the angle between the sample excitation point and the centre of the detector crystal line (Fig. 1.3).
2. The solid angle; which is the three dimensional, cone shaped angle between the X-ray point source and the surface of the detector crystal (Fig. 1.4).

If the take off angle is small, collected X-rays travel a longer distance through the sample to reach the detector (Fig. 1.5). This can produce the following undesired effects:

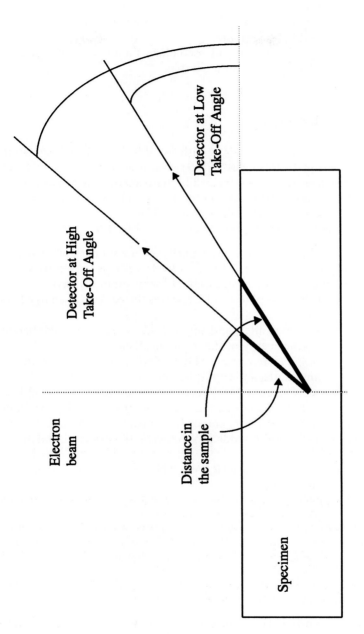

Fig. 1.5. Difference in the X-ray path within the sample when using a detector at a low and high take-off angle.

(*a*) absorption by the sample and therefore reduction of the amount of X-rays detected;

(*b*) reduced sensitivity for light elements because low energies are more likely to be absorbed by the sample;

(*c*) higher X-ray absorption and secondary fluorescence because X-rays travel a greater distance in the sample.

C   X-ray detection

The analysis of X-rays can be done with two different kinds of spectrometer. Scanning electron microscopes can be equipped with wavelength and/or energy dispersive spectrometers while transmission microscopes are in general, due to geometry, space limitation and relatively low beam current, only equipped with an energy dispersive system.

A general ED system has six primary parts (Fig. 1.6).

- *Crystal.* This can be either high purity germanium (HpGe) or lithium drifted silicon material (Si(Li)). It converts X-rays to electronic charge packages proportional to the energy of the incoming X-rays.
- *Preamplifier*, which amplifies and integrates the packets of charge from the detector.
- *Pulse processor*, which filters and amplifies the pulses. The pulse heights are proportional to the energy of the incoming X-rays.
- *Analogue to digital converter* (ADC). This converter sorts the pulses according to amplitude, and counts the number of pulses that fall into each energy interval. When an X-ray analyser is calibrated, it is important to know the voltage level representing 0 eV. Because of slight instabilities in the electronics, the zero level can drift and cause inaccuracy. Therefore a circuitry (ZERO DAC) is added to the ADC to correct automatically for this drift.
- *Random access data memory* (RAM). This contains the storage locations for X-ray count totals.
- *Spectral display* which shows the number of X-ray counts for each channel.

The ED detector has three major variable parts, namely: the type of crystal (Si(Li), or HpGe), crystal area (10 mm^2 or 30 mm^2), and the window material (beryllium, 1 atmosphere window, ultrathin, or windowless).

1   Si(Li) versus HpGe crystal

In the past, the use of HpGe detectors for EDS at photon energies below about 2 keV was impossible due to severe distortions of the peak shapes (Fig. 1.7) and large shifts in peak positions. These effects were the result of

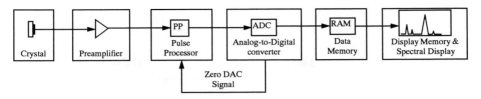

Fig. 1.6. Flowchart of an EDS detector.

incomplete charge collection and were most pronounced at energies just above the energy of germanium L absorption edges (1.2 to 1.4 keV) (Llacer, Haller & Cordi, 1977; Barbi & Lister, 1981; Cox, Lowe & Sareen, 1988). Using new techniques, HpGe detectors are now manufactured that do not exhibit significant spectral peak distortion from incomplete charge collection. The major causes of incomplete charge collection are trapping of charge carriers in the bulk material of the sensitive volume or at the surfaces of the detector. In particular the front surface, with its associated 'dead layer', was thought to be the limiting factor in previous HpGe detectors.

A number of measurements have been adopted which indicate quality of charge collection in a detector. These measurements include the ratio of Full-Width-Tenth-Maximum (FWTM) to the Full-Width-Half-Maximum (FWHM), the statistical broadening or dispersion (D) of spectral peaks (measured peak resolution (FWHM) minus the electronic noise in quadrature), peak-to-background ratio, and estimates of the dead layer thickness. The resolution of a peak is determined by the dispersion D and the electronic noise of the entire spectrometer. D is defined by the equation:

$$D = \sqrt{(5.52F\varepsilon E)}$$

where E is the energy of the peak, F is the Fano factor, and ε is the mean number of electron–hole pairs produced per unit energy. Given the fact that the Fano factor for Ge and Si is roughly equal and that ε_{Ge} is 27% less than ε_{Si} (Fink, 1981), the resolution of an HpGe is better than a Si(Li) detector. The difference in ε also improves the signal-to-noise ratio for HpGe detectors. The high stopping power of germanium leads to increased detection efficiency relative to silicon at X-ray energies above 20 keV (Fig. 1.8) (McCarthy, Ales & McMillan, 1990). In addition, because of increased absorption, the HpGe detector has a much smaller background from Compton-scattered photons of high energies. This leads to improved detection sensitivity at low energies. These properties have been compared for HpGe and Si(Li) by using an Am[241] source (Fig. 1.9) (McCarthy *et al.*, 1990).

Another advantage of using the HpGe detector for X-ray microanalysis is that, due to high energy photon detection efficiency and the improved

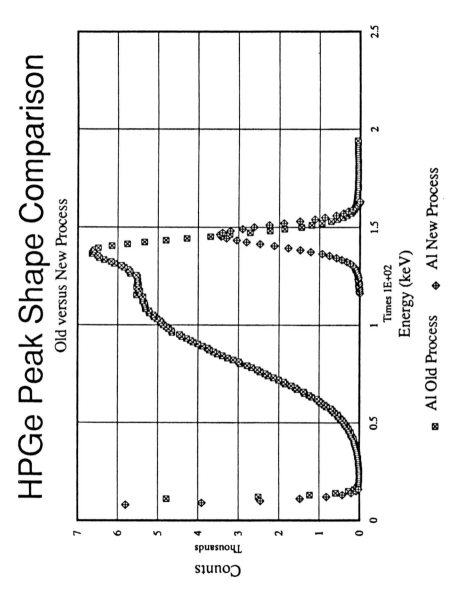

Fig. 1.7. Peak shapes of aluminium acquired with an HpGe detector manufactured by the old and new processes.

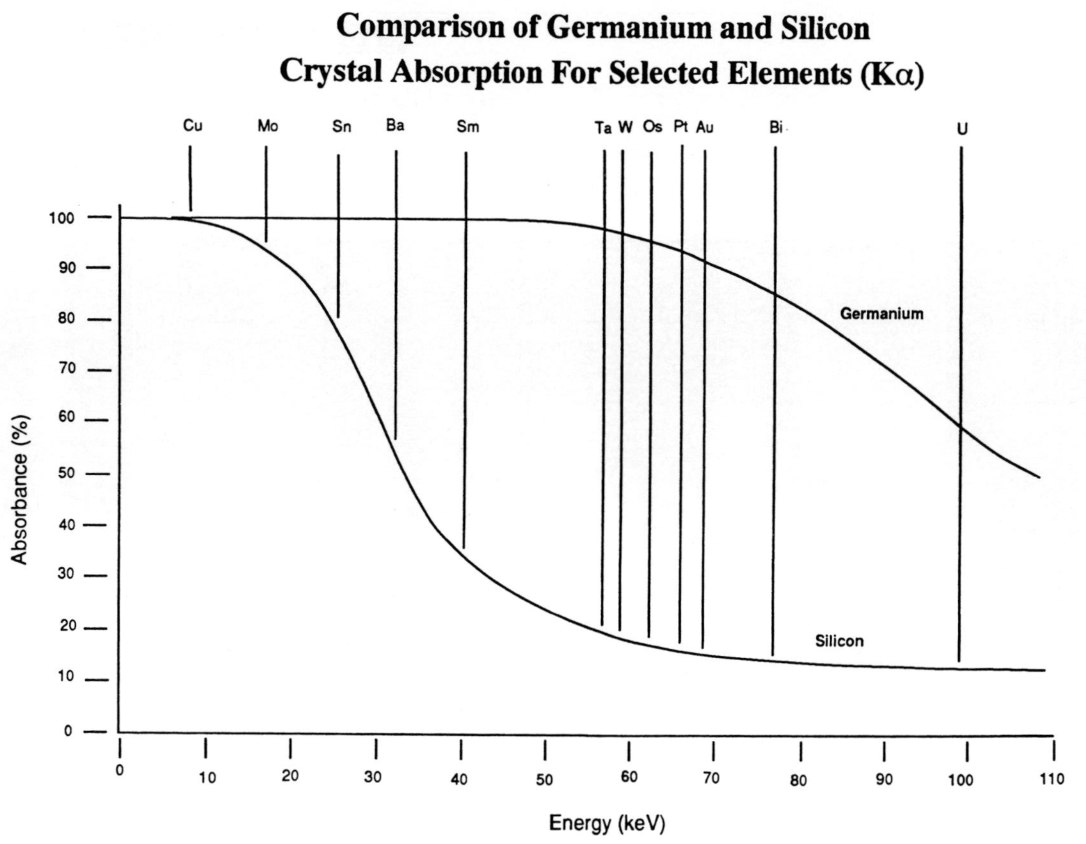

Comparison of Germanium and Silicon Crystal Absorption For Selected Elements (Kα)

Fig. 1.8. Comparative photon absorption, expressed as a percentage, measured in 33 mm thick slices of Si and Ge.

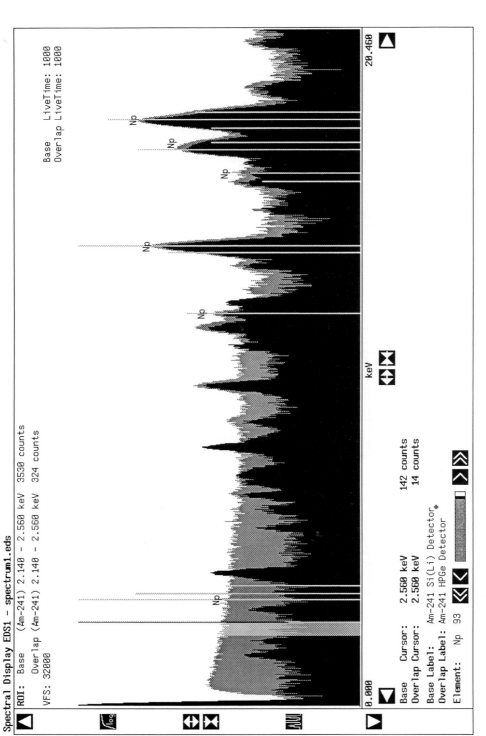

Fig. 1.9. Comparison of low-energy spectrum from an HpGe detector (black) and Si(Li) detector (grey), analysed under identical conditions. Note that the background signal is lower for HpGe than for Si(Li) at the low-energy region.

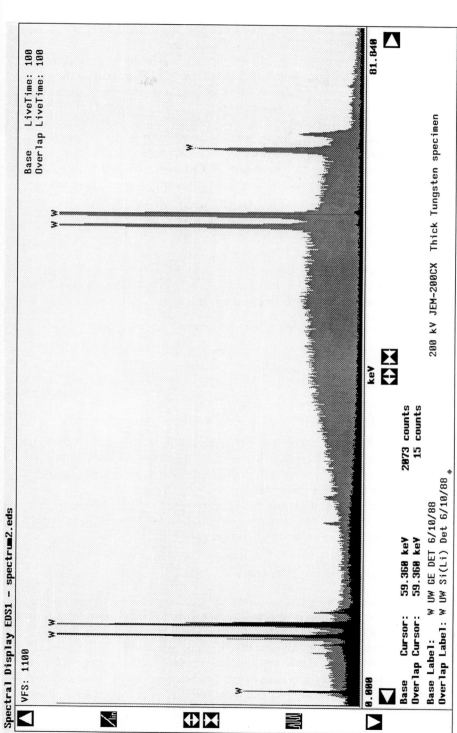

Fig. 1.10. Comparison of tungsten spectrum obtained from HpGe (grey) and Si(Li) (black) detectors collected under identical conditions. The specimen was relatively thick; this is confirmed in the HpGe spectrum, where a broad background below the W-K (energy: 59.318 keV) line is visible.

dispersion, the whole range of energies released from the specimen can be observed. This is useful in TEM studies (Fig. 1.10).

The lower dispersion of the HpGe detector leads to improved peak resolution values over those from a Si(Li) detector. This is especially the case for elements with a low atomic number, in which the dispersion of the Si(Li) detector is increased due to the dead layer. It has been established that improved resolution leads directly to lower detection limits, which result in enhanced detection sensitivity in the case of overlapped peaks, or when the relative peak heights are significantly different. This correspondence is easily established by considering the expression of Ziebold (1967) which estimates the minimum detectable concentration of an element in a given specimen as:

$$C = 3.29 \frac{a}{\sqrt{[(P/B)\,Pt]}}$$

where P/B is the peak-to-background ratio, P is the peak intensity in counts per second per nA, t is the total counting time in seconds, and a is the parameter depending on the specimen composition. For a given beam current and counting time the factor:

$$F = \frac{1}{\sqrt{[(P/B)\,P]}}$$

can be used to compare two detectors.

Via this expression we can come to the following conclusions.

1. Since the P/B ratio increases and the peak resolution decreases (FWHM), the detection limit decreases when using the HpGe crystal. This is particularly valid for light and ultralight elements which are predominantly of interest in biological research.
2. Due to the high resolution and better sensitivity of the HpGe compared with the Si(Li) crystal the beam current can now be reduced on sensitive specimens without any loss on analytical sensitivity.

2   Crystal size

For Si(Li) crystals the highest resolution can be obtained with a 10 mm² active crystal area. The limitation of this relative small size is that a long counting time or high beam current is required to get enough statistical data. This will result in a damage of radiation-sensitive organic specimens (von Zglinicki, this volume). With the HpGe crystal the standard active crystal area is 30 mm². This means that the beam current can be reduced by at least a factor of three. This guarantees a longer chemical stability of the specimen.

3   *Window choice*

Depending on the biological material to be investigated, a choice can be made between different windows.

(i) BERYLLIUM WINDOW

A choice of the Be-window can be made with studies limited to the elements with atomic number above ten (sodium).

Window characteristics: it is strong and reliable.

(ii) ULTRATHIN WINDOW AND WINDOWLESS

Depending on the type of microscope (transmission or scanning) two different forms are available. On the TEM, detectors are delivered windowless or with an ultrathin window which is protected against vacuum changes via a gate valve.

On the SEM, detectors are supplied windowless as well as with a beryllium and an ultrathin window. When vacuum changes occur (for example, during a specimen change) the Be-window must be put in front of the crystal. When light elements have to be measured the ultrathin window can be selected. The advantage of using an ultrathin window or windowless detector is that it offers the highest sensitivity of light element analysis (i.e. boron, carbon, nitrogen and oxygen). The ultrathin windows are recommended to prevent both contamination of the crystal and the detection of light, which causes an increment in the background signals.

Characteristics of an ultrathin window: it is fragile and especially vulnerable when changing vacuum.

(iii) ONE ATMOSPHERE WINDOW

This single window detector is also able to detect light elements (i.e. boron, carbon, nitrogen and oxygen) although the sensitivity for some of the elements is lower compared with the ultrathin window. This detection limit depends on the absorption of the elements by the window material. For example, if the material is pure carbon, nitrogen is less easy to detect. The microscope operation related to specimen change with this window is easy and quick both on TEM and SEM and results in a good light element detection capability.

Window characteristics: it can withstand the pressure of one atmosphere.

D   X-ray analysis in biological research

The difficulties of using X-ray microanalysis on biological specimens are mainly caused by specimen preparation. In addition, there is a risk of damaging the specimen in the microscope by the electron beam (von Zglinicki, this volume). The main problem during specimen preparation is the leaching and displacement of elements of interest which can lead to wrong or misleading conclusions (Morgan, Davis & Erasmus, 1978; Morgan, 1980). Two methods are available for overcoming these important constraints: (1) wet preparative methods, involving aqueous media for fixation, staining, the precipitation of diffusible constituents and embedding; and (2) cryopreparation avoiding liquid media.

The wet preparation method can be used in the context of X-ray microanalysis where the element of interest is insoluble or bound by the tissue. Hall (1979) gave examples of the research areas where aqueous method have been successfully used.

(*a*) Characterisation of unidentified electron-dense regions seen in chemically prepared EM sections.

(*b*) Characterisation of foreign bodies in tissues, for example, asbestos fibres in lung tissue (Champness, Cliff & Lorimer, 1976).

(*c*) Confirmation of histochemical 'stains' and precipitating agents, for example, silver stain to localise chlorine and bromine (van Stevininck *et al.*, 1976), and pyroantimonate for the presence of calcium (Yarom & Chandler, 1974). (See the chapter by Sumner in this volume.)

(*d*) Localisation and assay of enzyme activity, e.g. phosphatases (Hale, 1962).

(*e*) Comparison of pathological with normal tissue (Maunder, Yarom & Dubowitz, 1977; Janssens, 1983).

Cryopreparation (see Zierold, this volume) is used in studies where the elements of interest are in a chemically free, diffusible state.

With a specimen present in the electron microscope, practical pitfalls of the energy dispersive technique become apparent. Comparatively minor peaks in the spectrum can be masked by adjacent peaks or background. The peak-to-background ratios can be improved by increasing the peak or decreasing the background. The proper choice of the microscope parameters, type of support grid, and coating, can improve the P/B ratio. With respect to microscope parameters, a small probe diameter is a great advantage in the determination of trace elements in small areas of the specimen. This arises because in thin specimens at least the actual resolution is improved and also because, with the reduced total mass being analysed, background readings are lower and sensitivity is thus enhanced. However, most of the background arises from scattered electrons striking parts of the microscope and specimen support. This may be reduced by a correct choice of analytical conditions (accelerating voltage, current, specimen tilt) and the use of carbon and beryllium grids (and/or microscope coating) to create a light element specimen environment.

E   Conclusion

Electron microscopy, combined with X-ray microanalysis as a technique for tissue analysis, provides a number of extra advantages when using the HpGe detector instead of an ordinary Si(Li) detector.

1. Good sensitivity for light and ultralight elements.
2. Low background readout at the lower energies.
3. Improved resolution.
4. Better detection limits due to higher P/B ratios.
5. High efficiency for energies up to 80 keV.

References

Barbi, N.C. & Lister, D.B. (1981). A comparison of silicon and germanium X-ray detectors. *NBS Special Publication*, **604**, 35–44.

Castaing, R. & Guinier, A. (1949). Application des Sondes Electroniques a l'Analyse Metallographique. *Proc. 1st Int. Conf. in Electron Microscopy*, Delft, pp. 60–3.

Champness, P.E., Cliff, G. & Lorimer, G.W. (1976). The identification of asbestos. *Journal of Microscopy*, **108**, 231–49.

Cox, C.E., Lowe, B.G. & Sareen, R.A. (1988). Small area high purity germanium detectors for use in the energy range of 100 eV to 100 keV. *IEEE Trans. on Nucl. Sci.*, **35**, 28–32.

Duncumb, P. (1962). An electron optical bench for microscopy, diffraction and X-ray microanalysis. *Proc. 5th Int. Congress on Electron Microscopy*. New York: Academic Press.

Fink, R.W. (1981). Properties of silicon and germanium semi-conductor detectors for X-ray spectrometry. *NBS Special Publication*, **604**, 5–34.

Fitzgerald, R., Keil, K. & Heinrich, K.F.J. (1968). Solid state energy dispersion spectrometer for electron microscope X-ray analysis. *Science*, **159**, 528–30.

Goodhew, P.J. & Chescoe, D. (1980). Microanalysis in the Transmission Electron Microscope. *Micron*, **11**, 153–81.

Hale, A.J. (1962). Identification of cytochemical reaction products by scanning X-ray emission microanalysis. *Journal of Cell Biology*, **15**, 427–35.

Hall, T.A. (1979). Biological X-ray microanalysis. *Journal of Microscopy*, **117**, 145–63.

Janssens, A.R. (1983). *Copper Metabolism in Primary Biliary Cirrhosis*. Thesis, University of Leiden, The Netherlands.

Llacer, J., Haller, E.E. & Cordi, R.C. (1977). Entrance windows in Germanium low-energy X-ray detectors. *IEEE Trans. on Nucl. Sci.* **NS-24**, 53–60.

Maunder, C.A., Yarom, R. & Dubowitz, V. (1977). Electron microscopic X-ray microanalysis of normal and diseased human muscle. *Journal of Neurological Science*, **33**, 323–6.

McCarthy, J.J., Ales, M.W. & McMillan, D.J. (1990). High purity germanium detectors for EDS. *Microbeam Analysis*, pp. 79–84.

Morgan, A.J., Davis, T.W. & Erasmus, D.A. (1978). Specimen preparation. In *Electron probe microanalysis in biology*, ed. D.A. Erasmus, pp. 94–147. London: Chapman and Hall.

18   *Detection and quantification of X-rays*

Morgan, A.J. (1980). Preparation of specimens. Changes in chemical integrity. In *X-ray microanalysis in biology*, ed. M.A. Hayat, pp. 65–165. Baltimore: University Park Press.

van Steveninck, R.F.M., van Steveninck, M.E., Peters, P.D. & Hall, T.A. (1976). Ultrastructural localization of ions. IV. Localization of chloride and bromide in *Nitella translucens* and X-ray energy spectroscopy of silver precipitation products. *Journal of Experimental Botany*, **27**, 1291–6.

Yarom, R. & Chandler, J.A. (1974). Electron probe microanalysis of skeletal muscle. *Journal of Histochemistry and Cytochemistry*, **22**, 149–54.

Ziebold, T.O. (1967). Precision and sensitivity in electron microprobe analysis. *Analytical Chemistry*, **39**, 858–61.

2 Quantitative software for biological applications of X-ray microanalysis

Robert B. Shen

A Introduction

Quantitation routines which have been developed for use in materials science cannot usually be applied to biological specimens because biological material has special characteristics.

(a) It usually consists of elements of interest such as sodium, magnesium, potassium etc., which are present at low concentrations (less than 5 %) in an organic matrix of the low atomic number elements carbon, hydrogen, oxygen and nitrogen. These matrix elements are not detected by the usual beryllium window detector. Because of this composition, standards are required when biological material is analysed.

(b) The major component of living tissue is water. This is generally removed during specimen preparation, though hydrated tissue can be studied if a suitable cold stage is available.

(c) Biological specimens often present a rough surface and are of uneven thickness.

Because of these factors, software for the application of quantitation to biological specimens must include routines which are able to cope with the differences in specimen thickness that are encountered (e.g. using the Hall method) and also offer routines for the estimation of local water content. In addition, the programme should be easy to use and should be capable of processing spectra automatically when large amounts of data are being collected (as is the case with biological specimens).

B Application of the Hall method (continuum normalisation)

The continuum normalisation method developed by Hall (see Hall, 1979) is independent of specimen thickness, since the concentration of an element is determined from the ratio of the characteristic intensity to the continuum intensity (2.1):

$$C_u/C_s = (R_u/R_s) \times (G_u/G_s) \tag{2.1}$$

where C = concentration of a given element
R = peak to continuum ratio (Ip/Ic)
subscripts s and u refer to the standard and the unknown respectively
G = matrix correction factor, which is the mean of Z^2/A (Z = atomic number, A = atomic weight) for all the elements present.

From equation (2.1):

$$C_u = F_s \times R_u \times G_u \tag{2.2}$$

where

$$F_s = C_s/(R_s \times G_s). \tag{2.3}$$

F is a constant correction factor (called the 'S factor' in the new version of EDAX thin film software called PBTHIN) which differs for each element, and is obtained from the analysis of a standard of known composition. Details about standards for biological specimens can be found in Warley (1990). Once the F factors have been determined, equation (2.2) can be solved by iteration (Roomans, 1988) to determine C_u and G_u (2.4).

$$G_u = C_m(Z^2/A)_m + \text{Sum}\,(C_i \times (Z^2/A)]_i \tag{2.4}$$

where   m = organic matrix
i = all elements which can be analysed in the EDS system.

C   Corrections for continuum intensity

The most important consideration in regard to continuum normalisation is that the continuum intensity must be that part of the overall level that is derived solely from the specimen. When a thin specimen is analysed, therefore, the extraneous contributions from the grid and film must be subtracted from the total reading to give the continuum derived solely from the specimen. This subtraction is carried out by use of the following equation (2.5):

$$Ic = [B_1 - B_2 - B_3] \times [P_1 - P_2)/P_3] \tag{2.5}$$

where Ic = corrected continuum
P = characteristic peak of a reference element in the grid material
B = measured continuum
subscripts 1, 2 and 3 denote measurements made on the sample, the film (without sample) and the grid (without film).

Further details of this correction procedure can be found in Roomans

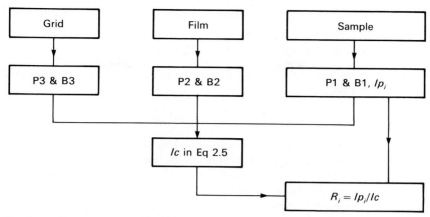

Fig. 2.1. Correction procedure for the determination of specimen peak/continuum ratios. The flow diagram outlines the sequence of data required for derivation of a characteristic peak for an element of interest in the sample (Ip_i), corrected continuum for the sample (Ic) and the resulting corrected peak to continuum ratio (R_i). Other symbols are as defined for text equation (2.5).

(1988). A flow diagram of the steps which are necessary for the determination of peak to continuum ratios in a specimen is shown in Fig. 2.1.

D Measurement of local water concentrations

Several methods are available for the measurement of local water concentrations, some of which are given below.

1 Comparison of frozen–hydrated and frozen–dried specimens

The pioneers with this method were Hall and Gupta (see, for example, Gupta & Hall, 1979). They used the differences in peak to continuum ratios in a specimen measured in the frozen–hydrated and the frozen–dried state to determine the local water content in their specimens (2.6). This method is only suitable for semi-thick specimens (0.5 μm or thicker). The rapid loss of mass which occurs when hydrated sections are analysed precludes the use of this technique with ultrathin sections (Zierold, 1988).

$$Fw = (R_h - R_d)/(R_d + K \times R_h) \tag{2.6}$$

where Fw = the mass fraction of water in the specimen
 R = peak to continuum ratio (Ip/Ic)
 subscripts h and d refer to the frozen–hydrated and frozen–dried state respectively.

The constant K is determined from G factors in the hydrated and dried states (2.7)

$$K = (G_h - G_d)/G_d \tag{2.7}$$

where $G_d = 3.28$ and $G_w = 3.6667$. G_d can be adjusted for different organic matrices (see Table 2.1).

2   Measurement of oxygen concentration

If an ultrathin windowed or a windowless detector is available, the water content can be determined by measurement of the oxygen concentration in the frozen–hydrated and frozen–dried states (2.8). This method is again only applicable to bulk specimens. For further details, see Roomans (1988) and Marshall & Condron (1985).

$$Fw = (O_h - O_d)/(O_d - 0.8889) \tag{2.8}$$

where O = oxygen concentration.

3   Use of peripheral standard of known water content

Water content can be measured in frozen–dried sections provided that the specimen is surrounded by a peripheral standard of known water content (2.9). Further discussion of this is found in Warner (1986) and Roomans (1988).

$$Fd_u/Fd_s = (Ic_u/Ic_s)*(G_s/G_u)*(D_s/D_u) \tag{2.9}$$

where Fd = dry mass fraction
 Ic = corrected continuum
 G = matrix correction factor
 D = density.

E   Worked examples

The new version of software produced by EDAX INTERNATIONAL (PBTHIN) for biological applications is based on the Roomans 1988 paper. Quantitation is based on the Hall continuum normalisation method as outlined above, and the programme also includes routines for the measurement of water concentration. When the programme is used, standards are first analysed and S-factors are stored in a parameter file. A second parameter file contains the G-factors for a number of different matrices (see Table 2.1), and can be edited by the user to enter his own matrices. The programme has been tested, and examples of operation and results are given below.

Table 2.1. *Some values for* Z^2/A *for biological specimens*

Epoxy resin	= 3.060
Gelatin	= 3.310
Albumin	= 3.200
Heart (freeze dried)	= 3.140
Liver (freeze dried)	= 3.190
Heart (frozen–hydrated)	= 3.540
Liver (frozen–hydrated)	= 3.530

From Roomans (1988).

Table 2.2. *Worked example of quantitation programme: determination of levels of KCl (mmoles/kg dry wt) in gelatin standards of known composition*

Sample no.	Si	S	Cl	K	Ca	Given KCl
STD02	442.5	122.3	763.9	787.3	94.7	757.0
STD03	375.9	135.4	736.1	769.0	55.1	757.0
STD09	291.7	123.1	468.3	487.9	53.4	475.0
STD18	221.9	140.1	307.9	312.1	53.4	272.0
STD19	214.8	149.9	313.7	308.2	52.7	272.0
STD24	586.2	145.3	681.7	629.6	41.1	693.0
STD24	642.4	139.4	634.7	696.3	38.2	693.0

Standards are listed STD02–STD24, with X-ray microanalytical results in the main part of the table. Known levels of KCl are given in the right-hand column. The elements Si, S and Ca are either extraneous or are present in the gelatin base.

The accuracy of the quantitation procedure is indicated by the close similarity in the XRMA values of K and Cl with the known molarities of KCl in respective standards.

1 Operation of programme – calibration and S-factors

Calibration constants for K and Cl were obtained by analysis of standards consisting of known amounts of KCl in gelatin, calibration constants for the remaining elements were determined by analysis of isoatomic droplets as outlined by Morgan elsewhere in this volume. Analysis was carried out at 80 kV for 100 s livetime. The *S*-factors were:

$$Na = 372.8 \quad Mg = 149.1 \quad Si = 60.0 \quad P = 55.53$$
$$S = 52.5 \quad Cl = 44.9 \quad K = 42.37 \quad Ca = 40.6$$

Values for *S*-factors will differ depending on the conditions used for analysis, and the region of the spectrum which is selected for measuring the continuum.

2   Results

Some results obtained by testing the programme against gelatin standards of known concentration using the S-factors are presented in Table 2.2.

Acknowledgement

The author would like to thank Dr A. Warley, St Thomas's Hospital Medical School, London, for providing and analysing the gelatin standards.

References

Gupta, B.L. & Hall, T.A. (1979). Quantitative electron probe X-ray microanalysis of electrolyte elements within epithelial tissue compartments. *Federation Proceedings*, **38**, 144–53.

Hall, T.A. (1979). Biological X-ray microanalysis. *Journal of Microscopy*, **117**, 145–63.

Marshall, A.T. & Condron, R.J. (1985). Normalisation of light element X-ray intensities for surface topography effects in frozen hydrated bulk biological samples. *Journal of Microscopy*, **140**, 99–108.

Roomans, G.M. (1988). Quantitative X-ray microanalysis of biological specimens. *Journal of Electron Microscopy Technique*, **9**, 19–43.

Warley, A. (1990). Standards for the application of X-ray microanalysis to biological specimens. *Journal of Microscopy*, **157**, 135–47.

Warner, R.R. (1986). Water content from analysis of freeze-dried thin sections. *Journal of Microscopy*, **142**, 363–9.

Zierold, K. (1988). X-ray microanalysis of freeze dried and frozen hydrated cryosections. *Journal of Electron Microscopy Technique*, **9**, 65–82.

3 *X-ray mapping techniques in biology*

Janet Wood

X-ray mapping is now an important qualitative and quantitative aspect of X-ray microanalysis of biological specimens, and has been the subject of a number of recent research papers and review articles (Chang, Shuman & Somlyo, 1986; Fiori, 1986; Fiori *et al.*, 1988; Ingram *et al.*, 1987; Lamvik *et al.*, 1989; Sauberman & Heyman, 1987). The related technique of electron energy loss mapping has also been the subject of much recent interest (Jeanguillaume *et al.*, 1983; Leapman & Ornberg, 1988) and is discussed separately by de Bruijn *et al.* in this volume.

X-ray mapping is the technique whereby an image is formed using the X-ray signal from a specified energy range, in order to show elemental distribution within the sample. X-ray mapping in its simplest form has been available since the early days of X-ray microanalysis. The technique has evolved in sophistication over the past 20 years, the main milestones being the development of digital mapping in the late 1970s and quantitative mapping in the late 1980s. The methods currently available are:

(*a*) analogue dot mapping,
(*b*) qualitative digital mapping,
(*c*) quantitative digital mapping.

A   Analogue dot mapping

This procedure was developed in the mid 1960s. Dot maps are formed when X-rays generated within a specified energy range appear as bright dots on the electron microscope display as the electron beam scans the sample. The advantage of the technique is that maps are very quick to set up and acquire. However, the simplicity of the technique also infers a number of limitations.

The main disadvantage relates to the fact that the data can only be 'stored' on photographic film and therefore no post processing of the data is possible. This is a particular problem when mapping elements present in low concentrations and samples with a rough surface. A further problem for low concentration mapping is the inability on most microscopes to integrate successive frames. In consequence, the dot map is formed from a single scan giving a poor signal/noise ratio. The resultant map can be difficult to interpret in terms of element distribution. This is illustrated in Fig. 3.1.

25

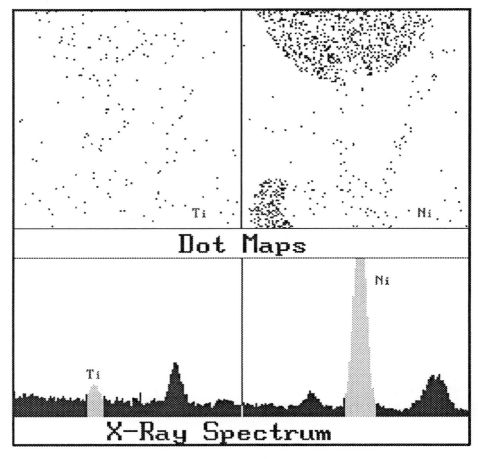

Fig. 3.1.   Examples of dot maps from low and high concentration elements. The Ti
and Ni have exactly the same distribution in the sample, but the poor
signal/noise in the Ti map degrades the distribution information.

Dot mapping is also inconvenient in that the number of elements that can
be simultaneously mapped is extremely limited. With the majority of electron
microscopes dot acquisition is limited to one element, although some
microscopes can map up to four elements simultaneously.

1   Dot mapping low concentrations

To specify which X-ray energies are to be used for mapping, a window is
defined on the spectrum over the required energy range. As we can see from
Fig. 3.2, this window includes both continuum and characteristic (peak and
background) X-rays. With dot mapping there is no reliable way of separating

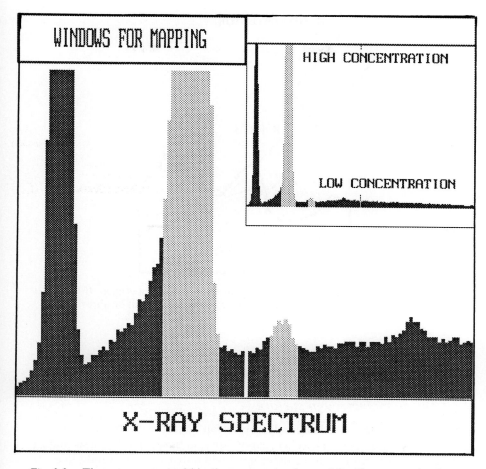

Fig. 3.2.   The energy range within the spectrum to be used for X-ray mapping is
defined by drawing a window over the region of interest. This is shown
as the light area in the above spectrum. This figure also shows the
relative background contributions as a proportion of total signal, for low
and high concentration elements.

these two signals. This presents a problem because continuum X-rays are
generated over the whole specimen and serve to degrade the information from
the characteristic X-rays which gives the actual element distribution.

As Fig. 3.2 shows, if an element is present in high concentrations, the
continuum contribution is small and is usually insignificant to the clarity of
the distribution information in the image. However, with low concentration
elements, the continuum contribution can be more than 50%. In such cases
it is impossible to unambiguously distinguish between the true elemental

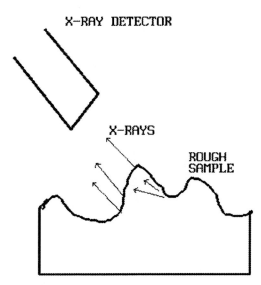

Fig. 3.3.   Schematic diagram of X-rays produced in a rough specimen. If there is no clear line of sight to the X-ray detector, X-rays will be absorbed and not contribute to the map.

distribution and the random continuum distribution. This problem is severe enough to render dot mapping of most biological samples almost useless.

2   Dot mapping rough surfaces

The ability of a biased secondary electron detector to bend secondary electrons, gives rise to the excellent topographic detail available in a secondary electron image. X-rays on the other hand will only travel in straight lines. In rough samples this leads to the dichotomy that areas of the sample imaged by the secondary electron detector are not necessary imaged by the X-ray detector. The shadowing effect of sample roughness is shown schematically in Fig. 3.3. The effect on a dot map for a sample of homogeneous composition would be an apparent inhomogeneous distribution. X-rays from areas with an obstruction in the path to the X-ray detector will be absorbed and not contribute to the X-ray image.

B   Qualitative digital mapping

Digital mapping overcomes many of the limitations of dot mapping. In a digital map the X-ray microanalysis computer takes control of the microscope scan coils and steps the beam across the sample, usually in a 128 × 128 matrix. Data are acquired at each pixel and stored to disk. With the data stored in the

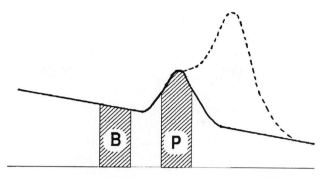

Fig. 3.4. X-ray spectrum showing the energy regions or windows used for mapping. The dashed lines show the potential of overlap from a nearby peak.

computer it is easy to apply image processing in order to extract useful information from the maps.

Digital mapping has many advantages. Frame integration is available over multiple frames which allows the characteristic X-ray signal to build up giving improved signal/noise images. Multiple images can be acquired simultaneously, up to 32 in advanced systems. This allows all elements plus electron images to be acquired at the same time. The use of colour to distinguish between elements is available and most systems have access to graphics packages to enhance data presentation. In fact digital mapping in combination with colour is amenable to a whole range of data presentation methods.

X-ray maps can be superimposed in colour over the electron image better to show distribution. In some systems X-ray maps can be superimposed with colour mixing to denote the overlap of element distributions. At its simplest this involves superimposing binary maps where each pixel shows whether an element is present or not, and mixed colours are used where two or more elements coexist. More sophisticated methods allow the superimposition of multi-level images so that several concentration levels can be shown for more than one element.

1   Processing low concentration X-ray maps

As described above, the problem with mapping low concentration elements is the interference of the proportionately large continuum contribution with the characteristic X-ray image. This can be minimised by acquiring a background map and subtracting this from the peak + background map. The relative positions of the two maps with respect to energy range on the spectrum is shown in Fig. 3.4.

Fig. 3.4 also illustrates how overlapping peaks from other elements

Table 3.1. *Comparison of peak/background ratio on a flat and rough specimen*

Flat specimen	
Peak – background counts = 500	
Background counts	= 10
Peak/background	= 50
Rough specimen	
Assume that X-rays are absorbed by 10%	
Peak – background counts = 450	
Background counts	= 9
Peak/background	= 50

Table 3.2. *Processing in quantitative mapping*

Escape peaks	*Background*	*Overlapped peaks*
Removed from the spectrum.	Removed by a digital filter. A specially developed filter is available for peaks with less than 300 counts.	Deconvoluted by a least squares fit routine between standard profiles and spectrum.
Peak area	*Continuum area*	
Accurate peak area calculated.	For thin biological specimens the Hall quantitative method requires a continuum map. This map can be acquired from any energy range within the spectrum. It is not necessary to specify a peak free area as the software will automatically remove any peaks within the range.	
Statistics	*Spectrometer drift*	*Specimen drift*
Standard deviation is calculated at each pixel and stored in a separate map.	In the case of long acquisition times it can be useful to correct for drift in the EDS spectrometer. This is particularly useful if trying to deconvolute strong overlaps, e.g. PmB/SK.	With thin samples in the TEM it is almost inevitable that some sample drift will occur over a long acquisition period. Drift compensation software must be available to overcome this problem.

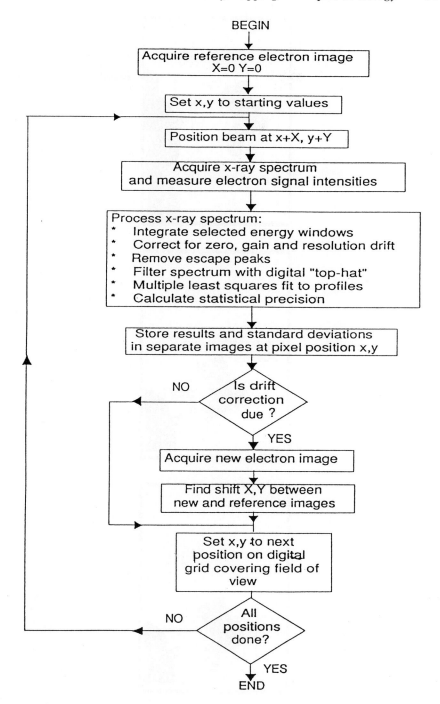

Fig. 3.5. Flow diagram to show the sequence of stages involved in quantitative mapping, with drift correction.

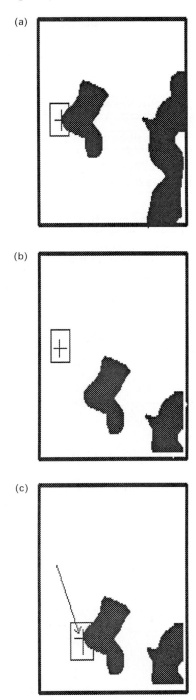

Fig. 3.6. For legend see facing page.

represents a further complication. In this case it is difficult to distinguish between X-rays from the two elements. Processing methods are available to minimise the problem. In general, a map will be collected from another spectral region, for example from Kb, L or M lines, then scaled and subtracted from the convoluted image.

In both these cases, although the processing described offers a distinct improvement over dot mapping, these methods of background subtraction and peak deconvolution are often too crude to allow accurate quantitative information to be derived. In addition, background subtracted maps of low concentration elements are still plagued by the low grade image structure introduced by statistical artefacts.

2   *Processing maps from rough specimens*

Topographical artefacts in X-ray maps can be minimised by exploiting one simple assumption. This is, that continuum and characteristic X-rays of approximately the same energy, generated in approximately the same region, are absorbed to approximately the same extent in the sample. If we therefore take a background map using an energy band close to that for the characteristic map and divide one by the other the resultant map will be independent of topography. The simple calculation in Table 3.1 illustrates this.

Errors in this method originate from two main sources: (i) inaccuracies in the peak minus background map; and (ii) the fact that the background map is not of exactly the same energy as the peak (assuming the relative energy ranges shown in Fig. 3.4). Obviously the limit at which processing can improve data is exceeded if sample roughness is so great that all generated X-rays are absorbed before reaching the X-ray detector. This limit is equal to the range of X-rays in the sample.

C   Quantitative mapping

An advanced form of digital mapping, quantitative mapping, involves full spectrum processing at every pixel within the image. This results in accurate background removal, accurate peak deconvolution and statistical data from which quantitative information can be derived. Although acquisition times

---

Fig. 3.6.   Principle of drift correction. (a) Cursor is placed on a feature in the reference image. (b) Subsequent electron image exhibits drift. (c) Computer searches to find point corresponding to original cursor position and deduces the drift-correction vector.

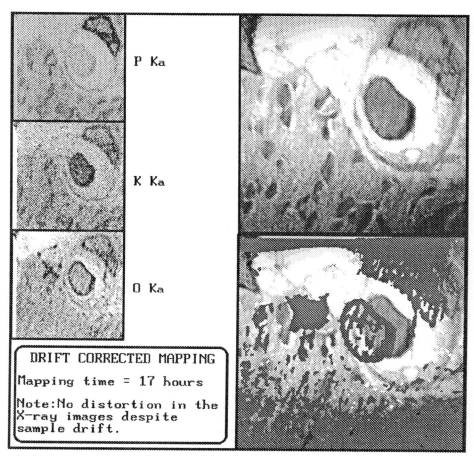

Fig. 3.7. Examples of drift correction applied to X-ray mapping of rat's heart. The map was acquired over a period of 17 hours and although the sample has drifted (as shown by the superimposed binary image in the left-hand quadrant), the X-ray maps show no distortion. (Sample courtesy of Dr Alice Warley, St Thomas's Hospital, London.)

are usually longer than for qualitative digital mapping (of the order of 1 s per pixel or more), the benefits in terms of improved information are great.

Table 3.2 shows the possible processing involved at each pixel in a quantitative map.

A flow chart of how the computer implements quantitative mapping is shown in Fig. 3.5. Processing at each pixel is complete within 1 second and this therefore dictates the minimum suitable acquisition time. For maximum efficiency, the computer processes data from the previous pixel simultaneously with data acquisition. Nevertheless the minimum time for a 128×128 map is around 4 hours. Time can be saved by acquiring only a 64×64 map but at the expense of image detail. In practice, it is most convenient to acquire these maps overnight. This allows for a longer acquisition per pixel (useful in biological specimens) without tying up instrument time during the day. Obviously some automatic form of specimen drift compensation is required for this to be practical.

With acquisition times of 10 hours and longer one might expect a greater problem with radiation damage than may occur with qualitative mapping or point analysis. This is not the case. The specimen is only scanned once to form the map. Therefore the beam only rests for a few seconds at each pixel resulting in much less damage than may be induced during acquisition for a standard point analysis.

1   Coping with drift

Long acquisition times bring the added complications of specimen drift and possible spectrometer drift. Routines are available though to compensate for both types of drift.

(i) SPECIMEN DRIFT

For TEM analysis in particular, correction for specimen drift is essential to quantitative mapping. Routines available for correction are based on relatively simple principles as illustrated in Fig. 3.6. An electron image is acquired immediately prior to mapping. A reference feature is marked in the image. At periodic intervals during the map, acquisition is suspended and a new electron image acquired. The computer then searches for the original reference feature and deduces the drift correction vector. This vector is used to modify the scanned area and mapping restarts. The result is that despite specimen drift, no smearing of the X-ray maps occurs. An example is shown in Fig. 3.7. The whole procedure is fully automatic and after set up does not require any operator intervention.

(ii) SPECTROMETER DRIFT

Peak overlap is dealt with by the least squares fit method (Statham, 1977). This involves matching standard peak shapes to the acquired spectrum. In order to be effective the matching process requires that there be no drift in peak position (energy) between the standard peaks and acquired spectrum. As all electronic systems undergo some degree of drift, for maximum accuracy in deconvolution, a correction for the drift is required. Various methods are available to correct for spectrometer drift. Some more advanced systems incorporate special hardware (known as a Zero Strobe) to monitor drift

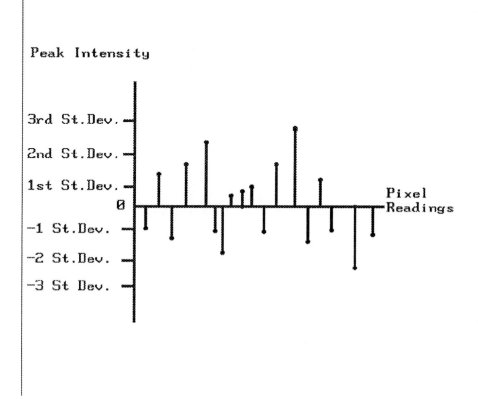

Fig. 3.8. Statistical distribution of the X-ray intensity within a window for an element below the detection limit. The average value is zero.

which is then exploited in the compensation. Another popular method to compensate for spectrometer drift is to use derivative profiles in the fitting procedure. One important disadvantage of using 2nd derivatives is that they can degrade precision by up to 60% (Statham, 1988). To put this in perspective, in biological samples with low peak counts a 30% improvement in precision is equivalent to having twice the number of counts in the peak.

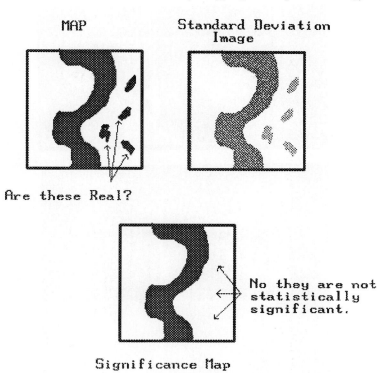

Fig. 3.9. Principles of significance maps. Pixels in the map are tested against the corresponding standard deviation image and set to zero if the intensity is less than a specified significance level.

2 Information generated from quantitative maps

Many different types of image can be acquired with quantitative mapping. Software allows the analyst to choose the exact combination. For example, one may choose to acquire quantitative maps just for elements present in low concentrations or where peaks overlap, and use window integrals for the remaining elements. Alternatively, quantitative data for all elements may be generated. Different types of specimen maps that are of particular interest to biologists are listed (*a*)–(*f*) below.

(*a*) Characteristic X-ray map with accurate pixel intensities,
(*b*) accurate continuum map,

(a)
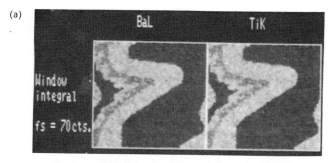

(b)
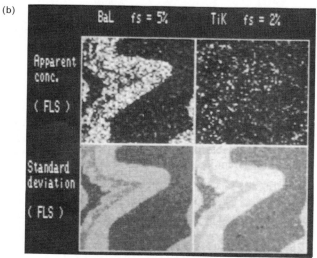

(c)
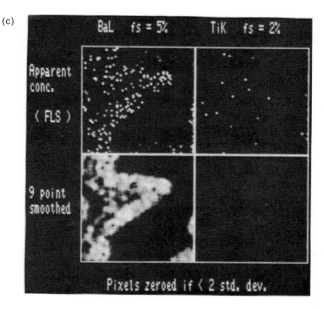

Fig. 3.10.   For legend see facing page.

(c) standard deviation map,
(d) fit index map,
(e) window integral map,
(f) electron image.

(i) STANDARD DEVIATION MAPS

When a quantitative map is acquired, the standard deviation at every pixel may be stored as a separate image. So, every elemental map has a corresponding standard deviation map. These statistical data are extremely important, particularly in the case of low concentration maps where they can be used to unambiguously verify the presence of an element. In fact this aspect of quantitative mapping is useful in its own right, irrespective of whether there is a need for quantitative information.

To understand why a standard deviation map is useful we need to understand the role that statistics play in producing artefacts in X-ray maps. In an X-ray spectrum, as a rule of thumb, a peak is said to be statistically significant if its intensity is more than three times the standard deviation of the background. If the intensity is less than this, the element may still be present but it is below the detection limit of the EDS detector. To illustrate the effect of statistics let us assume that we take a series of peak intensity readings for an element that is below the detection limit. It is statistically impossible that every reading will show a peak intensity of zero. However, we will find that readings are statistically distributed such that the *average* value is zero. This is shown schematically in Fig. 3.8.

This statistical distribution leads to the presence of a low grade (low intensity) structure in the image. With high concentration maps, the characteristic intensities are so much stronger than the low grade structure that they are rendered insignificant. As discussed before, with low concentration images the intensity of the characteristic peaks is close to the intensity in the low grade structure, making identification of the true elemental distribution somewhat subjective.

The standard deviation image can be exploited to solve this problem. The principle is shown in Fig. 3.9. What is required to remove statistical artefacts

---

Fig. 3.10. Example of significance maps as applied to mapping Ti in a Barium glass. (a) BaL and TiK are completely overlapped and simple window mapping will always show the presence of Ti with Ba. (b) Same sample but with quantitative mapping applying spectrum processing at each pixel. Barium and titanium are separated. Titanium may be present in low concentrations. (c) Same sample with pixel values set to zero if less than twice the corresponding standard deviation. Titanium, if present is below the detection limit of EDS.

(a)

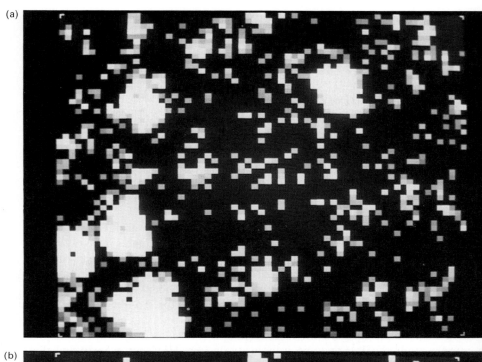

(b)

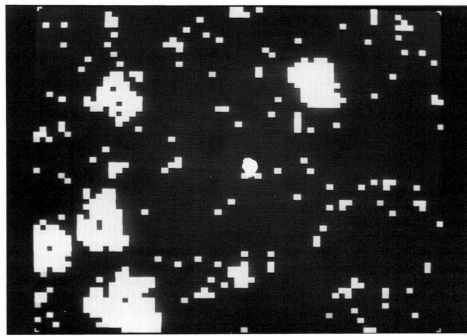

Fig. 3.11.   For legend see facing page.

from the image is to suppress all pixel intensities below three standard deviations. The computer can do this automatically by comparing the elemental map with its own standard deviation map. At each pixel where the intensity in the element map is below three times the corresponding pixel in the standard deviation map, this pixel is suppressed to zero. The result is a significance map where we can be 99 % confident that every single pixel in the image corresponds to the presence of the element in the sample.

This technique is an important analytical tool. It allows the analyst to use qualitative X-ray mapping safe in the knowledge that if there are any ambiguities in a low concentration map they can be objectively assessed by using quantitative mapping to produce a significance map. Examples of significance maps are shown in Fig. 3.10 and 3.11.

One point to note is that, equally, pixels below two standard deviations or one standard deviation can be suppressed. The analyst has the choice.

(ii) FIT INDEX MAPS

The fit index is a parameter that shows how good a match has been achieved between standard peak shapes and the acquired spectrum during the least squares fit routine. If there is a good match the fit index would be 1. The poorer the match the higher the fit index. Generating a fit index map by storing this parameter at each pixel provides a useful diagnostic tool by which the success of peak deconvolution can be judged. If there are no problems with the least squares fit a noisy map with pixel intensities distributed around 1 would be expected. Any high intensity structure in the image indicates immediately that, for some reason, accurate peak deconvolution has not been achieved. The most common reason for poor deconvolution is the presence of a peak in the acquired spectrum that the computer has not been asked to fit. An example of this is shown in Fig. 3.12. In this example, the sample contains Fe, Cr and Mn but the computer has only been asked to fit Fe and Cr. The bright areas in the fit index map correspond to the presence of Mn in the sample. An example where a fit index map verifies successful peak deconvolution is shown in Fig. 3.13. In this case the sample is a cocoa bean and a K map with the associated fit index map is shown. The overlap being separated is K and Ca. As would be expected with successful deconvolution, the fit index map is low intensity and completely random.

Fig. 3.11. Significance mapping of potassium in the presence of calcium in a cocoa bean. (a) Window map of potassium. (b) Same sample with pixel values set to zero if less than twice the corresponding standard deviation. The combination of FLS and significance testing at each pixel guarantees there are no distortions in the map from the K/Ca overlap or statistics.

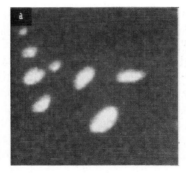

Fig. 3.12.   Example of a fit index map of a Fe/Cr/Mn steel. Most of the pixel values are close to unity (a satisfactory fit), but the bright areas indicate a region of poor fit. In this experiment profiles for Cr and Fe were included but ignored for Mn in the FLS processing. The bright regions effectively indicate where Mn is present.

(a)

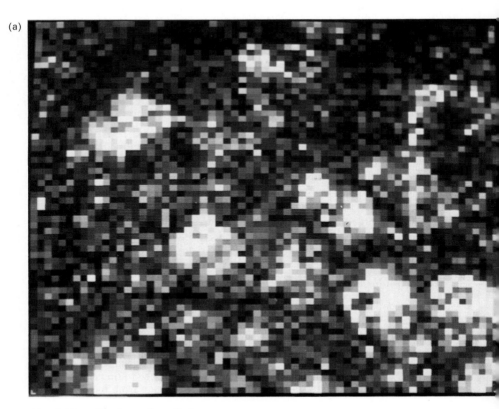

Fig. 3.13a.   For legend see facing page.

(b)

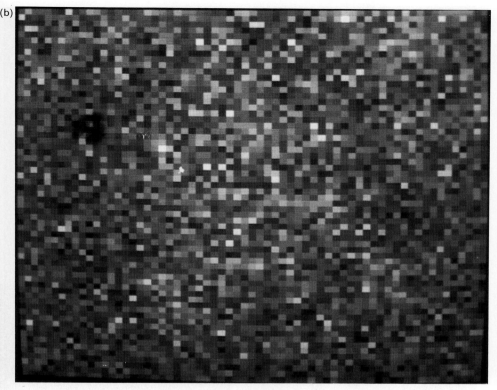

Fig. 3.13.   Use of a fit index map to verify successful deconvolution of the K/Ca overlap in the X-ray mapping of K in a cocoa bean. (a) Quantitative map of K. (b) Same sample, fit index map exhibits a low intensity random distribution indicative of a good fit. (Data courtesy of Dr B. Brooker, NIRD Reading.)

(iii) QUANTITATIVE MAPS FROM BIOLOGICAL THIN SECTIONS

Quantification of thin biological sections is achieved via the Hall technique which utilises a ratio of the characteristic peak to continuum intensity. A quantitative map uses exactly the same principles as quantifying a single spectrum. The difference being that with a map the process occurs over a matrix of points rather than a single point.

To produce a quantitative map the following components are required.

(*a*) Background subtracted elemental map,
(*b*) grid and film corrected continuum map,
(*c*) element proportionality constant from a standard.

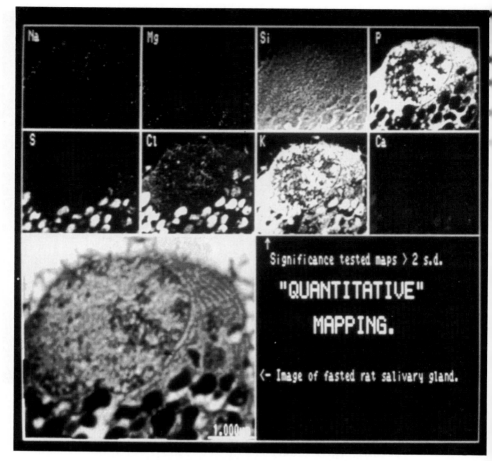

Fig. 3.14.   Quantitative maps from a thin section of rat salivary gland. Quantification is achieved via the Hall technique. Drift correction was applied. (Sample courtesy of University of Washington, Seattle.)

These components combine as shown below, to produce a map calibrated to elemental mass fractions of mm/kg

$$\text{where Quantitative map} = \frac{\text{Characteristic peak map}}{\text{Continuum image}} \times \text{constant from standard.}$$

Examples of quantitative maps are shown in Fig. 3.14.

Table 3.3. *Comparisons of element concentrations in human labial gland secretory granules from image and spot analyses*

	Na	Mg	K	Ca
Image	89	89	210	193
Spot	85	94	179	208

(iv) QUANTIFICATION OF SELECTED REGIONS

In the case of low concentrations, it may be preferable to use quantitative values averaged over a region (image analysis) rather than single pixel (spot analysis) concentrations. This can easily be achieved by using a mask to isolate the region of interest and then determining the average pixel value within the isolated area. This task can be achieved by a computer in seconds. Comparisons have been made by Wong *et al.* (1989) between quantitative results determined by the above method and averaging the results from 19 individual spot spectra. The data shown in Table 3.3 for separate elemental determinations by image and spot analysis show that the quantitative information obtained by these two analytical procedures is closely comparable.

D Conclusions

Quantitative mapping offers a powerful tool to the biologist in terms of providing both statistically reliable and quantitative data. Even if the experiment does not require quantitative information, the ability to remove statistical artefacts from maps is of significant benefit.

References

Chang, C.-F., Shuman, H. & Somlyo, A.P. (1986). Electron probe analysis, X-ray mapping and electron energy loss spectroscopy of Ca, Mg and monovalent ions in log phase and in dividing *Escherichia coli* B cells. *J. Bacteriol.*, **167**, 935–9.

Fiori, C.E. (1986). Quantitative compositional mapping of biological cryosections. In *Microbeam analysis*, ed. A.D. Romig & W.F. Chambers, pp. 183–6. San Francisco: San Francisco Press.

Fiori, C.E., Leapman, R.D., Swyt, C.R. & Andrews, S.B. (1988). Quantitative X-ray mapping of biological cryosections. *Ultramicroscopy*, **24**, 237–50.

Jeanguillaume, C., Tence, M., Trebbia, P. & Colliex, C. (1983). Electron energy loss chemical mapping of low Z elements in biological sections. *Scanning Electron Microscopy*, **2**, 745–56.

Ingram, P., LeFurgey, A., Davilla, S.D., Lamvik, M.K., Kopf, D.A., Mandel, L.J. & Lieberman, M. (1987). Real-time quantitative elemental analysis and imaging in cells. In *Analytical electron microscopy*, ed. D.C. Joy, pp. 179–83. San Francisco: San Francisco Press.

Lamvik, M.K., Ingram, P., Menon, R.G., Beese, L.S., Davilla, S.D. & LeFurgey, A. (1989). Correction for specimen movement after acquisition of element-specific electron microprobe images. *J. Microscopy*, **156**, 183–90.

Leapman, R.D. & Ornberg, R.L. (1988). Quantitative electron energy loss spectroscopy in biology. *Ultramicroscopy*, **24**, 251–68.

Statham, P.J. (1977). Deconvolution and background subtraction by least-squares fitting with prefiltering of spectra. *Analytical Chemistry*, **49**, 2149–54.

Saubermann, A.J. & Heyman, R.V. (1987). Quantitative digital X-ray imaging using frozen hydrated and frozen dried tissue sections. *J. Microscopy*, **146**, 169–82.

Statham, P.J. (1988). Pitfalls and advances in quantitative elemental mapping. *Scanning*, **10**, 245–52.

Wong, J.G. *et al.* (1989). Quantitative elemental imaging in the analytical electron microscope with biological applications. *Scanning*, **11**, 12–19.

4 Quantitative X-ray microanalysis of thin sections in biology: appraisal and interpretation of results

Alice Warley

There is considerable interest in the application of X-ray microanalysis to the study of biological specimens since this is one of the few techniques which allows the study of diffusible elements at a subcellular level. Although it is easy to obtain element ratios from spectra produced by X-ray microanalysis, fully quantitative data are required for the results to take their place alongside those obtained by use of other physiological techniques. Quantitative data are still published routinely from only a small number of laboratories, implying that some difficulties may be encountered in the application of quantification to biological specimens. Here, the two main methods used for obtaining quantitative results from the analysis of thin sections will be outlined briefly; details of other methods can be found in Hall (1989a). The discussion will concentrate on the appraisal and interpretation of results as this is an area where problems may arise.

The preparative steps involved in the study of diffusible elements are, firstly, cryofixation (reviewed by Zierold, this volume), which is usually followed by cryosectioning and analysis of the sections in the frozen–dried state (the sections can be analysed frozen–hydrated, but this is technically much more difficult – see Warley & Gupta, 1991). As an alternative the cryofixed specimen may be freeze dried and embedded, or embedded using low temperature techniques. The initial part of this discussion will focus on the quantification of data obtained from the analysis of frozen–dried cryosections; in a later section the problems encountered when applying quantitation to embedded material will be highlighted.

A Obtaining quantitative results from the analysis of frozen–hydrated or frozen–dried specimens

1 Measurement of elemental mass per unit mass (*continuum normalisation*)

Continuum normalisation, which measures the mass of an element per unit mass of specimen, is the method most commonly used for obtaining quantitative data from the analysis of thin biological specimens. This method was first proposed by Hall (see Hall, 1989b) and has been refined and developed by Hall and co-workers; full details of the derivation of the equations and their use can be found in the many publications from this group (for example: Gupta & Hall, 1979; Hall, 1979; Hall & Gupta, 1982; Hall & Gupta, 1986). A validation of the technique can be found in Shuman, Somlyo & Somlyo (1976). The principle of this method is that the amount of a given element in the area of analysis is determined from the net number of counts under its characteristic peak, whilst the continuum radiation gives a measure of the total mass in the area of analysis. Thus the ratio of characteristic peak counts to the continuum counts gives a measure of the amount of x per unit mass of the specimen. The equation used for quantification is:

$$Cx = Cx_s \cdot \frac{(Px/W)}{(Px/W)_s} \cdot \frac{\overline{Z^2/A}}{\overline{Z^2/A}_s}$$

(4.1)

where Cx is the concentration of x, Px the net counts in the characteristic peak, and W the continuum reading taken in a fixed region of the spectrum. The subscript s denotes values which are determined for a thin standard. The term $\overline{Z^2/A}$, sometimes called the G-factor, is the mean value, for all elements of the specimen, of atomic number squared divided by atomic weight, and corrects for differences in efficiency of continuum generation between the specimen and the standard.

Continuum normalisation measures the concentration of an element in the specimen as it exists at the time of analysis. When frozen–dried cryosections are analysed data is obtained in terms of mmoles/kg dry weight, and when frozen–hydrated material is analysed the results are in terms of mmoles/kg wet weight. When measuring the diffusible elements considerable care must be taken to ensure that movement of elements does not occur before cryofixation or during the cryopreparation steps, since this will mean the composition of the specimen is changed before analysis takes place.

The advantages of continuum normalisation are that the method is independent of specimen thickness, and of the beam current used for analysis, since alteration of these parameters affects both characteristic and continuum radiation to the same extent. Because of this there is no requirement to analyse a standard alongside each specimen.

The major disadvantage of the technique is that the continuum reading (W in equation (4.1)) must be that emitted solely from the specimen. Thus when a thin section supported on a film is analysed there is a requirement to correct for the continuum produced by the film, the grid, and other surrounding material. Problems may arise when carrying out these corrections and this will be discussed in a later section.

2   Measurement of mass per unit volume (characteristic alone)

The second method which is used for determining element concentrations in spectra from biological specimens depends solely on the measurement of characteristic radiation, and is variously called the characteristic alone method, the peripheral standard method, or the Rick method (after its principal users from the Physiology group in Munich; see, for example, Rick *et al.*, 1979). This method requires the use of a peripheral standard. To provide the standard the specimen is immersed in albumin, or a similar solution, of known composition immediately before quench freezing. The adherent layer forms the peripheral standard which is an integral part of the specimen and is frozen and sectioned along with it. When the specimen is analysed, the peripheral standard is also analysed, ideally using the same probe current, and for the same length of time, although changes in these parameters can be corrected for. Quantification is achieved by direct comparison of peak heights between the specimen and the standard using the equation:

$$C'x = (Px/Px_s)\, C'x_s. \tag{4.2}$$

Again Px is the number of counts in the characteristic peak, in this case the subscript s denotes the peripheral standard. $C'x$ is the concentration of x which in this case is measured in mmoles/l of specimen. Further details about the derivation of this equation can be found in Hall & Gupta (1986).

Three points should be noted when using this method.

1. Even though the specimen is usually analysed in a dried state the units are mmoles/l of wet specimen (Hall & Gupta, 1986).
2. The method depends on the assumptions that the specimen and standard are of equal thickness, and that when freeze dried, the specimen and standard experience the same amount of shrinkage. Since both the specimen and the standard usually are of similar composition, and are sectioned and freeze dried under the same conditions, it is likely that these assumptions hold. However, assumptions about even sectioning and shrinkage may not hold true for specimens with large extracellular spaces which do not contain an organic matrix. In this case the growth of large ice crystals during quench freezing will lead to the displacement of elements and also make the cutting of sections of identical thickness difficult.
3. It is not necessary for every element of interest to be included in the peripheral standard. Provided that one element is present in the standard in known concentration, the concentration of the other elements can be determined from relative sensitivity factors. Details about the use of relative sensitivity factors can be found in Morgan & Winters (this volume).

The major advantage of this technique is that there is no requirement for the measurement of the specimen continuum, with its attendant difficulties.

The disadvantage is the need for a peripheral standard which must be shown not to alter the element composition of the specimen.

B   Standards

A standard of known composition is required whatever method is used to obtain quantitative results from the analysis of biological specimens. The usual approach is to use a standard of similar composition to the specimen, the reason being to compensate for the loss of mass from the specimen matrix during the course of analysis. The general assumption is made that mass loss will be similar between the specimen and standard, but this assumption has been tested in only a very few cases (the problem of mass loss is discussed in the chapter by von Zglinicki, this volume). Details about the different types of standard which have been used have been reviewed recently (Warley, 1990). Problems may arise in the choice of a standard when analysing resin embedded tissue; this is discussed in a later section.

C   Appraisal and interpretation of results

It is common for results from X-ray microanalysis to be obtained using commercial software. This can be easy and rapid means of producing a large quantity of data. However, the ease of producing results should not lead to a false sense of security. Problems can and do occur. An example of this is shown in Table 4.1. Discussion of these results will help to show how to recognise when problems occur, and highlight the source of the problems.

The results shown in Table 4.1 were obtained from bronchial smooth muscle cells grown in culture. The cells were prepared for analysis either as a whole cell preparation in which the external medium was removed by washing in 0.3 M sucrose (Wroblewski *et al.*, 1983) before freeze drying, or as sectioned cells. In the latter case a drop of 20 % PVP in medium was added to the surface of the cells before cryofixation. The monolayer of cells was then cryofixed, sectioned at low temperature, and freeze dried before analysis. In each case areas of the cytoplasm were analysed.

The results for potassium from the whole cell preparations are in agreement with those published by James-Kracke *et al.* (1980) for cultured aortic smooth muscle; sodium and chloride concentrations are lower as might be expected in cells which have been washed to remove the overlying medium. The results obtained by the continuum normalisation method from the sectioned cells for all elements are at least double those from the whole cell preparations. The concentrations obtained from analysis of the peripheral standard using continuum normalisation also are increased to approximately twice the

Table 4.1. *Element concentrations (mmoles/kg dry weight ± SE) in bronchial smooth muscle cells in culture and a peripheral standard estimated using either continuum normalisation (CN) or the continuum alone (CA) procedure*

	Na	Cl	K
Whole cells (CN)	52 ± 6	126 ± 13	441 ± 12
Sectioned cells (CN)	286 ± 78	328 ± 45	960 ± 141
Sectioned cells (CA)	108	124	429
Peripheral standard	1089 ± 129	696 ± 77	52 ± 7
Calculated conc.	643	501	24

Contributions to the continuum

	Film	Grid	Specimen
Whole cells	3964	100	14885 ± 2094
Sectioned cells	2477	2464 ± 78	2743 ± 501
Standard	2477	2528 ± 28	3766 ± 557

known (calculated) values. Thus analysis of the peripheral standard confirms that the results from the sectioned cells are in error. If the characteristic alone method of quantification is applied to the data from the sectioned cells, and an intracellular water concentration of 80% is assumed, the concentrations of potassium and chloride are in agreement with those obtained from the analysis of the whole cell preparation. The sodium concentration is higher, but the differences in sodium concentration might well reflect differences in specimen preparation.

The probable source of the error when continuum normalisation was applied to the sectioned cells (and standard) is shown when the values of the individual contributions to the continuum are examined. The contributions from the surroundings to the total continuum reading is in excess of the contribution from the specimen. This is likely to lead to errors in the estimation of the continuum produced by the specimen. When the whole cells were analysed the extraneous contributions are a much lower fraction of the continuum.

Several pertinent points emerge from considering these results.

1. The results produced using X-ray microanalysis should not be taken at face value, but should always be considered both in the light of what is already known about the system under study, and how the results agree with general knowledge about cell physiology and cell biology. The effects of both the method of specimen preparation and the method of quantitation on the final results should always be considered.
2. When carrying out quantitative analysis, if possible, both the continuum normalisation method of quantitation and the characteristic alone

methods should be applied to the same set of results. This proposal is not new, and is the procedure that was adopted by Gupta & Hall (see, for example, Gupta & Hall, 1979). When the results from the two different methods concur this gives confidence in accepting the results. On the other hand, discrepancies between the two procedures indicate that some problems are occurring. More recently, Warner, Myers & Taylor (1985) and Steinbrecht & Zierold (1989) have also suggested that a peripheral standard should be included as an internal monitor for quantitative analysis.

3. When the continuum normalisation method is used for quantitation, extraneous contributions to the continuum should be monitored. The majority of commercial software includes routines for the subtraction of extraneous continuum based on the methods described in Hall (1979) and Roomans (1988). However, problems with background subtraction and variability in continuum production occur (Warner *et al.*, 1985; Roomans, 1988; Steinbrecht & Zierold, 1989). In the worst case the corrections to the continuum exceed the total continuum count giving a zero or negative value for the specimen continuum (Roomans, 1988), but inaccuracy in the estimation of the continuum can also lead to considerable variability in results (Warner *et al.*, 1985; Steinbrecht & Zierold, 1989) or even to results which are in error as reported here. If the continuum produced from extraneous sources is monitored and found to be in excess of that produced by the specimen, steps should be taken to reduce the external continuum. The various sources of extraneous continuum in the microscope have been documented (Nicholson *et al.*, 1982; Morgan, 1985): they include the grid, the specimen holder and surfaces in the specimen chamber. Practical ways of reducing the problem are: ensuring that the electron beam is correctly aligned before carrying out analysis; using thick apertures to prevent X-rays being generated by sources higher in the column; and confining analysis to the central area of the grid. Full practical details can be found in Nicholson *et al.* (1982), Morgan (1985), and Warley & Gupta (1991).

D   Measurement of water content

In the majority of studies, the continuum normalisation method for quantification is carried out on frozen–dried sections and the results are expressed in terms of mmoles/kg dry weight. Although dry weight concentrations are relevant for elements such as Mg or Ca, which are predominantly bound to cellular components, such units are not relevant for the elements Na, K or Cl which are free within the aqueous phase. In this case the relevant units are mmoles/l of cell water. In order to express results obtained from X-ray microanalysis in this form, an estimation of local water concentration is required. A second reason why some attempt should be made to estimate

Table 4.2. *Element concentrations and water content in isolated cardiac myocytes*

	Na	Mg	Cl	K
	(mmoles/kg dry weight $\pm$ SE)			
myofibrils	76 ± 15	86 ± 10	319 ± 18	685 ± 30
mitochondria	32 ± 6	36 ± 5	62 ± 19	331 ± 12
		H_2O		
myofibrils		82%		
mitochondria		66%		
	Na	Mg	Cl	K
	(mmoles/l H_2O)			
myofibrils	16.7	19.0	70.0	171.0
mitochondria	16.3	18.3	31.0	168.0

Element concentrations were estimated in freeze dried isolated cardiac myocytes in mmoles/kg dry weight (upper values). Local water content was estimated by comparison with a protein standard, and values were then converted to mmoles/l H_2O (lower values).

the water content is that total dry mass differs between different subcellular compartments. This can lead to apparent differences in element concentrations between organelles when results are expressed on a dry weight basis, but these gradients disappear when the results are expressed in terms of mmol/l water (see Table 4.2 and Dorge *et al.*, 1974; Zierold, 1986). The different methods for estimating water concentration are given briefly in the chapter by Shen in this volume and a detailed description of the techniques can be found in von Zglinicki (1991).

E   Quantification of results from resin-embedded specimens

A situation in which interpretation of the results from X-ray microanalysis becomes difficult is when quantitative results are obtained from resin-embedded specimens. Recently there has been a resurgence of interest in the use of freeze drying and low temperature embedding techniques, so a discussion of the problems encountered is appropriate here.

One problem is that of specimen preparation. When specimens are freeze dried (or freeze substituted) and embedded before analysis the possibility arises that redistribution of elements might occur during the preparation steps. This aspect remains controversial: some authors maintain that diffusible elements remain at their subcellular sites throughout drying and embedding (Elder *et al.*, 1988; Wroblewski & Wroblewski, 1986), or freeze substitution and embedding (Edelmann, 1991), whereas others claim that movement of elements occurs (Roos & Barnard, 1985). Clearly, when using these techniques some attempt should be made to monitor whether re-distribution occurs. This can be achieved simply by estimating the K/Na ratio, which should be high, and comparing it with values obtained from other sources.

The second problem is encountered when trying to interpret the biological significance of data obtained, and trying to compare the results with data obtained from other techniques. When continuum normalisation is applied to frozen–dried sections the results are obtained in terms of mmoles/kg of dried specimen, which is relevant to known cell physiology, whereas when the specimen is embedded the results are in terms of mmoles/kg of specimen plus resin, which has dubious biological significance. To be able to evaluate the results from resin-embedded tissue, some account needs to be taken of the contribution of the resin to the spectrum. This problem was recognised by earlier workers (see, e.g., Ingram & Ingram, 1986; Nicaise *et al.*, 1989) and different approaches were used to cope with it. Both groups made the assumption that specimen water is replaced exactly by the resin. Nicaise *et al.* (1989) used the known water content of the tissue, and the mass of the polymerised resin per unit volume to estimate the weight of resin present in the analysed area. Ingram & Ingram (1986) embedded the freeze dried tissue in a resin (Epon 826) which had been doped with dibromoacetophenone (this resin will give a characteristic signal when analysed). Wet weight concentrations were measured, and the original water content was determined by comparing the bromine signal in the resin-embedded specimen with that obtained from resin-embedded albumin standards of known water content.

Recently, the problem of quantitation of embedded material has been reappraised (Hall, 1989a; Hall, 1991). Hall points out that it is highly unlikely that resin would totally replace specimen water, especially in tissues which have been freeze dried before embedding, because shrinkage of the tissue during freeze drying would reduce the space available for the resin. In addition, full penetration of the resin is unlikely to occur due to barriers in the tissue. Quantification in resin embedded tissue is also complicated by loss of mass from the resin. Hall (1991) points out that the continuum signal from the resin can be considered to be an extraneous contribution to the total continuum signal, and that if its contribution is subtracted, along with other extraneous contributions, then a true dry weight concentration will be obtained. He proposes that the contribution of the resin can be estimated provided that a tagged resin which emits a characteristic signal (such as Epon doped with dibromoacetophenone suggested by Ingram & Ingram) is used. The ratio of continuum to characteristic radiation can then be obtained from analysis of the resin alone. This ratio can then be used to determine the contribution of the resin to the continuum when the embedded specimen is analysed. This method is identical to that used for determining the contribution of other extraneous sources of radiation to the total continuum reading as outlined by Shen in this volume. The results obtained by such a method would be unaffected by specimen shrinkage, since both the characteristic and continuum signal are derived solely from the specimen in the area of analysis. Hall (1991) points out that if this approach is used, the relevant standard would not be one based on embedding resin; since the continuum reading of interest is generated solely by the mass of specimen, a protein based standard would be more appropriate.

The peripheral standard method can also be used to obtain quantitative results from the analysis of resin embedded tissue (Hall, 1991) in terms of mmol/unit volume. In this case the resin itself can serve as the peripheral standard, using either an element which occurs naturally in the resin or a deliberately added tag element as the element for standardisation. The advantages of the peripheral standard method are, as before, the lack of requirement for the estimation of the continuum, and the fact that loss of mass from the matrix does not pose a problem. However, there is a disadvantage in the measurement of element concentrations per unit volume when freeze dried and embedded sections are studied. Specimens are known to shrink during freeze drying, causing a reduction in their initial volume. Since the amount of elements present is not changed, there is a consequent increase in concentration of any element per unit volume after freeze drying. This change in concentration occurs before the specimen is embedded and sectioned, that is before the section is exposed to the external standard. Since there is no further shrinkage during embedding and sectioning the specimen and standard are not exposed to the same degree of shrinkage during specimen preparation. Consequently the volume concentrations obtained from analysis of the freeze dried embedded sections are likely to be in error.

The procedures outlined by Hall (1991) for quantitation of analytical data obtained from freeze dried and embedded specimens offer advantages over the methods used by either Ingram & Ingram (1986) or Nicaise *et al.* (1989), since they do not depend on any assumptions about the extent of replacement of water by the embedding medium. In addition, if continuum normalisation is used in conjunction with a tagged resin it should be possible to carry out the necessary calculations without the need to modify commercially available software.

F   Conclusions

Quantitative X-ray microanalysis is now a well established technique for the study of elemental concentrations in biological specimens, and many of the problems and pitfalls which may be encountered have been documented. This chapter was not meant to convey the impression that the method is exceptionally difficult but to show that, as with any technique, the results obtained should be evaluated carefully. Provided that such critical appraisal is carried out there can be no doubt that X-ray microanalysis will help in our understanding of tissue, cell and subcellular physiology.

Acknowledgements

I would like to thank Dr T.A. Hall for reading the manuscript critically and making helpful comments. I would also like to thank The Garfield Weston Foundation for financial support.

References

Dorge, A., Gehring, K., Nagel, W. & Thurau, K. (1974). Intracellular Na–K concentration of frog skin at different states of Na-transport. In *Microprobe analysis as applied to cells and tissues*, ed. T.A. Hall, P. Echlin & R. Kaufmann, pp. 337–49. London: Academic Press.

Edelmann, L. (1991). Freeze-substitution and the preservation of diffusible ions. *Journal of Microscopy*, **161**, 217–28.

Elder, H.Y., Bovell, D.L., Pediani, J.D., Wilson, S.M., McWilliams, S.A. & Jenkinson, D.M. (1988). On the validity of block freeze drying and vacuum resin embedding of cryoquenched tissues for quantitative intracellular X-ray microanalysis. *Institute of Physics Conference Series*, Number 93, (EUREM 88) 3, pp. 575–6.

Gupta, B.L. & Hall, T.A. (1979). Quantitative electron probe X-ray microanalysis of electrolyte elements within epithelial tissue compartments. *Federation Proceedings*, **38**, 144–53.

Hall, T.A. (1979). Problems of the continuum-normalisation method for the quantitative analysis of sections of soft tissue. In *Microbeam analysis in biology*, ed. C.P. Lechene & R. Warner, pp. 185–208. New York: Academic Press.

(1989a). Quantitative electron probe X-ray microanalysis in biology. *Scanning Microscopy*, **3**, 461–6.

(1989b). The history of electron microprobe analysis in biology. In *Electron probe microanalysis in biology. Applications in biology and medicine*, ed. K. Zierold & H.K. Hagler, Springer Series in Biophysics 4, pp. 3–15. Berlin: Springer-Verlag.

(1991). Suggestions for the quantitative X-ray microanalysis of thin sections of frozen–dried and embedded biological tissues. *Journal of Microscopy*, **164**, 67–79.

Hall, T.A. & Gupta, B.L. (1982). Quantification for the X-ray microanalysis of cryosections. *Journal of Microscopy*, **126**, 333–45.

(1986). EDS quantitation and applications to biology. In *Principles of analytical electron microscopy*, ed. D.C. Joy, A.D. Romig Jr. & J.I. Goldstein, pp. 219–48, New York: Plenum Press.

Ingram, M.J. & Ingram, F.D. (1986). Cell volume regulation studies with the electron microprobe. In *The science of biological specimen preparation for microscopy and microanalysis 1985*, ed. M. Muller, R.P. Becker, A. Boyde & J.J. Wolosewick, pp. 43–9. Chicago: Scanning Microscopy.

James-Kracke, M.R., Sloane, B.F., Shuman, H., Karp, R. & Somlyo, A.P. (1980). Electron probe analysis of cultured vascular smooth muscle. *Journal of Cell Physiology*, **103**, 313–22.

Morgan, A.J. (1985). X-ray microanalysis in electron microscopy for biologists. *Royal Microscopical Society Handbook 5*. Oxford: Oxford University Press.

Nicaise, G., Gillot, I., Julliard, A.K., Keicher, E., Blainue, S., Amsellem, J., Heyran, J.C., Hernandez-Nicaise, M.L., Crapa, B. & Gleyzal, C. (1989). X-ray micro-analysis of calcium containing organelles in resin embedded tissue. *Scanning Microscopy*, **3**, 199–220.

Nicholson, W.A.P., Gray, C.C., Chapman, J.N. & Robertson, B.W. (1982). Optimising thin film X-ray spectra for quantitative analysis. *Journal of Microscopy*, **125**, 25–40.

Rick, R., Dorge, A., Bauer, R., Gehring, K. & Thurau, K. (1979). Quantification of electrolytes in freeze dried cryosections by electron microprobe analysis. *Scanning Electron Microscopy*, **1979/II**, 619–26.

Roomans, G.M. (1988). The correction for extraneous background in quantitative X-ray microanalysis of biological thin sections: some practical aspects. *Scanning Microscopy*, **2**, 311–18.

Roos, N. & Barnard, T. (1985). A comparison of subcellular element concentrations in frozen–dried, plastic embedded, dry cut sections and frozen–dried cryosections. *Ultramicroscopy*, **17**, 335–44.

Shuman, H., Somlyo, A.V. & Somlyo, A.P. (1976). Quantitative electron probe microanalysis of biological thin sections: methods and validity. *Ultramicroscopy*, **1**, 317–39.

Steinbrecht, R.A. & Zierold, K. (1989). Electron probe X-ray microanalysis in the silkworm moth antennae – Problems with quantification in ultrathin sections. In *Electron probe microanalysis in biology. Applications in biology and medicine*, ed. K. Zierold, & H.K. Hagler, Springer Series in Biophysics 4, pp. 87–97. Berlin: Springer-Verlag.

Warley, A. (1990). Standards for the application of X-ray microanalysis to biological specimens. *Journal of Microscopy*, **157**, 135–47.

Warley, A. & Gupta, B.L. (1991). Quantitative biological X-ray microanalysis. In *Electron microscopy in biology: a practical approach*, ed. J.R. Harris, pp. 243–81. Oxford: IRL Press.

Warner, R.R., Myers, M.C. & Taylor, D.A. (1985). Inaccuracies with the Hall technique due to continuum variation in the electron microscope. *Journal of Microscopy*, **138**, 48–52.

Wroblewski, J., Roomans, G.M., Madsen, K. & Friberg, U. (1983). X-ray microanalysis of cultured chondrocytes. *Scanning Electron Microscopy*, **1983/II**, 777–84.

Wroblewski, J. & Wroblewski, R. (1986). Why low temperature embedding for X-ray microanalytical investigations? A comparison of recently used preparation methods. *Journal of Microscopy*, **142**, 351–62.

Zglinicki, T. von. (1991). The measurement of water distribution in frozen specimens. *Journal of Microscopy*, **161**, 149–58.

Zierold, K. (1986). The determination of wet weight concentrations of elements in freeze dried cryosections from biological cells. *Scanning Electron Microscopy*, **1986/II**, 713–24.

SECTION B
ASSOCIATED TECHNIQUES

Electron probe X-ray microanalysis (EPXRMA) is only one of a large number of techniques that are available for investigating the chemical composition of cells and tissues. These techniques use a wide variety of physical and chemical principles, and each has its own advantages and disadvantages (see Table B1). Some are non-microscopical and, at best, give information only at a rather coarse level of spatial resolution, although their chemical sensitivity may be very high. Among the microscopical techniques, many should be regarded as highly specialised, in the sense that the equipment is not easily available, and in some cases is not produced commercially. Other methods, such as histochemistry and autoradiography, may be regarded as routine, but do not, on the other hand, yield the same type of information as XRMA and other techniques. Indeed, while all the techniques listed in Table B1 can be used to study the chemical composition of cells or tissues, it seems that they are complementary rather than competing. For example, histochemical procedures can identify compounds, such as enzymes, which are not generally accessible by the other methods. As well as autoradiography, techniques such as SIMS (Thellier, Ripoll & Berry, 1991) and LAMMA, which can identify isotopes, are well adapted for the study of dynamic processes using tracers. Of the techniques considered in the following two chapters, particle-induced X-ray emission (PIXE) is most similar to EPXRMA, yet nevertheless shows important differences in specimen penetration, resolution and sensitivity. Unfortunately, PIXE equipment remains, so far, the preserve of a few laboratories who have the expertise and money to build and maintain it. Electron energy loss spectroscopy (EELS) (Leapman & Ornberg, 1988) is, on the other hand, commercially available at a comparatively modest price; spatial resolution and analytical sensitivity are both very high, but quantification is difficult, and elemental analysis is restricted to very thin specimens. It is clear, therefore, that considerations of cost and availability apart, the techniques listed in Table B1 must not be thought of as rivals, but as complementary, the most appropriate one depending on the problem being investigated.

References

Leapman, R.D. & Ornberg, R.L. (1988). Quantitative electron energy loss spectroscopy in biology. *Ultramicroscopy*, **24**, 251–68.

59

Table B1. *Comparison of biological microanalytical techniques*

Technique	Analytical sensitivity	Elemental range	Spatial resolution	Information given	Quantification	Availability	Comments
Microscopical							
X-ray microanalysis (XRMA)	Moderate	$Z > 4$	Subcellular	Elemental	Easy	Widely available	Thick or thin specimens
Particle-induced X-ray emission (PIXE)	High	$Z > 4$	Subcellular	Elemental	Easy (?)	Very specialised	Thick or thin specimens
Electron energy loss spectroscopy (EELS)	High	All	Ultrastructural	Elemental	Difficult	Very specialised	Ultrathin specimens only
Secondary ion mass spectrometry (SIMS)	High	All	Subcellular	Elemental/isotopic	Difficult	Very specialised	Surface analysis – destructive
X-ray fluorescence	High	$Z > 4$	Histological	Elemental	?	Very specialised	
X-ray absorption			Ultrastructural	Elemental	?	Very specialised	
LM histochemistry			Subcellular	Molecular	Generally not simple	Routine	
EM histochemistry			Ultrastructural	Molecular	Difficult	Specialised	
Autoradiography		All	Subcellular	Molecular/elemental	Fairly easy	Routine	
Laser microprobe mass analysis (LAMMA)	High	All	Subcellular	Elemental/isotopic/molecular	Difficult	Very specialised	Destructive
Non-microscopical							
Ion-selective electrodes	? Moderate		Cellular	Free ions	Easy	Specialised	
Atomic absorption spectroscopy	High	Most metals	Histological		Easy	Widely available	Destructive

Modified from Sumner (1988).

Sumner, A.T. (1983). X-ray microanalysis: a histochemical tool for elemental analysis. *Histochemical Journal*, **14**, 501–41.
Thellier, M., Ripoll, C. & Berry, J.-P. (1991). Biological applications of secondary ion mass spectrometry. *Microscopy and Analysis* no. 23, May 1991, pp. 13–15.

5 Proton probe microanalysis in biology: general principles and applications for the study of Alzheimer's Disease

F. Watt, J.P. Landsberg, G.W. Grime and B. McDonald

A Introduction

The use of a focused beam of high energy (MeV) protons to image and analyse thin samples offers many unique advantages over existing techniques. As shown in Fig. 5.1, the interaction of the proton beam with matter can take many forms. Non-nuclear reaction products such as characteristic X-rays, secondary electrons, backscattered protons and nuclear reaction products such as gamma rays and nuclear particles, can provide a wealth of analytical information on the sample under investigation. If the sample is thin enough to allow the protons to pass through, then the subsequent detection of the transmitted protons allows structural information to be obtained.

Described here is an example of the use of three of the techniques associated with the proton beam interaction, as applied to Alzheimer's Disease tissue. These techniques, utilising the detection of characteristic X-rays, back-scattered protons and transmitted protons, can be simultaneously applied to form a powerful array of nuclear particle based techniques which we have called nuclear microscopy.

B Nuclear microscopy: techniques for analysis and features of the scanning proton microprobe

1 Proton induced X-ray emission (PIXE)

When a particle collides with an atom, the probability of knocking out an inner core electron is optimised when the velocity of the incoming particle matches the velocity of the inner core electron. This occurs at keV energies when electrons are used as the impinging particle, and at MeV energies when protons are used. Following the removal of an inner core electron, the newly created vacancy in the inner shell is quickly filled by an electron falling from a higher shell. The energy made available by this re-arrangement is either emitted as a photon (X-ray emission) or used to eject an electron (Auger emission), and is characteristic of the parent atom. It is the measurement of

62

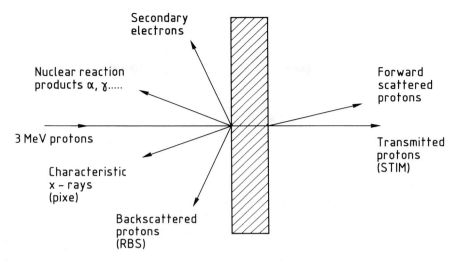

Fig. 5.1. High energy ion beam interactions with matter.

the energy of this characteristic X-ray radiation which forms the basis of X-ray microanalysis.

Electron induced X-ray emission (or electron probe X-ray microanalysis), by far the most common and well used technique, suffers from relatively high background levels (bremsstrahlung) caused by the rapid change in momentum of the incident electrons as they undergo collisions with the electrons in the target atoms. In proton induced X-ray emission, however, the incident proton with a mass of around 1800 times that of the electron does not suffer the same degree of deceleration during the collision process, and consequently the bremsstrahlung background caused by the incident proton beam is very low. Fig. 5.2 shows a study carried out in Heidelberg on a pollen grain using the two techniques, and shows that the background associated with the proton beam is 100 to 1000 times less than that observed by using electrons. A large body of work has shown that with PIXE, parts per million levels of analytical sensitivity can be achieved for elements from Na and above in the periodic table, particularly on biological specimens (see, for example, Johansson & Campbell, 1988, or Watt & Grime, 1987).

A second important advantage of utilising high energy protons rather than electrons is that the range in matter is greater. Fig. 5.3 shows a graph of proton range against proton energy, and shows that for a typical proton energy of 3 MeV, its range in biological tissue is about 100 μm. A further advantage is that as the proton beam traverses the sample the beam does not broaden appreciably, since the proton beam does not suffer large angle scattering from electron collisions. As an example, a beam of 3 MeV protons

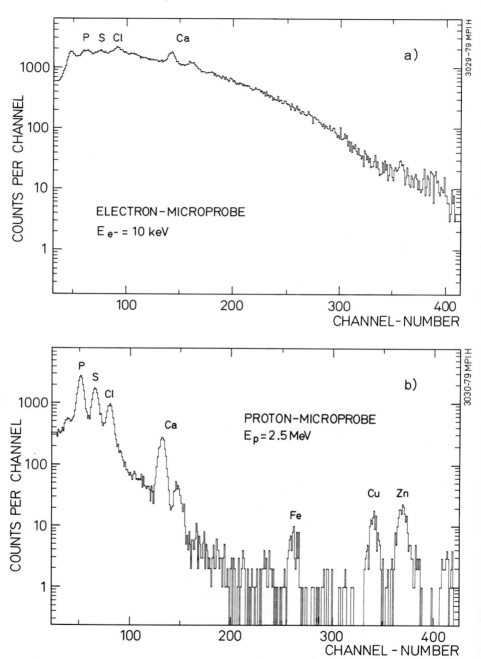

Fig. 5.2. For legend see facing page.

passing through a carbon matrix 4 μm thick will broaden only by 50 nm (Watt & Grime, 1987).

2 Rutherford backscattering spectrometry (RBS)

As the high energy proton traverses the specimen inducing characteristic X-rays, there is a smaller probability that a nuclear collision will also occur. If the proton penetrates the nucleus then a nuclear reaction may occur, with a corresponding release of a reaction product such as a gamma ray or an alpha particle etc. Frequently, however, the nuclear collision is elastic and the dynamics of the collision is based on the conservation of momentum. In this case the energy lost by the incident particle is dependent on the mass of the target nucleus. By accurately measuring the energy of a proton which has been elastically scattered back out of the sample, the target atom can be identified. Energy is also lost by the proton as it travels through the sample, and if scattered at a backward angle will lose further energy as it is reflected back out of the sample. In principle therefore not only can the target atom be identified but also the depth at which the interaction took place. This is the basis of the technique of Rutherford backscattering spectrometry (RBS) (Chu, Mayer & Nicolet, 1978). Although RBS is a technique which has found wide usage in materials research for depth profiling, its use in biological microanalysis is not so widespread. This technique when used on thin biological specimens complements PIXE in that elemental analysis of the light elements can be carried out, and in particular C, N and O profiles can be obtained.

3 Scanning transmission ion microscopy (STIM)

If the specimen under investigation is sufficiently thin (see Fig. 5.3), then the proton beam will pass through the sample. The energy loss of each proton in the transmitted beam measured at 0° to the initial beam direction will be dependent on the number of electron collisions made and therefore the transmitted proton beam contains information on the electron density within the sample.

By displaying the average energy loss of the protons at each pixel in the STIM image on a grey scale where darker colours represent higher energy loss and therefore higher density, a density map (mass per unit area) can be assembled. By imaging transmitted protons which have suffered a specific energy loss, then effective density contour maps (mass per unit area) of the sample can be constructed. This technique, analogous to STEM in electron

---

Fig. 5.2. X-ray energy spectrum of a pollen tube (CTC treated, fixed with glutaraldehyde and air dried) taken with (a) an electron probe operating at an energy of 10 keV, and (b) the Heidelberg proton microprobe at an excitation of 2.5 MeV. Reproduced with the kind permission of K. Traxel.

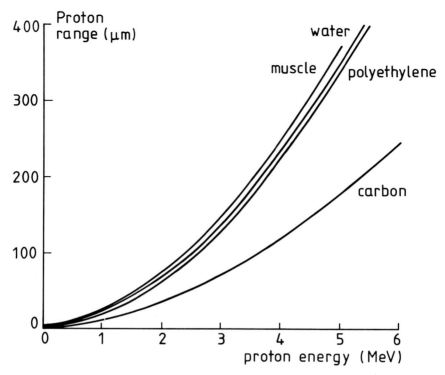

Fig. 5.3.   Proton range versus energy for a variety of target materials (Watt & Grime, 1987).

microscopy, has only been utilised in a few laboratories throughout the world (Overley *et al.* 1983; Sealock, Mazzolini & Legge, 1983), and has the potential for imaging the structure of relatively thick biological samples at high resolution (Bench & Legge 1989).

If the transmitted protons are detected off axis (dark field/scattered beam) then similar images can be produced, although the energy loss mechanism is determined by a more complex mixture of electron collisions and small angle nuclear elastic scattering.

The use of STIM is not limited to high energy protons. The use of other energetic nuclear particles, e.g. alpha particles (or helium ions), enables greater structural depth resolution to be attained as a consequence of the greater energy loss suffered by these heavier ions as they pass through matter.

4   Specimen preparation

The sample preparation procedures necessary for analysis using nuclear microscopical techniques are as stringent as those required for analysis using the electron probe, i.e. we must ensure that the sample retains its physical structure and elemental make-up as far as possible. It is clear that the nuclear microscopist must adopt the techniques extensively developed by the electron probe analyst, and adapt the specimen preparation to suit the problem. For example, for the analysis of mobile elements such as Ca, Na and K, flash freezing followed by cryomicrotome sectioning is necessary, whereas for more stable elements such as S and P, freeze drying may be satisfactory.

A minor difference in specimen preparation (in relation to thickness) arises as a consequence of the physical properties of the proton beam, which because of its higher momentum penetrates further into the sample compared with electrons. In addition, the proton beam tends to maintain its shape, and therefore its spatial resolution, as it traverses the specimen. Criteria for determining the maximal acceptable thickness of sample for transmission analysis, which for high resolution electron probe studies is determined by the amount of beam broadening which can be tolerated within the sample, can now be relaxed, and samples of 10 µm or more can be scanned at the sub-micron resolutions currently available with the nuclear microscope.

5 The nuclear microscope

The potential of the nuclear microscope (or the scanning proton microprobe) is in the utilisation of fast-moving massive particles. Unfortunately this makes the technology of the nuclear microscope more complex than that of the ubiquitous electron microscope. The production of a beam of nuclear particles (protons, deuterons, alpha particles etc.) at MeV energies requires a small nuclear accelerator with a high brightness ion source, and although these can be purchased commercially they require a high level of expertise for operation and maintenance. The focusing system required for producing a finely focused beam of protons is also complex, since we are unable to make use of conventional electron microscope technology. Most nuclear probe installations throughout the world use a system of strong focusing magnetic quadrupole lenses (Grime & Watt, 1984) which again require a high level of expertise and are more difficult to use than electron focusing devices. Since as yet there is not a full nuclear microscope package commercially available, it is not surprising that the nuclear microscope tends to be limited to physics or materials science laboratories within Universities and Government Establishments.

The Oxford Nuclear Microscope (scanning proton microprobe) is based around a 1.7 MV tandem Van de Graaff accelerator (NEC 5SDH2) capable of producing a beam of 3 MeV protons (Grime *et al.*, 1991). A triplet of magnetic quadrupole lenses focuses this beam of 3 MeV protons down to a spot size of 300 nm with sufficient beam intensity (100 pA) to carry out PIXE, RBS and dark field (off axis) STIM. Spatial resolutions are estimated to be below 100 nm using bright field (0°) STIM as a consequence of the reduction

Fig. 5.4.   Photograph of the Oxford Nuclear Microscope focusing system and
target chamber. The proton beam enters the system through the beam
tube passing through the wall, is focused by the three cylindrical
magnetic quadrupole lenses, and scans across the sample positioned at
the centre of the target chamber. The sample can be observed by optical
microscopes for positioning purposes and the X-ray detector is situated
out of view on the far side of the chamber.

in beam current used in the analysis. The target chamber used in the analyses
(see Fig. 5.4), has a Link Systems Si(Li) 80 mm² X-ray detector positioned at
45 degrees to the sample (PIXE), with silicon surface barrier particle detectors
at backward angles (RBS) and forward angles (STIM). The STIM detector is
moveable to allow either bright field or dark field STIM to be carried out. The
focusing system and target chamber is shown in Fig. 5.4. The beam is scanned
over the sample using pre-lens magnetic coils and scan sizes of up to
2.5 mm × 2.5 mm square can be achieved.

The data acquisition system is based around a sophisticated computer
interface (CAMAC-ADC, VME bus incorporating a colour display and a
68020 microprocessor). When a signal from the X-ray, STIM or RBS
detectors is received by the ADC interface the spatial coordinates of the
instantaneous beam position are also registered. These incoming data are

then sorted on-line into as many as 40 256 × 256 pixel maps, which can be monitored during the experiment by a dedicated microVAX workstation. In this way both elemental (PIXE and RBS) and density contour (STIM) maps can be collected and monitored simultaneously. The beam can also be positioned at particular points of interest within the scan, and elemental analyses carried out on-line using PIXE analysis and RBS simulation software. Typical scans take of the order of one half to one hour to accumulate, with point analyses taking about 5 to 15 minutes.

C   The use of nuclear microscopy in Alzheimer's Disease

1   General features of the disease

Alzheimer's Disease (AD) is a neurodegenerative disease that affects a significant proportion of the population over the age of 60. Although the cause of the disease is still unknown there is research that suggests that increased levels of aluminium and silicon may be associated with the disease.

The senile plaques and neurofibrillary tangles which form the major pathology of AD have been the focus of much attention, in particular the trace metal content of these features. Many of the groups working in this field, however, have published conflicting results. Nikaido *et al.* (1972), using electron probe microanalysis, observed an increase in the Si content of the rims and cores of senile plaques. Duckett & Galle (1976) also using an electron probe reported possible increases in Al in some plaque cores, whereas more recent LAMMA (laser microprobe mass analysis) investigations have revealed aluminium in the neurofibrillary tangles (Perl & Good, 1988), but not in the senile plaque (Stern *et al.* 1986). Candy *et al.* (1985), using energy dispersive X-ray microanalysis (EDX) and solid state NMR (nuclear magnetic resonance), claimed that most of the senile plaques they investigated contained focal deposits of alumino-silicates, whereas more recent electron probe work (Jacobs *et al.* 1989) failed to demonstrate any aluminium in the plaque cores at a claimed sensitivity of 20–25 ppm.

We have recently analysed the cores of many plaques with the Oxford nuclear probe using the techniques of PIXE and RBS (Landsberg *et al.*, 1991) and our results are also inconclusive. Although aluminium and silicon were observed in approximately 20 % of senile plaques at a level of 50 ppm or greater, the results were complicated by problems of contamination by alumino-silicates, which are universally present as airborne contamination. The situation was complicated even further by the observed presence of this type of contamination in high purity aqueous and organic reagents used for staining the tissue.

AD plaques, largely composed of β/A4-amyloid protein, are traditionally

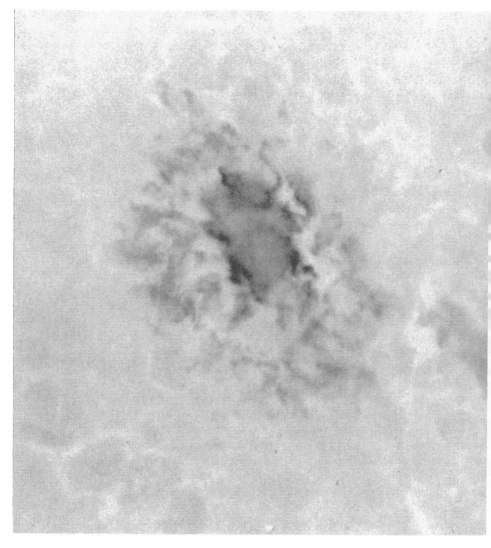

Fig. 5.5.   Photomicrograph of an immunohistochemically stained senile plaque
(see text) showing the characteristic dense core and a more diffuse halo.
The size of the plaque (including halo) is 50 to 80 μm.

located and identified in tissue using a suitable staining technique. Fig. 5.5
is a photomicrograph of a typical plaque, immunohistochemically stained
using antisera raised against complement component C3d, showing the
characteristic densely stained core with the lighter neuritic halo. Identification

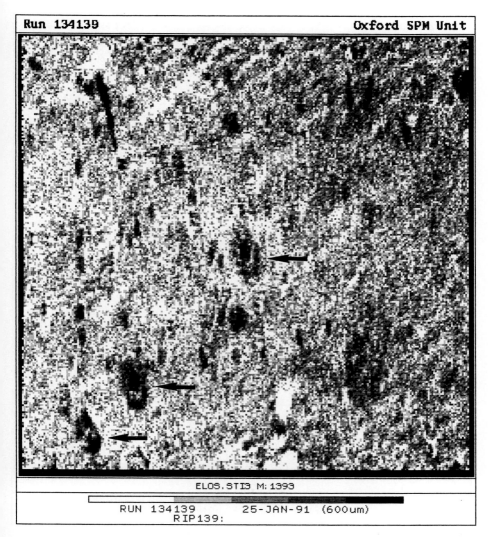

ELOS.STI3 M: 1393

RUN 134139   25-JAN-91 (600um)
RIP139:

Fig. 5.6.   STIM map of a 600 × 600 μm scan over stained tissue showing three well-
defined plaques (see arrows). The STIM map has been processed to
display the mean energy loss at each pixel and therefore represents the
density distribution (mass per unit area) in the sample. Black represents
high energy loss (i.e. high density) and lighter colours represent lower
energy loss (i.e. lower densities).

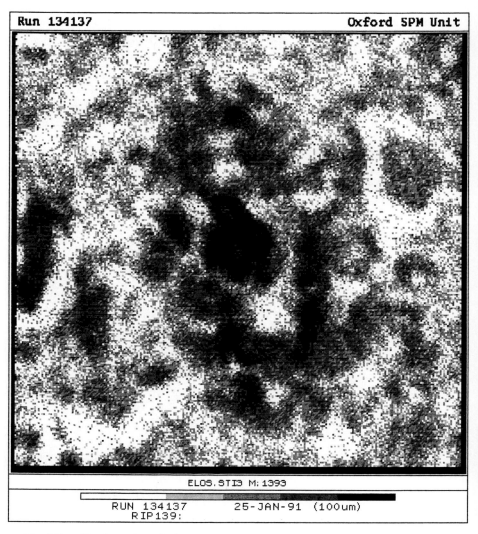

Fig. 5.7a.   For legend see facing page.

and analysis of senile plaques without recourse to chemical staining would be a significant step forward in the microanalysis of such features.

2   *Identification of senile plaques using STIM*

Identification of plaques using STIM relies on the β/A4-amyloid protein which makes up the bulk of the plaque having a greater density than the

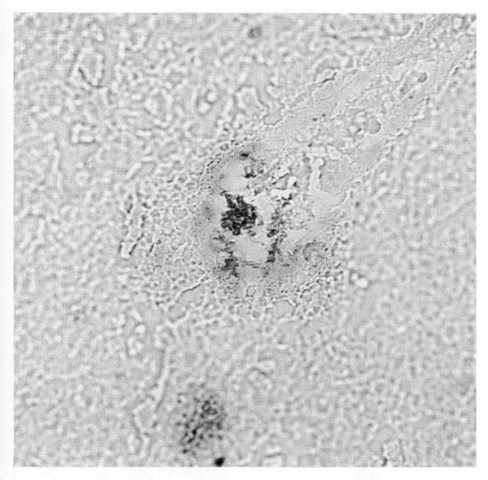

Fig. 5.7b. (a) Higher resolution STIM map (100 × 100 μm) of the stained plaque at
the centre of the image shown in Fig. 5.6 processed to represent the
density distribution and (b) corresponding photomicrograph of the same
region.

surrounding tissue. Evidence for this was obtained from our previous studies
using RBS, where increased C and N levels were observed in the stained
plaque (Landsberg *et al.*, 1991). As a first step to test whether STIM can be
used to identify AD features, stained tissue with a known distribution of
plaques was scanned.

Blocks of tissue were taken at autopsy from the hippocampus and temporal
cortex (regions of the brain most affected by AD). The blocks were mounted
on cork discs and snap frozen in liquid nitrogen before storage at −70 °C.

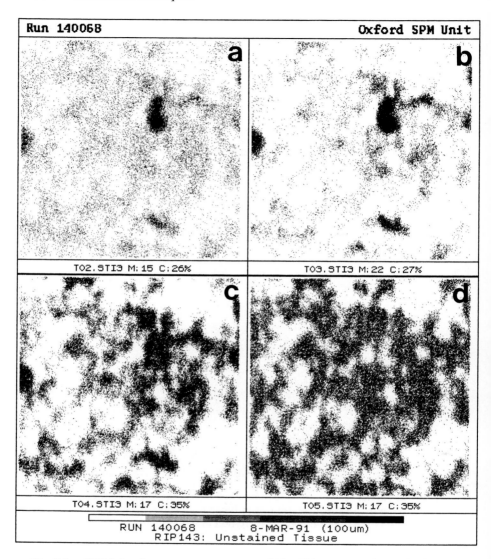

Fig. 5.8.   STIM density contour maps over a 100 × 100 μm area of unstained AD
tissue. The maps represent (a) high energy loss, (b) medium–high energy
loss, (c) medium–low energy loss, and (d) low energy loss.

Sections of 30 μm thickness were cut using a cryostat microtome, picked up
onto pioloform coated slides and stained using antisera raised against
complement component C3d. Polyclonal rabbit C3d antibody (Dakopatts) at
a dilution of 1 in 100 in 50 mM tris–HCl pH 7.4 containing 200 mM NaCl

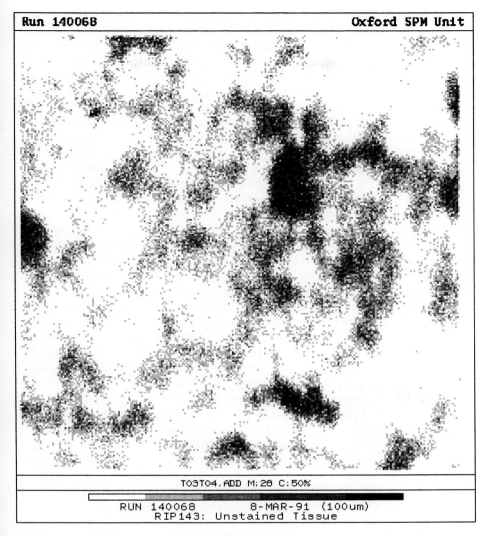

Fig. 5.9. STIM density contour map assembled by using an energy loss window corresponding to the sum of the energy windows used in Figs 5.8b and 5.8c.

was incubated for 60 minutes at room temperature with the sections, followed by three 5 minute washes in tris buffer. The sections were then overlaid with peroxidase conjugated goat anti-rabbit antiserum (Dakopatts) at a dilution of 1 in 50 in tris buffer for 60 minutes at room temperature, washed three times with tris buffer and the bound antibody complex visualised using diaminobenzidene (DAB, Sigma, 60 mg/100 ml in tris containing 40 µl 30 % vol. hydrogen peroxide). The substrate was applied to the sections and incubated for 3–5 minutes resulting in a brown reaction product. The reaction was terminated by washing the sections in milli-Q water. The pioloform membrane was floated off the slides by immersion in milli-Q water and then laid down across the 8 mm diameter hole in the nuclear microscope target holders.

A 600 µm square region of stained tissue containing several well defined plaques with cores, as well as several diffuse plaques, was scanned by the nuclear microscope. Results of the STIM scan are shown in Fig. 5.6, where at least three well defined plaques can be observed. A second scan of higher magnification (Fig. 5.7a) was carried out over the plaque located centrally in Fig. 5.6. This STIM image reproduces extremely well the shape of the senile plaque, a photomicrograph of which is shown in Fig. 5.7b.

STIM was then applied to a section of unstained tissue, known from optical observations on stained serial sections to contain a small quantity of well defined senile plaques. A 30 µm thick section was cut from a tissue block as described above, and fixed with acetone vapour for 10 minutes before mounting across the target holder. A sequence of adjacent scans was carried out over the unstained tissue in order to search for possible plaque type structures, and STIM density contour maps were produced for each scan. Fig. 5.8 shows the results of one particular scan, where four density contour maps are shown. These maps correspond to areas of high energy loss (high density) in Fig. 5.8a, to low energy loss (low density) in Fig. 5.8d. A convincing interpretation of these maps is that Fig. 5.8a shows up the plaque core, Fig. 5.8b shows up the plaque core with a trace of corona, with Figs 5.8c and 5.8d emphasising the plaque corona and the surrounding tissue. By choosing a suitable energy loss window (in this case by summing the energy windows corresponding to Figs 5.8b and 5.8c), the characteristic shape of a senile plaque can be identified (Fig. 5.9). It is perhaps worth mentioning at this stage that the results above were produced using off-axis STIM, and that bright field STIM yielded very similar results. In addition, although the STIM density maps obtained with stained tissue showed sufficient contrast to enable the plaques to be readily identified, in the case of the plaque imaged in the unstained tissue it was the density contour maps which produced the optimum contrast. It is not expected that this was a consequence of the staining procedure.

3   Analysis of senile plaques using PIXE and RBS

If off-axis STIM is used in the identification of the senile plaque, and sufficient beam current (100 pA) is used in this identification procedure, then PIXE and RBS elemental mapping can be carried out simultaneously. Fig. 5.10 shows some of the results of the elemental mapping aspect of nuclear microscopy,

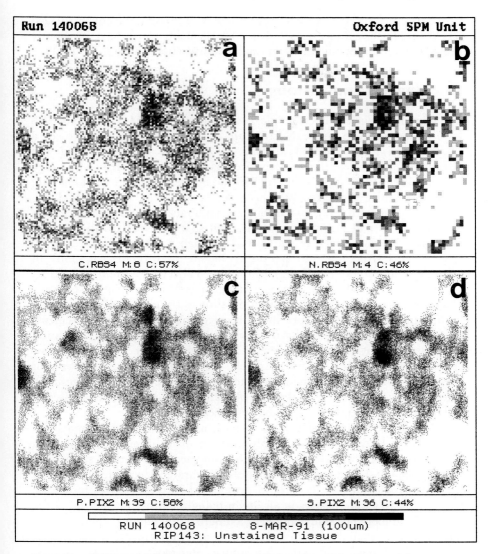

Fig. 5.10. PIXE and RBS elemental maps of (a) carbon, (b) nitrogen, (c) phosphorus, and (d) sulphur. The scanned region is the same as that depicted by the STIM image in Fig. 5.9.

with maps of C, N, P and S correlating well with the simultaneously imaged STIM profile shown in Fig. 5.9. These are elements which have previously been shown to be characteristic of senile plaques (Landsberg *et al.*, 1991).

D   General conclusions

The combined nuclear microscopical techniques of STIM, PIXE and RBS have the potential to image the structure of unstained tissue and provide a simultaneous multielemental analysis of any part of the sample. The advantages of nuclear microscopy are many fold, and include (*a*) high analytical sensitivity (ppm levels for Na and above in the periodic table), (*b*) high quantitative accuracy (10% accuracy is considered routine, and 5% can be achieved with care), (*c*) the ability to simultaneously measure the major elements C, N and O, (*d*) a relatively high spatial resolution for analysis (currently 300 nm) which is maintained through thick specimens, and (*e*) simultaneous structural information at 300 nm or with the capability of higher spatial resolutions (50 nm) using direct STIM (low current) operation.

We have shown that this can be a valuable asset in the analysis of senile plaques in Alzheimer's Disease. Determination of the elemental composition of the senile plaque core has previously been complicated by the requirements of chemical staining for identification, which leads to possible contamination of the sample by alumino-silicates. The use of nuclear microscopy now enables senile plaques to be positively identified and analysed without chemical staining, thereby enabling us to provide a more conclusive result in the controversy surrounding the role of aluminium in Alzheimer's Disease. This work is in progress and will be published elsewhere.

The nuclear microscope is in its infancy, and further developments will undoubtedly lead to a major involvement in many disciplines. One such development area, which may benefit the biological and medical sciences, is that of the high spatial resolution analysis of hydrated specimens. Because of the high momentum of the proton beam used in the investigations, it is feasible to carry out analyses in air, by bringing the proton beam out of the vacuum chamber via a thin window or a differentially pumped aperture. The quest for better spatial resolutions is a matter of improvements in technology, through developments in lenses, ion sources and more stable accelerators. This is in progress and the next few years should see an improvement in spatial resolutions to 100 nm for analysis and 10 nm for STIM.

Acknowledgements

The authors wish to acknowledge the support of the Wellcome Trust in funding this programme, and to Jean Roberts for the expert preparation of the samples.

References

Bench, G.S. & Legge, G.J.F. (1989). High resolution STIM. *Nucl. Instr. and Meths*, **B40/41**, 655–8.

Candy, J.M., Edwardson, J.A., Klinowski, J., Oakley, A.E., Perry, E.K. & Perry, R.H. (1985). Co-localisation of aluminium and silicon in senile plaques: implications for the neurochemical pathology of Alzheimers disease. In *Senile dementia of the Alzheimers type*, ed. J. Traber & W.H. Gispen. Berlin: Springer-Verlag.

Chu, W.-K., Mayer, J.W. & Nicolet, M.-A. (1978). *Backscattering spectrometry*. New York: Academic.

Duckett, S. & Galle, P. (1976). Mise en evidence de l'aluminium dans les plaques de la maladie d'Alzheimer: etudie a la microsonde de Castaing. *C.R. Seances Acad. Sci.* D, **282**, 393–5.

Grime, G.W. & Watt, F.W. (1984). *Beam optics of quadrupole probe forming systems*. Bristol: Adam Hilger.

Grime, G.W., Dawson, M., Marsh, M., McArthur, I.C. & Watt, F. (1991). The Oxford submicron nuclear microscopy facility. *Nucl. Instr. and Meths*, **B54**, 52–63.

Jacobs, R.W., Duong, T., Jones, R.E., Trapp, G.A. and Scheibel, A.B. (1989). A reexamination of aluminium in Alzheimers Disease: analysis by energy dispersive X-ray Microprobe and flameless atomic absorption spectrophotometry. *Can. J. Neurol. Sci.*, **16**, 498–503.

Johansson, S.A.E. & Campbell, J.L. (1988). *PIXE: a novel technique for elemental analysis*. John Wiley and Sons.

Landsberg, J.L., McDonald, B., Roberts, J.M., Grime, G.W. & Watt, F. (1991). Identification and analysis of senile plaques using nuclear microscopy. *Nucl. Instr. and Meths* **B54**, 180–5.

Nikaido, T., Austin, J., Trueb, L. & Rivehart, R. (1972). Studies in aging in the brain II. Microchemical analysis of the nervous system in Alzheimers patients. *Arch. Neurol.*, pp. 549–54.

Overley, J.C., Connolly, R.C., Seiger, G.E., MacDonald, J.D. & Lefevre, H.W. (1983). Energy loss radiography with a scanning MeV ion probe. *Nucl. Instr. and Meths*, **218**, 43–6.

Perl, D.P. & Good, P.F. (1988). Laser Microprobe Mass Analysis evidence that aluminium selectively accumulates in the neurofibrillary tangle. *J. Neuropath. Exp. Neurol.*, **47**, 318.

Sealock, R.M., Mazzolini, A.P. & Legge, G.J.F. (1983). The use of He microbeams for light element X-ray analysis of biological tissue. *Nucl. Instr. and Meths*, **218**, 217–20.

Stern, A.J., Perl, D.P. & Munoz-Garcia, X. (1986). Investigation of silicon and aluminium content in isolated senile plaque cores by LAMMA. *J. Neuropath. Exp. Neurol.*, **45**, 361.

Watt, F. & Grime, G.W. (1987). (eds.) *Principles and applications of high energy ion microbeams*. Bristol: Adam Hilger.

6 Electron energy-loss spectroscopy and electron probe X-ray microanalysis of biological material: a comparative quantitative analysis of electron microscopical images

W.C. de Bruijn, C.W.J. Sorber, G.A.M. Trommelen-Ketelaars, J.F. Jongkind, A.L. Beckers and E.S. Gelsema

A Summary

Electron energy-loss spectroscopical (EELS) data and those acquired by electron probe X-ray microanalysis (XRMA) add chemical information to morphological structures in electron microscopical images. Recent instrumental developments lead to the need for a comparison of their various microanalytical potentials in relation to biological materials, especially their possibilities for quantitative image analysis. The aim of this chapter is to describe the present state of the art in electron energy-loss image analysis and to compare the results with those from previous studies on images of the same material with electron probe X-ray microanalysis. The acquisition of digitised electron spectroscopic images (ESI) from ultrathin sectioned cells and tissues allows the simultaneous morphometric and chemical analysis of such material. A recurring problem in the analysis is the segmentation of the images. It is shown that segmentation can be achieved by thresholding, using the first derivative of the grey-value frequency histograms constructed from such images. This method applies to electron energy-loss images as well as to images acquired by X-ray microanalysis. The influence of the image contrast and of the image averaging procedure during acquisition on the morphometrical parameters and perimeter is indicated. It is demonstrated that quantification of element-related spectra and net-intensity images is possible. One of the advantages of digitised images, to show the co-localisation of several elements within a single organelle, is illustrated. An example of element concentration determination using co-embedded Bio-standards present in the same section as the 'unknown' is given. Some of the restrictions of the morphometrical methods, and some of the problems to acquire element-concentration distribution images, are described in detail.

B   Introduction

1   *General comparison between electron energy-loss and X-ray microanalysis of biological materials*

The predominantly morphologic information obtained from electron microscopic images has in recent years increasingly been supplemented with chemical data. Multi-detector instruments have allowed simultaneously acquired signals to be related to items of interest present in the image under observation.

Two types of signals, containing either X-ray microanalytical or electron energy-loss spectroscopical data, have been obtained in the form of spectra from recognised points in the image (point analyses). The selection of these points was mostly guided by pre-existing knowledge of the biological specimen or by the operator's curiosity. However, afterwards the spectral data recorded had to be related, in one way or the other, to (the same ?) points in the acquired micrograph of the area.

In this respect, there is no difference in principle between X-ray or electron energy-loss microanalytical spectral data. Both comprise relative intensity tracings (on the abscissa), recorded versus an energy scale (on the ordinate; in eV for EELS and in KeV for XRMA spectra). In both types of spectrum a specific element-related portion (peak or edge) is riding on an unspecific part called background. Both types of spectrum have many aspects in common (like spectral resolution, peak to background ratio, peaks or edges that overlap each other) but other criteria may favour the choice of one of them.

Initially, spectra and images were recorded in an analogue way (as print and micrograph), which seriously hampered the integration of the two data streams. Gradually, computer-assisted spectral processing and off-line analogue to digital conversion of micrographs eased the path to integration (Gravenkamp *et al.*, 1982).

2   *Elemental mapping and image analysis by electron energy-loss and electron probe X-ray microanalysis*

It soon became apparent that elemental-distribution images, either acquired by the X-ray or electron energy-loss signal, were to be preferred. Not only because qualitatively, such images could register objectively elements at unpredicted sites in the images, but also as they allowed element-related pictures to be mutually compared and to be morphometrically quantitated.

In recent years, new means of image acquisition have opened the way to digital image processing. There are generally two strategies to acquire digitised images:

(*a*) pre-acquisition digitisation, in which the specimen area is covered by a matrix of spots of selectable diameter from which signals are recorded, like in a scanning electron microscope (SEM);

(*b*) post-acquisition digitisation in which a complete image is obtained, which is subsequently digitised by dividing it into a matrix of pixels.

The strategy mentioned under (*a*) is used in all sorts of scanning instruments. The second strategy has come into use since the advent of new image acquisition techniques such as TV-cameras and solid state CCD-cameras.

In multi-detector instruments and in electron microscopes with an electron spectrometer, multiple images of the same specimen, but showing different characteristics, can be acquired and subsequently related to each other by the use of such a matrix. The digital information per matrix/image point is a mean value, irrespective of the signal recorded. The acquired digital information can be stored, preserving both the matrix position and the value acquired therein. The presence of this matrix allows one not only to relate objectively images acquired sequentially or serially from the same area, but also to obtain morphometric quantification of the images.

In this chapter, we will describe some aspects of the quantitative analysis of digitised images acquired from ultrathin-sectioned biological material. A comparison will be made between the analysis of images acquired by transmission EM/electron energy-loss (TEM/EELS) and by scanning transmission EM/electron probe microanalysis (STEM/EPMA). Both methods have aspects in common which are illustrated in the following scheme.

$$
\left.\begin{array}{l}
\text{XRMA signal} \\
\text{EELS signal}
\end{array}\right\}
\left.\begin{array}{l}
\text{Image analysis} \\
\text{Signal analysis}
\end{array}\right\}
\text{Digitisation}
\left\{\begin{array}{l}
\text{Segmentation} \\
\text{Deconvolution}
\end{array}\right.
\left.\begin{array}{l}
\text{Integration} \\
\text{Co-localisation}
\end{array}\right.
$$

In the images, the items of interest have to be separated from the remainder of the cell by a process called image segmentation. Similarly in both types of spectra the element-related peak or edge information has to be separated from the unspecific background, a process called deconvolution. Once processed element- or structure-related images are acquired, the co-localisation of two or more images has to be established, a step called image integration. For all images it is assumed that morphometric quantification of the area taken by a certain element must be possible. So, additional aspects, those of reproducibility and accuracy of the measurements, have to be considered. Three aspects will be treated in particular:

1. the morphometric quantification of items of interest inside a cell, with the aim to establish their diameter, area or area fraction;
2. the conversion of energy-related spectral quantitation to the quantitation of energy-selected images;

3. the use of ultrathin-sectioned standard material to convert relative concentration images into concentration distribution images.

C   Materials and methods

Analysis was performed using images of ultrathin-sectioned cells and tissues. The items to be analysed are those that have survived a classical aldehyde fixation procedure. In some cases cytochemical contrast-differentiating reactions have been induced in order to detect enzyme activity in some of the organelles. The exact reaction conditions are given elsewhere (van Dort *et al.*, 1987, 1989). In most cases a classical osmium tetroxide postfixation was omitted and the ultrathin sections were observed unstained. Fig. 6.1 shows electron and X-ray populations around a thin section after point irradiation by a small beam of high energy (80 keV) electrons. For STEM/imaging, the annular backscattered electron detector is used, which collects the electrons emitted by each irradiated point. Subsequently, X-rays emitted from the same matrix of points are acquired with a much longer dwell time per point. In Fig. 6.1, the detectors for elastically backscattered electrons and the X-rays are combined with those for the transmitted (in)elastically forward scattered electrons on their way through the in-column spectrometer to the energy dispersive plane. Here the energy-selected populations are separated by the slit diaphragm and guided to the screen to make an image or to the photomultiplier to make a spectrum.

The instrument used for STEM/EPMA is a Philips EM 400 (Philips Nederland, Eindhoven, The Netherlands), connected to a Tracor Northern 2000 (Tracor Europe, Bilthoven, The Netherlands) for X-ray analysis. For TEM/EELS analysis, the instrument used is a Zeiss EM 902 (Zeiss, Oberkochen, Germany), connected to an IBAS 2000 (Zeiss, Kontron) and an Olivetti M280 computer. A detailed description of the system is given elsewhere (Sorber *et al.*, 1990a, b).

The in-column spectrometer and energy slit in the energy-dispersive plane allow energy spectra and energy-selected images to be collected at energies between 0 and 2000 eV. Historically, images formed by 0–50 eV electrons have been named zero-loss images, whereas images formed by electrons with an energy just below the carbon edge are called Δ 250 eV, or dark field images (Adamson-Sharpe & Ottensmeyer, 1981; Sorber *et al.*, 1990a). Element-related images are acquired at the corresponding ionisation edges. However, since the spectra and related images are composed of a complex of electron populations, the population of interest has first to be separated from the background. This enables the acquisition of net-intensity elemental spectra and/or distribution images. Electron spectroscopic images and energy-loss images are collected sequentially, using different parts of the energy band in the energy-dispersive plane.

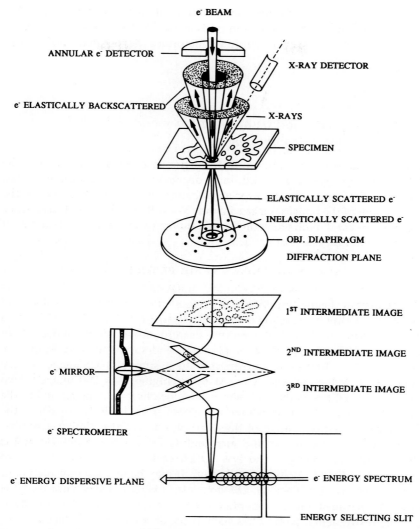

Fig. 6.1. Instrumental cross-section combining the various X-ray and electron populations in use for image and signal formation in STEM/electron probe X-ray microanalytical and TEM/electron energy-loss spectroscopical instruments.

For element quantitation, Bio-standard material containing an exactly externally determined mean concentration of a certain element is co-embedded with the tissue and present in the ultrathin section at the periphery of the tissue to be analysed. In some cases, ultrathin-sectioned Bio-standard material was also used for the optimisation of instrumental parameters. Details of the use of such Bio-standards both for X-ray and for electron energy-loss analyses can be found in the literature (Sorber *et al.*, 1991a, b).

D   Results

1   *Morphometric analysis of ultrathin sectioned cells by electron spectroscopic imaging*

As compared with a conventional EM, the presence of a spectrometer in the EM 902 gives the operator an additional degree of freedom in optimising the contrast of the images. In the process of image analysis, the problem of the segmentation of the images is a crucial one. We will analyse here the influence of the instrumental parameters on the measurements of area and perimeter of an item of interest, e.g. a cell.

(i) THE ACQUISITION OF AREA AND/OR PERIMETER

The grey-values of a digitised 512×512 image may be arranged in a grey-value frequency histogram. For our images, the digitised grey-values are expressed in eight bits, running from 0 (black) to 255 (white). In Fig. 6.2, two such grey-value frequency histograms corresponding to an image of an erythrocyte contrasted by 46 wt% platinum and surrounded by Epon, are shown. The first histogram corresponds to a simple $(1 \times)$ image; in the second, noise reduction was achieved by averaging over 200 images. Two Gaussian-shaped grey-value populations can be observed, one related to the erythrocyte, one to the surrounding Epon. For both histograms, the corresponding first derivative of the distributions are also shown. Contrast between the erythrocyte and the surrounding Epon may be defined as the difference (Δ GV) between the two peaks, each corresponding to a zero crossing in the first derivative distribution. It may be noticed that the image averaging has no influence on the contrast, whereas it has improved the separation between the two populations. Contrast may be increased by changing instrumental parameters such as the diameter of the objective-lens diaphragm or the accelerating voltage. Also, the contrast depends on the type of image, i.e. zero-loss or Δ 250 eV image. Such parameters have a corresponding influence on contrast and hence on the value of (Δ GV). Whatever the mode of operation, the value applied as a threshold between the two populations, i.e. the item of interest and the background, determines the number of pixels supposed to be generated by the erythrocyte. This number may be converted to a measure of the area by the application of EM-magnification calibration techniques.

Once the points belonging to the erythrocyte have been identified, the boundary of the erythrocyte may be traced. The number of pixels constituting

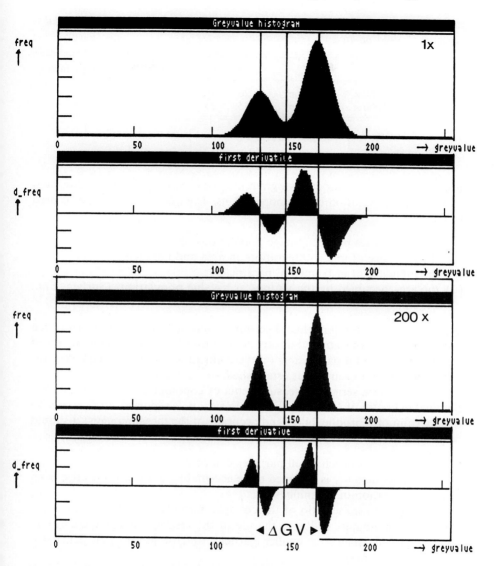

Fig. 6.2.   Grey-value frequency histogram and its first derivative from an erythrocyte.

the boundary gives an estimate of the perimeter of the item. The influence of the instrumental parameters and image averaging on the perimeter value can also be established.

In summary, it has been shown that in some instances, the first derivative

Table 6.1. *Morphometric analysis from nominal 1 nm colloidal gold particles*

M	D_{circle}	σ	n	$'N_{pix/part.}$	Shape-factor
225 000 ×	1.67 nm	0.26	48	25	0.57
400 000 ×	1.32 nm	0.35	25	50	0.38

80 KeV, $\alpha = 90 \, \mu m$, zero loss ($E = 0$ eV), 20 × integration.

of the grey-value frequency histogram may be used to determine a threshold value for the separation of two populations. The first derivative can be used to find:

(*a*) the boundaries between image populations objectively;
(*b*) the influence of image-integration on area and perimeter;
(*c*) the influence of instrumental parameters;
(*d*) the area occupied by the various grey-value populations (Sorber *et al.*, 1990a).

The perimeter of an item thus determined depends on the extent to which image averaging was applied during acquisition. Moreover, the influence of various instrumental parameters on the contrast and their consequences for the determination of area was established.

In Table 6.1, a series of measurements of colloidal gold particles with a nominal diameter of 1 nm in an ultrathin section is reported. In reference to the question of reproducability and accuracy of the measurements, it should be pointed out that:

(*a*) the diameter of such small particles can be measured;
(*b*) the reproducibility is assumed to be acceptable, but problems arise from the calibration of magnification;
(*c*) the signal to noise ratio of the colloidal particles is just sufficient for a good discrimination due to the use of the zero-loss imaging condition. For global images or smaller particles and/or lower Z-elements discrimination may fail;
(*d*) the accuracy of 25 or 50 points per particle is extremely low and according to Young in the order of 7–8 % (Young, 1988).

(ii) THE ACQUISITION OF THE AREA FRACTION PER CELL CROSS-SECTION

When items of interest are present in single cells their area fraction can be determined from the grey-value frequency histogram. Now three grey-value populations (Epon, cytoplasm, items of interest) have to be separated by the first derivative. The area fraction is obtained as the ratio of the area of the items to that of the cell area. The restrictions mentioned before can play an important role here too, such as when the items of interest do not show up as a separate population in the grey-value frequency histogram.

(iii) ACQUISITION OF THE AREA FRACTION PER STANDARD TISSUE AREA

When individual cell boundaries cannot be detected as a cross-over between two populations in the first derivative of a grey-value frequency histogram, as in tissue sections from homogeneously distributed cells, the analysed cell area can be related to the total frame area by histogram analysis techniques. This was demonstrated for liver ferritin particles (Sober *et al.*, 1990b). For a population of liver parenchymal cells we determined a mean ferritin iron core area fraction of 2.758×10^{-3} in 19 randomly chosen frames of $0.54 \ \mu m^2$. The mean iron core diameter was 5.93 ± 0.46 nm ($n = 2400$, 54 pixels per iron core).

2   Quantitative spectral and image analysis

(i) QUANTITATIVE SPECTRAL ANALYSIS

In electron energy-loss spectra of cellular material, the presence of elements is shown as an elevation beyond the element ionisation edge on a gradually decreasing background. The background is produced by unspecific energy losses and by tails from elements with their ionisation edges before the ionisation edge of the element of interest. For quantitative measurements of the elemental concentration, the elevation has to be separated from the background by extrapolation of the latter. For most of the elements a power law for the background may be assumed:

$$I(E) = A \, E^{-r} \qquad\qquad (6.1)$$

where A and r are constants. A and r may be determined by curve fitting techniques, for which various models have been proposed (Sorber *et al.*, 1991a, b) and their accuracy is discussed (de Bruijn *et al.*, 1991).
Fig. 6.3 illustrates the procedure of background subtraction in a spectrum containing barium and cerium edges in a lysosome and indicates the place and the width of the slit in the energy dispersive plane.

In X-ray microanalytical spectra of cellular material, the presence of the elements is observed as peaks superimposed on a continuum at the absorption edge. The peaks and the continuum may be separated by interpolation techniques.

Composed example spectrum

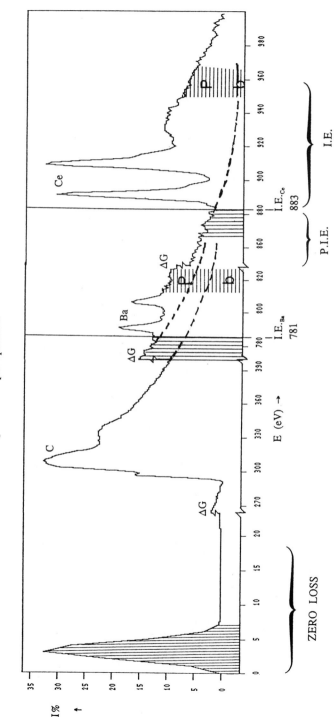

Fig. 6.3. Composed spectrum to show the zero-loss electrons between 0–20 eV, the low electron energy range just before the carbon edge (C) in the text indicated as Δ 250 eV images. At the right-hand side, extrapolations are shown to separate the peak (P) from the background (b) after the ionisation edge. The striated zones indicate the parts of the spectrum where (about 15 eV wide) electron populations are selected to form the various element-related images (for Ba and Ce). ΔG, intensity changes.

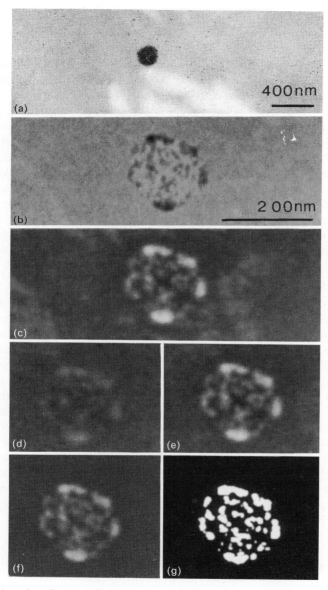

Fig. 6.4.   A series of images from the same cerium-containing lysosome. (a) Low
power CTEM-image recorded on photographic plate; (b) zero-loss
image; (c) Δ 250 eV image (= darkfield image); (d) pre-ionisation edge
image from cerium (see vertical shading in Fig. 6.3); (e) ionisation edge
image from cerium (see horizontal shading in Fig. 6.3); (f) resultant
image acquired by subtraction of 'b' image (after extrapolation) from
ionisation edge (e) – image; (g) is the binary image from (f).

(ii) QUANTITATIVE IMAGE ANALYSIS

In TEM/electron energy-loss images, background subtraction may be performed in a way similar to the procedure applied to spectra. The curve fitting and extrapolation procedure has to be performed for each pixel in the image. When the initial computer storage capacity is sufficient, this type of image processing can be done off-line.

In Fig. 6.4, a set of seven images is shown as obtained from a cerium-containing lysosome in which the presence of acid phosphatase is demonstrated by a cerium phosphate precipitate. In this case, the background image is calculated as a linear extrapolation of two pre-ionisation edge images subsequently subtracted from the ionization edge image to yield the net-intensity image (Sorber *et al.*, 1991a,b). Note the improved resolution as compared to the TEM (zero-loss) image of the same area (Fig. 6.4g vs. 6.4b). To acquire X-ray microanalytical element distribution images, the spectral deconvolution procedure has to be performed on-line for each point in the matrix, for each element of interest.

(iii) IMAGE INTEGRATION AND CO-LOCALISATION

The goal of a particular study may be to observe qualitatively the localisation of one element within a grey-value image acquired simultaneously from the same area or to relate multiple element-distribution images to each other. Examples can be found in the literature for electron probe X-ray micro-analytical images (de Bruijn & Cleton-Soeteman, 1985) and for electron energy-loss images (Sorber *et al.*, 1990a,b). The morphometric procedure described above can be performed to acquire the areas taken by the various element-containing items. Objective segmentation is not a problem as the subtracted image (Fig. 6.4f) is already a binary image. The final resolution and the localisation sharpness in such TEM/electron energy-loss images is improved over those previously acquired by STEM/electron probe X-ray microanalysis images from similar material. This was the main preferential argument for the combined use of TEM/electron energy-loss over STEM/ electron probe X-ray microanalyses (de Bruijn, 1985; de Bruijn *et al.*, 1987; de Bruijn & van Miert, 1988) in such image analysis.

3   The use of ultrathin-sectioned Bio-standard material

(i) IN SPECTRAL ANALYSIS

Bio-standard material has been used in an effort to convert relative element concentration values into absolute values. In TEM/electron energy-loss spectra, the relative element concentration R_x^* may be determined by relating the net-intensity value acquired in an energy range (Δ) beyond the ionisation edge to the electron yield in an energy range of the same width at the extreme

Table 6.2. *Quantitative analysis as wt % of calcium oxalate crystals and primordia with the use of Ca-containing Bio-standards*

Ca-crystal	n	Stone-primordia	n	Bio-standard	n	COM
30.2	7	21.56	4	6.13	12	27.43

COM = calcium oxalate monohydrate. Standard deviation values were respectively: Ca-crystal = $\pm 30\%$, primordia = $\pm 14\%$ (electron energy-loss analysis), Bio-standard = 1% (= Neutron Activation Analysis).

low-energy end of the spectrum. Ratios of element concentrations may be established as well by relating net-intensity values from two elements.

Recently, we proposed the use of co-embedded Bio-standards for quantitation of TEM/electron energy-loss spectra (Sorber *et al.*, 1991b). In such preparations, the relative concentration of homogeneously distributed elements in the item of interest (R_x^*) may be related to the relative concentration of the Bio-standard $(R_{x,\,\mathrm{st}}^*)$. Since for the latter, the absolute mean concentration $(C_{x,\,\mathrm{st}})$ is known as well, the absolute element concentration in the item (C_x) may be estimated as:

$$C_x = (R_x^* / R_{x,\,\mathrm{st}}^*) \, C_{x,\,\mathrm{st}}. \tag{6.2}$$

Table 6.2 shows measurements of calcium concentration for a small series of calcium oxalate-containing crystals and crystal primordia in rat kidney cells as determined using this method (Sorber *et al.*, 1991b).

(ii) IN IMAGE ANALYSIS

For relative concentration-distribution and absolute concentration-distribution images, the situation is more complicated. A few instrumental parameters have to be set, for which the optimal values are as yet unclear.

1. The first parameter is the width of the energy selecting slit in the energy dispersive plane. EM 902 spectra are recorded with a slit of 1 eV wide and R_x^*-values are based on an integration range with a width (Δ) of 50 or 100 eV both at the ionization edge of the element and at the zero- and low-loss end (0–100 eV) of the spectrum. For images, e.g. of iron at an ionisation edge of 708 eV, however, the energy selecting slit must at least be 5 to 15 eV wide in order to achieve a sufficient electron yield per image point.
2. The second (set of) parameter(s) concerns the position(s) of the energy selecting slit in the 50 eV (or 100 eV) (Δ) area. In Table 6.3, values of R_x^* for

Table 6.3. *Changes in relative spectral concentration values* (R_{Fe}^*) *for iron in relation to the position of a 15 eV wide slit between 728 and 833 eV*

E (eV)	R_{Fe}^* ($\times 10^{-2}$)
728	3.43
743	3.33
758	3.67
773	3.82
788	4.38
803	4.72
818	4.89
833	4.92

iron are shown as a function of the position starting at 728 eV (excluding the white lines between 708–728). The R_{Fe}^*-values calculated in a 15 eV wide slit depend systematically on this position.

E   Discussion

1   Morphometry

An image segmentation method based on the grey-value frequency histogram for contrast-related electron spectroscopic-images is shown to result in an appreciably accurate determination of the area of 1 nm nominal colloidal gold particles in ultrathin sections. In element-related images, segmentation and area determination may be performed in essentially the same way.

It has been shown previously (de Bruijn & Cleton-Soeteman, 1985; de Bruijn & van Miert, 1988) that a similar type of image segmentation may also be applied to STEM/electron probe X-ray microanalytical images, acquired with the Philips EM 400 both by the X-ray detector and the annular backscattered-electron detector. However, in the experiment described in de Bruijn *et al.* (1987), both detectors could not be active simultaneously. The images had therefore to be acquired sequentially. Also, the X-ray information was limited to four spectral regions only. These drawbacks have been removed in more recent instruments of this type. This results now in up to 16 simultaneously-acquired, background subtracted images, which can be acquired and stored within one run, plus the (up to 2048²) 16 bit deep backscattered-electron image.

However, in scanning transmission electron microscopic/backscattered-electron images, there is a time delay between the acquisition of the first and the last point. This delay increases with the total number of points and the dwell time per point. The dwell time in turn depends on the beam diameter and the energy collected in the beam, which must be determined in relation to the counting statistics required.

TEM/electron energy-loss spectroscopic images are acquired in sequence, resulting in a time delay between the various energy-related images taken in one run. Averaging over 100 images, the total acquisition time for one 512×512 image is 1 to 2 seconds.

Recently, Trebbia and co-workers (Trebbia & Bonnet, 1990; Trebbia & Mory, 1990) proposed a different way to acquire objectively a set of images from a large population of grey values acquired by a dedicated STEM/ electron energy-loss spectroscopic instrument. Although the way TEM/ electron energy-loss spectroscopic images are collected differs from their way of acquisition (post-acquisition vs pre-acquisition digitisation) it seems interesting to compare both methods of segmentation.

2   Quantitative spectral analysis

All methods of background correction discussed above are based upon mathematical modelling of the background. In both types of spectrum, in the case of multiple elements, peak or edge overlap may cause problems in separating the contributions of the different elements. In STEM/electron probe X-ray microanalysis, we encountered problems separating barium and cerium. In TEM/electron energy-loss spectroscopy, similar problems with cerium and phosphorus occurred. There the detectability of phosphorus and sulphur using electron energy-loss spectroscopic imaging methods has been questioned due to the low signal to background ratio, in spite of fairly convincing images (Adamson-Sharpe & Ottensmeyer, 1981; Arsenault & Ottensmeyer, 1983).

3   Background correction in element-related images

Background correction in STEM/electron probe X-ray microanalytical images has been described previously (de Bruijn *et al.*, 1987; de Bruijn & van Miert, 1988). In TEM/electron energy-loss spectroscopic images, background correction is more complicated. It has been shown that in images of ferritin-loaded lysosomes, the subtraction of a background image obtained by (linear or exponential) extrapolation leads to a better determination of the area of the ferritin particles than the simple procedure by which one pre-ionisation edge image is subtracted from the ionisation-edge image (Sorber *et al.*, 1990b). The images resulting after appropriate background correction have a high spatial resolution so that co-localisation of elements may be demonstrated with high accuracy. The question of the optimal position(s) of the energy selecting slit in the ionisation edge-related zone remains as yet unsolved. The optimal position(s) may depend on the type of edge of a particular element (hydrogen-type, with white lines, delayed).

4   Co-embedding of Bio-standards

The use of Bio-standards co-embedded in the vicinity of the tissue had been proposed some years ago for electron probe X-ray microanalysis and very recently for electron energy-loss spectroscopic analysis (de Bruijn, 1981a,b; Sorber *et al.*, 1991a,b). It was shown that the relative carbon concentration distribution in the Bio-standard matrix is different from that in an erythrocyte and in a macrophage cytoplasm. There are still unsolved problems in the conversion of elemental net-intensity images into concentration distribution images. We intend to use Bio-standard material, in which the carbon concentration has been determined, to solve some of these problems.

5   Comparison of energy-dispersive X-ray and electron energy-loss spectroscopic spectra

There is only a limited amount of information available, allowing comparisons of simultaneous electron energy-loss spectroscopic and energy dispersive X-ray-microanalyses of the same specimen. Our limited experience indicates that an appreciable electron energy-loss spectroscopic spectrum can be acquired from a 10 wt% Ni Bio-standard (using a 1 nm spot, maximal field emission-gun intensity and a parallel electron energy-loss spectroscopic energy detection system), whereas in the same time virtually no counts were acquired in the (EDAX) X-ray detector.

6   Pre- versus post-acquisition digitation

(i) RESOLUTION

In the past (de Bruijn *et al.*, 1987), we have proposed an equation (6.3) expressing the influence of the various experimental conditions on the distance between points ($= dmin =$ resolution) in STEM/electron probe X-ray microanalytical instruments:

$$dmin = IPDsp = dsp = \frac{dsc \cdot F^* \cdot \sqrt{2} \cdot f}{M \cdot N \cdot F}. \tag{6.3}$$

In which *dsp* and *dsc* are distances on specimen and screen respectively, *IPD* is the inter pixel distance, *M* the microscope magnification, *N* the number of points (pixels) per line (here assumed to be 1024) and *F* and *F** factors when less than the full screen is analysed or the N-value is different from the assumed value 1024. The value $\sqrt{2}$ comes from the degree of coverage one can acquire by completely filling the item of interest with points or pixels. When an M_{max} of the microscope is assumed to be 5×10^5 times, the STEM-image resolution for an accuracy of 25 pixels per particle ($f = 5$) is: 0.64 and 0.38 nm for $N = 2098^2$ and 4096^2 respectively.

(ii) TOTAL ACQUISITION TIME

In STEM/electron probe X-ray microanalysis, the total duration of exposure for one image increases with N^2. For a 256×256 image, assuming a dwell time per point of 1 s, the total duration of one exposure is 18.2 hours. For larger N-values, at the same dwell time, the total duration soon becomes unacceptably long. Reduction of the dwell time per point reduces this duration with factors of 10^{-3} when changing from seconds to milliseconds. In TEM/electron spectroscopic imaging, the total acquisition time will similarly be multiplied by four when it is assumed that increasing the value of N requires enlargement of the photon-sensitive plate in the camera.

(iii) X-RAY YIELD VERSUS ELECTRON YIELD

In STEM/electron probe X-ray microanalysis, assuming a detection quantum efficiency of unity and 100 % elements to be analysed, the X-ray yield is usually indicated by ω. When it is assumed that the electron yield is 1-ω, the efficiency boundary to choose between electron energy-loss and electron probe X-ray microanalysis is of the order of 0.5 (Cu or Zn). It should be realised that a detection quantum efficiency of unity is far from reality in both types of systems.

In conclusion: both types of acquisition modes have their advantages and disadvantages for quantitative spectral and image analysis of ultrathin sections from biological materials.

References

Adamson-Sharpe, K.M. & Ottensmeyer, F.P. (1981). Spatial resolution and detection sensitivity in microanalysis by electron energy loss selected imaging. *J. Microsc.*, **122**, 309.

Arsenault, A.L. & Ottensmeyer, F.P. (1983). Quantitative spatial distributions of calcium, phosphorus and sulfur in calcifying epiphysis by high resolution spectroscopic imaging. *Proc. nat. Acad. Sci. U.S.A.*, **80**, 1322.

de Bruijn, W.C. (1981a). Ideal standards for X-ray microanalysis of biological specimens. SEM. Inc. AMF O'Hare Il. USA, ed. O. Johari, *Scanning Electron Microsc.*, **1981/II**, 357–67.

(1981b). Ion exchange beads as standards for X-ray microanalysis of biological tissue. *Beitr. Elektronemikrosk. Direktabb. Oberfl.*, **16**, 369–72.

(1985). Integration of X-ray microanalysis and morphometry of biological material. SEM Inc. AMF O'Hare, Il. USA, ed. O. Johari, *Scanning Electron Microsc.*, **1985/II**, 697–712.

de Bruijn, W.C. & Cleton-Soeteman, M.I. (1985). Application of chelex standard beads in integrated morphometrical and X-ray micro-analysis. SEM. Inc. AMF O'Hare Il. USA, ed. O. Johari, *Scanning Electron Microsc.*, **1985/II**, 715–29.

de Bruijn, W.C., Ketelaars, G.A.M., Gelsema, E.S. & Sorber, C.W.J. (1991). Comparison of the Simplex method with several other methods for background-fitting for electron energy-loss spectral quantification of biological materials. *Microsc. microanal. microstruct.* **2**, 1–11.

de Bruijn, W.C., Koerten, H.K., Cleton-Soeteman, M.I. & Blok-van Hoek, C.J.G. (1987). Image analysis and X-ray microanalysis in cytochemistry. *Scan. Microsc.* **1**, 1651–67.

de Bruijn, W.C. & van Miert, M.P.C. (1988). Extraneous background-correction program for matrix bound multiple point X-ray microanalysis. *Scanning Microsc.*, **2**, 319–22.

Gravekamp, C., Koerten, H.K., Verwoerd, N.P., de Bruijn, W.C. & Daems, W.Th. (1982). Automatic image analysis applied to electron micrographs. *Intern. Rev.*, **6**, 656–7.

Sorber, C.W.J., de Jong, A.A.W., den Breejen, N.J. & de Bruijn, W.C. (1990a). Quantitative energy-filtered image analysis in cytochemistry I. Morphometric analysis of contrast related images. *Ultramicroscopy*, **32**, 55–68.

Sorber, C.W.J., van Dort, J.B., Ringeling, P.C., Cleton-Soeteman, M.I. & de Bruijn, W.C. (1990b). Quantitative energy-filtered image analysis in cytochemistry II. Morphometric analysis of element-distribution images. *Ultramicroscopy*, **32**, 69–79.

Sorber, C.W.J., Ketelaars, G.A.M., Gelsema, E.S., Jongkind, J.F. & de Bruijn, W.C. (1991a). Quantitative analysis of electron energy-loss spectra from ultrathin sectioned biological material. I. Optimization of the background-fit with the use of Bio-standards. *J. Microsc.*, **162**, 23–42.

(1991b). Quantitative analysis of electron energy-loss spectra from ultrathin sectioned biological material. II. The application of Bio-standards for quantitative analysis. *J. Microsc.*, **162**, 43–54.

Trebbia, P. & Bonnet, N. (1990). EELS elemental mapping with unconventional methods. I. Theoretical bases: image analysis with multivariate statistics and entropy concepts. *Ultramicroscopy*, **34**, 165–78.

Trebbia, P. & Mory, C. (1990). EELS elemental mapping with unconventional methods. II. Application to biological specimens. *Ultramicroscopy*, **34**, 179–203.

van Dort, J.B., Ketelaars, G.A.M., Daems, W.Th. & de Bruijn, W.C. (1989). Ultrastructural electron probe X-ray microanalytical reaction product identification of three different enzymes in the same mouse resident peritoneal macrophage. *Histochemistry*, **92**, 243–53.

van Dort, J.B., Zeelen, J.Ph. & de Bruijn, W.C. (1987). An improved procedure for the X-ray microanalysis of acid phosphatase activity in lysosomes. *Histochemistry*, **87**, 71–7.

Young, I.T. (1988). Quantitative image analysis. *Anal. quant. Cytol. histol.*, **10**, 269–79.

SECTION C
SPECIMEN PREPARATION

Specimen preparation is the cornerstone of successful biological X-ray microanalysis. A damaged or compromised specimen cannot be 'recovered' by sophisticated analytical hardware or software, or even by a skilled operator. But what constitutes a 'good specimen' or an 'acceptable degree of damage'? Is there such a thing as 'the ideal preparative procedure'? Answers to these fundamentally important questions are explicitly or implicitly offered in the chapters in this section of the book. However, two general observations should be made at this juncture.

First, there is no single ideal preparative procedure that meets all the requirements of the biological electron probe X-ray microanalyst (even if we confine our considerations to thin specimens of soft tissues). The starting point must be the nature of the biological problems that are to be tackled. Once this has been clearly defined then a suitable preparation procedure may be adopted, but even then the question (and thus the analytical objectives) may need to be modified in the light of the anticipated level of preparative damage. Measuring electrolyte concentrations in a small cohort of distinctive cells lying some distance beneath the surface of a complex tissue is a good example of a 'real microanalytical problem', as distinct from the analysis of well-chosen model systems. Electrolyte analysis requires cryofixation to limit redistribution, but the quality of freezing will not be high in deep cells. So the prospect of analysing compartments finer than 'nucleus' and 'cytoplasm' in such preparations may be unrealistic. Finding the cells that are the subject of the study may also present a severe problem, necessitating a preparation that can be trimmed and re-orientated with some ease. For such applications a resin-embedded cryopreparation should probably be a practical option. Thus, selecting a preparative procedure requires a pragmatic and flexible approach.

Second, a good method is the one that provides meaningful answers. But how are we to know that they are meaningful if, as is often claimed, the microprobe provides data not easily obtained by other means? Well, in the case, for example, of the electrolytes in deep-seated cells the data must initially make some sense to the physiologist. But of more general moment, the methods must be reproducible and transferable to other laboratories. A good method should not be so fickle that it prevents the analyst from meeting exacting statistical requirements. The very best methods are elegant if not simple; but given the crucial nature of preparative procedures, they all demand considerable operational vigilance.

Selected reading material

Morgan, A.J. (1980). Preparation of specimens. Changes in chemical integrity. In *X-ray microanalysis in biology*, ed. M.A. Hayat, pp. 65–165. Baltimore: University Park Press.

Roos, N. & Morgan, A.J. (1990). *Cryopreparation of thin biological specimens for electron microscopy: methods and applications.* Oxford: Oxford University Press, Royal Microscopical Society.

Steinbrecht, A. & Zierold, K. (1987). (ed.) *Cryotechniques in biological electron microscopy.* Heidelberg: Springer Verlag.

7 Rapid freezing techniques for biological electron probe microanalysis

Karl Zierold

A Introduction

Electron probe microanalysis is a sensitive tool to localise elements in biological cells and tissues. In comparison to most specimens studied by this method in materials sciences, cells are not static objects, but vary in their elemental composition, depending on their functional state. Therefore, intracellular components, in particular diffusible ions, have to be immobilised in the defined functional state to be investigated. Rapid freezing, also called cryofixation, is the most promising approach to meet this requirement. This chapter points to the crucial role of appropriate freezing techniques for biological electron probe microanalysis. This is in particular important for X-ray microanalytical studies of the following biological features: (1) intracellular element compartmentation, (2) cell viability and membrane damage, (3) ion transport systems and (4) ion shifts related to dynamic processes in cells.

B Preparation paths for electron probe microanalysis

Cells and tissues to be studied in an electron microscope have to be converted into a solid state specimen which is compatible with vacuum. Chemical preparation methods established for morphological investigations are based on fixation with aldehydes, staining by heavy metal salts, dehydration by alcohol and embedding in resin. Thereby, diffusible substances such as electrolyte ions are re-distributed and washed out (Zierold & Schäfer, 1988). As an alternative, low temperature preparation protocols as sketched in Fig. 7.1 were developed. They all start with cryofixation or with specimen sampling which means the appropriate handling of the specimen before rapid freezing to ensure the arrest of the functional state of interest. Then, there are essentially three different pathways to process the specimen.

1. Dehydration by freeze-drying or freeze-substitution, resin embedding and ultramicrotomy. The suitability of this protocol with respect to ion preservation in intracellular compartments is discussed controversially in

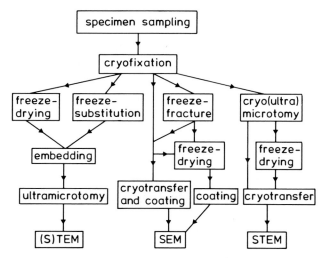

Fig. 7.1. Pathways of cryopreparation used for electron probe microanalysis of biological specimens. From Zierold & Steinbrecht (1987).

the literature (see e.g. Harvey, 1982; Roos & Barnard, 1985; Wroblewski & Wroblewski, 1986; Zierold & Steinbrecht, 1987; Condron & Marshall, 1990; Edelmann, 1991). At present, redistribution of diffusible substances by the substitution medium and the embedding material cannot be excluded. Therefore, in my opinion, this pathway can be recommended only for localisation of bound elements and qualitative analytical studies, for example to identify specific precipitates in histochemistry and cytochemistry (see Sumner, this volume).

2. Cryoprepration of the whole sample for X-ray microanalysis in the scanning electron microscope. Ion redistribution is unlikely to occur by using this pathway, but the spatial analytical resolution achievable in the bulk specimen is limited to several micrometres. If the specimen is not fractured, deposition of extracellular material on the specimen surface may interfere with the obtained X-ray spectrum of the specimen. Cleaning of cell surfaces of extracellular material by washing may affect the intracellular ion composition. Therefore, the washing medium has to be carefully checked for influences on the intracellular electrolyte balance before experiments are started (Zierold & Schäfer, 1988). This method is feasible for studying element contents on a whole-cell level but is not suitable to localise elements in intracellular compartments.

3. Cryoultramicrotomy and analysis of freeze-dried cryosections in a scanning transmission electron microscope. This pathway leads to the best results with respect to sensitivity and spatial analytical resolution. Therefore, the following examples of rapid freezing techniques are illustrated by results obtained from freeze-dried cryosections. In

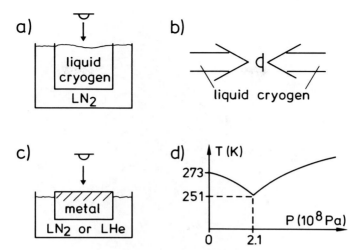

Fig. 7.2. Cryofixation techniques: (a) plunge freezing, (b) jet freezing, (c) impact freezing, (d) high pressure freezing. LN_2 = liquid nitrogen, LHe = liquid helium, T = temperature, P = pressure. From Zierold (1990).

addition to the element concentrations obtained by X-ray microanalysis in terms of mol per dry mass the dry mass portion, d, of the measured cell compartment is determined by darkfield intensity measurements in the scanning transmission electron microscope. Multiplication of the element concentration related to dry mass by $d(1-d)$ yields the physiologically often more significant element concentration related to the mass of water.

C   Rapid freezing techniques

Cryofixation can be defined as solidification of a biological specimen by cooling with the aim of minimal displacement of its components (Steinbrecht & Zierold, 1987). Cryofixation techniques and their use for biological electron microscopy were reviewed by, e.g., Plattner & Bachmann (1982), Robards & Sleytr (1985), Menco (1986), Sitte, Edelmann & Neumann (1987), Elder (1989). The physics of cryofixation was studied extensively by Bald (1987). As schematically depicted in Fig. 7.2, four physically different cryofixation techniques can be discerned: (1) plunging the specimen into a liquid cryogen, mostly propane or ethane cooled by liquid nitrogen, (2) jetting a liquid cryogen such as propane onto a specimen, (3) impacting the specimen against a cold polished metal surface, (4) high pressure cryofixation. By use of the first three methods small cell and tissue specimens up to a thickness of about 30 μm or a superficial region of 30 μm thickness in bulk specimens can be frozen without electron microscopically visible ice crystal segregations. This

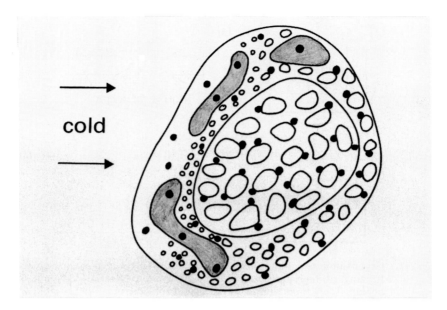

Fig. 7.3.   Ions in a frozen cell with ice crystal segregations depicted as holes of
varying size. The drawing shows three cellular compartments: cytoplasm,
nucleus and mitochondria. No segregations are found close to the cold
front and in compartments with high dry mass density such as
mitochondria (hatched areas). In the cytoplasm segregation size increases
with the distance from the cold front. Ice crystal segregation size is larger
in the nucleus than in the cytoplasm due to the lower dry mass portion
or higher water portion, respectively. Ions indicated by dark points are
displaced by the growing segregations and precipitate at the organic
matrix.

limit is due to the small thermal conductivity of water, the main component
of biological cells and tissues. Even an optimal cold sink on the specimen
surface would not be able to extend this region significantly because of the
slow flow of heat and ice crystallisation energy to the specimen surface.
During the time period of cooling water can crystallise until a temperature of
approximately 130 K is reached. In the temperature range between 273 K and
approximately 190 K small ice crystals can transform to larger ones thus
increasing the ice crystal segregations often visible in electron micrographs of
frozen specimens after dehydration by freeze-drying or freeze-substitution.
Ice crystal segregations appear larger in water-rich compartments (e.g.
extracellular space, vacuoles, nucleus) in comparison with compartments
rich in organic matrix (mitochondria, lipid or glycogen granules).

What happens with diffusible substances, in particular electrolyte ions,
dissolved in water during formation of ice crystal segregation? One has to
admit that this question cannot yet be answered accurately, but high
resolution X-ray microanalysis indicates that ions precipitate at the organic

Table 7.1. *X-ray microanalysis of erythrocytes*

	Plunge freezing	Impact freezing	High pressure freezing
n	24	26	18
d	0.33 ± 0.03	0.35 ± 0.02	0.37 ± 0.03
Na	21 ± 21	22 ± 34	20 ± 34
Mg	10 ± 8	13 ± 34	14 ± 10
P	48 ± 10	40 ± 9	38 ± 12
S	123 ± 12	134 ± 21	131 ± 13
Cl	79 ± 27	70 ± 39	77 ± 29
K	200 ± 29	192 ± 41	178 ± 35
Ca	3 ± 2	5 ± 4	5 ± 4
Fe	41 ± 5	46 ± 5	44 ± 5

Data obtained from cryosections of human erythrocytes in fresh blood after different cryofixation techniques. n = number of cells, d = dry mass portion. Element contents are given in mmol/kg dry mass $\pm$ standard deviation. Significance of statistical differences were calculated according to the t-test. Between impact freezing and high pressure freezing error probability for differences was higher than 20 %. Differences between plunge freezing and high pressure were significant with an error probability higher than 20 % for Na, Mg, Cl, Ca, and Fe and about 5 % for S and K. The difference for P vanishes after relating data to the intracellular water content. Data partly from Zierold *et al.* (1991).

matrix surrounding the segregations. This consideration is illustrated in Fig. 7.3. Ions probably do not move further than one diameter of a segregation hole. There are no indications for ion movement within the organic matrix after freezing and freeze-drying. Ion gradients over the border of compartments, for example potassium, sodium and chloride gradients at the cell membrane separating the extracellular from the intracellular space, remain steep provided the sections are kept free from humidity. It becomes obvious from Fig. 7.3 that the minimal microprobe area must be larger than the mean segregation hole to yield reasonable results.

The limited suitability of rapid freezing methods to dimensions considerably smaller than 1 mm suggest isolated and cultured cells to be the specimens of choice. Consequently, there are numerous reports on X-ray microanalysis of single and cultured cells, reviewed for example by Wroblewski & Wroblewski (this volume), Wroblewski & Roomans (1984), Zierold (1988, 1991).

Only high pressure freezing can provide specimens up to a thickness of 0.5 mm with electron microscopically invisible ice. The high pressure of 2.1 kbar during freezing causes an increase in the viscosity of water thus depressing the melting point to 251 K. For details of this technique see e.g. Moor (1987). In comparison to the other three mentioned freezing techniques,

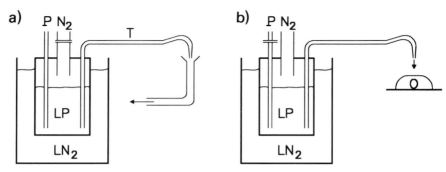

Fig. 7.4.   Propane jet freezing of biological objects *in situ*. Abbreviations: LN_2 = liquid nitrogen, LP = liquid propane, N_2 = pressurised gaseous nitrogen, O = biological object, P = propane, T = propane tube. (a) Propane is liquified in a steel container cooled by liquid nitrogen. The valve for gaseous propane is open, the valve for pressurised gaseous nitrogen is closed. Liquid propane flows slowly through the propane tube and the nozzle at the end. Propane flowing through the nozzle is taken up by a refill tube until it is sufficiently cold ($\sim$ 210 K). Then, the propane valve is closed. The elevation of the tube before the nozzle ensures that no liquid propane drops onto the biological object which is positioned now below the nozzle. (b) After opening the nitrogen valve pressurised gaseous nitrogen pushes the liquid propane through the nozzle onto the biological object. The rapidly frozen part of the biological object is cut out and transferred into liquid nitrogen.

the use of high pressure freezing for element localisation was considered with scepticism because of the rather low cooling rate and because of the high pressure supposed to cause ion shifts. Recent experiments with human erythrocytes do not indicate any ion distortions by high pressure freezing in comparison to impact and plunge freezing (Zierold, Tobler & Müller, 1991) as can be seen from the data compiled in Table 7.1. Measurements with a spot size of 50 nm in the centre of the cells and close to the membranes did not show any significant difference.

In spite of the advantage of high pressure freezing for large specimens, at present the first three mentioned methods (plunge freezing, jet freezing and impact freezing) are preferred for microanalytical purposes. One reason might be the high price of high pressure machines, but more important from the scientific point of view is that plunge freezing, jet freezing and impact freezing can be modified to arrest functional states and dynamic processes in cells and tissues. For example, jet freezing can be modified to an *in situ* method as shown by the drawing in Fig. 7.4. A similar jet-freezing device was combined with a binocular microscope to freeze streaming amoebae under light optical control (Zierold & Schäfer, 1988; Zierold, 1991). Hagler *et al.* (1983) used pliers with cooled copper jaws and in a more sophisticated version the 'cryosnapper' (Hagler, Morris & Buja, 1989) to freeze tissue *in situ*. Von Zglincki, Bimmler & Purz (1986) developed a precooled tubule-shaped steel

Table 7.2. *X-ray microanalysis of hepatocytes*

	Dissected tissue specimens	Cryopunched tissue	Isolated cells	Cultured cells
n	15	10	20	12
d	0.26 ± 0.05	0.29 ± 0.04	0.32 ± 0.03	0.30 ± 0.04
Na	300 ± 162	142 ± 48	23 ± 24	10 ± 14
Mg	26 ± 25	28 ± 20	26 ± 17	41 ± 14
P	387 ± 185	506 ± 86	265 ± 76	235 ± 54
S	201 ± 60	209 ± 30	211 ± 17	185 ± 24
Cl	403 ± 194	202 ± 21	93 ± 26	62 ± 9
K	226 ± 77	364 ± 47	137 ± 26	421 ± 48
K/Na	< 1	> 1	> 1	> 10

Data obtained from cryosections of rat liver cell cytoplasm after different preparation protocols for cryofixation. Dissected tissue specimens of 1–2 mm diameter were plunged into liquid propane. Cryopunched tissue was prepared from isolated whole liver by use of the device shown in Fig. 7.6. Isolated cells in suspension were plunged into liquid propane. Cultured cells were grown on Petriperm® foil and frozen by the technique shown in Fig. 7.7. n = number of cells, d = dry mass portion. Element contents are given in mmol/kg dry mass ± standard deviation. Data partly from Zierold & Schäfer (1988).

needle to take cryobiopsies from liver and heart muscle tissue. This *in situ* option at present seems hard to achieve with the high pressure freezing technique.

D   Rapid freezing of functional states

Electron probe microanalysis of biological specimens may show preparation-induced ion redistribution, in particular intracellular increase of sodium and chloride accompanied by a loss of potassium. This is illustrated by an experiment with rat liver. Microanalytical results from a dissected tissue sample of about 2 mm in diameter, plunge frozen and cryosectioned, are compiled in the first column of Table 7.2. Obviously, dissecting small pieces from an intact organ damages cell membranes and affects the original ion distribution before any distortion is seen in the ultrastructure. The morphology of the corresponding cryosection (Fig. 7.5) does not indicate any cell damage. Microanalysis requires *in situ* freezing in order to preserve the ion distribution related to the native functional state of the organ. This can be done for example by 'cryopunching' with a modified impact freezing device, schematically depicted in Fig. 7.6. Data obtained from cryopunched liver are shown in the second column of Table 7.2. The sodium concentration is not as low as expected for native liver cells. This might be due to the fact that this

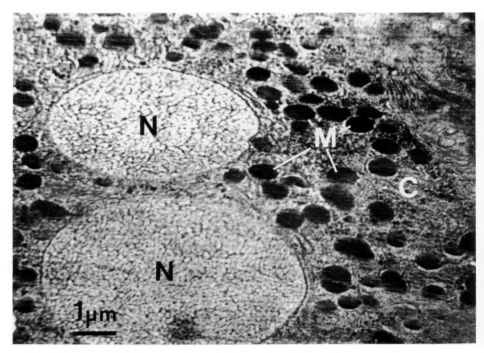

Fig. 7.5.   Freeze-dried cryosection of dissected and plunge frozen rat liver tissue.
C = cytoplasm, M = mitochondria, N = nucleus.

experiment was done with isolated liver and not with the animal. A further improvement can be expected by cryopunching specimens from an organ supplied by blood. Nevertheless, the intracellular K/Na ratio is improved in comparison to that from tissue samples dissected before freezing. The K/Na ratio can be considered as a sensitive viability criterion of animal cells. Healthy mammalian cells have a K/Na ratio significantly higher than 1. A K/Na ratio below 1 indicates cell damage or at least affected membrane functions.

Experiments with isolated cells provide an alternative to *in situ* cryofixation. This is illustrated by rat liver hepatocytes isolated by digestion with collagenase, washed and suspended in Tyrode buffer (Petzinger & Frimmer, 1988). A droplet of the cell suspension was pipetted onto a gold freeze-etch specimen support, frozen and cryosectioned for microanalysis. The results compiled in Table 7.2 indicate that cell suspensions are feasible systems for studying the action of extracellular parameters (osmolarity, drugs etc.) on the intracellular ion composition. However, hepatocytes usually do not exist as isolated cells but form a tight epithelium. For example, epithelial cells may lose their polarity by isolation, whereas other cellular features remain maintained in the nutrition solution. Petzinger *et al.* (1988) were able to grow

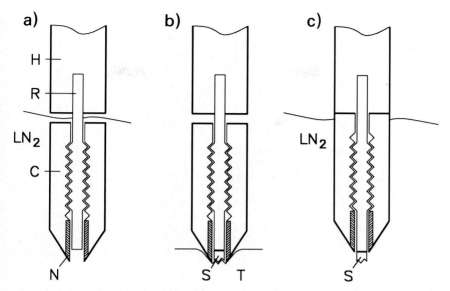

Fig. 7.6. Schematic drawing of a cryopunching device for rapid freezing of tissue samples *in situ*. Abbreviations: C = cooling cartridge, H = holder, LN_2 = liquid nitrogen, N = needle, R = cooling rod, S = specimen, T = tissue. (a) On the plastic holder the cooling rod (copper) is fastened. The cooling cartridge (copper) with a tube-shaped steel needle of 1 mm inner diameter at the lower end is screwed on the cooling rod. This device is precooled in liquid nitrogen. (b) The specimen of about 1 mm in diameter is punched out from tissue *in situ* and rapidly frozen by contact to the lower end of the cooling rod. (c) At low temperature the cooling rod is screwed against the surrounding cartridge until the specimen can be taken from the cooling rod by forceps.

hepatocytes on Petriperm®, a gas-permeable thin and flexible plastic foil. Flat growing cell monolayers are usually a difficult object for cryoultramicrotomy. By means of a modified plunge freezing technique, schematically depicted in Fig. 7.7, the cells grown on Petriperm could be cryosectioned, followed by microanalysis. The results compiled in Table 7.2 show a further increase in the intracellular K/Na ratio in comparison to the isolated cells in suspension. The data from the experiments with hepatocytes in Table 7.2 prove the sensitive relationship between the intracellular elemental composition and the physiological conditions immediately before cryofixation. The low K content in isolated cells in comparison to cultured cells is conceivable by the hypothesis of a K channel in the basolateral cell membrane. In isolated cells K leaks out by this channel. In the confluent epithelial layer formed in the culture dish the basolateral cell membrane is sealed from the extracellular medium by tight junctions as in intact tissue.

Papillary collecting ducts cells of rat kidney are not accessible to

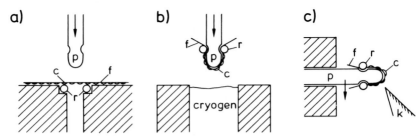

Fig. 7.7. Cryofixation of cells cultured on flexible gas-permeable foil for cryoultramicrotomy. Abbreviations: c = cell, f = foil, k = knife, p = Plexiglass rod, r = sealing ring. Arrows indicate direction of movement. (a) From the foil with cultured cells an area of 10 mm in diameter is punched out and placed down on the preparation block with a central hole. A 2 mm thick Plexiglass rod with a half-sphere like lower end and indentation above is placed into the central hole of the preparation block. Thereby, the foil with the cells is stretched over the rod and fixed by a sealing ring. (b) Cells are rapidly frozen by plunging the Plexiglass rod with the foil into liquid propane cooled by liquid nitrogen. (c) The Plexiglass rod is mounted onto the specimen holder of the cryoultramicrotome for cryosectioning with a dry glass knife. Cells can be seen by light reflected at the Plexiglass surface. The curved end of the Plexiglass rod permits approach of the knife edge to the frozen cell layer from different directions. From Zierold (1989).

cryofixation *in situ* without damaging the organ. Stokes, Grupp & Kinne (1987) developed a technique to isolate tubule fragments of the papillary collecting duct by enzymatic digestion. These still cohering cells maintained at least partly their epithelial properties. In addition, they could be frozen easily by plunging a suspension into liquid propane. Fig. 7.8 shows a cryosection from such a papillary collecting duct cell cluster. They exhibit K/Na ratios significantly higher than 1. More X-ray data from this model are reported by Pavenstädt-Grupp, Grupp & Kinne (1989) and Zierold (1991).

Small droplets of cells in suspension may dry during preparation before freezing. Drying of the droplet proceeds more quickly the smaller it is (Hall & Gupta, 1982). The resulting increase in extracellular osmolarity may influence the intracellular balance of water, ions and cell volume. The droplet can be protected from contact with air by a sandwich cryofixation device suitable for subsequent cryoultramicrotomy, illustrated in Fig. 7.9 (Zierold, 1988). Recently, environmental chambers have been used more and more to prepare small aqueous specimens for rapid freezing under controlled conditions of temperature and humidity (Bellare *et al.*, 1988; Somlyo, Shuman & Somlyo, 1989).

The above mentioned examples prove microanalysis of cells and tissues in defined functional states not to be as simple as supposed at first glance. The appropriate freezing technique has to be developed with respect to the experimental aim.

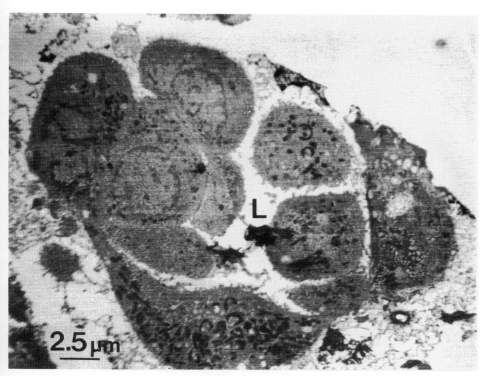

Fig. 7.8.   Freeze-dried cryosection from tubule fragments of isolated rat papillary collecting duct. The still cohering cells are arranged around the central lumen L.

E   Rapid freezing of dynamic processes

Investigating ion shifts related to dynamic processes in cells is perhaps the most attractive challenge for biological microanalysis. This purpose requires cryofixation techniques with controlled time resolution. By modifications of an air pressure driven plunge freezing device it was possible to freeze cells at defined time intervals after stimulating exocytotic processes with the accuracy of milliseconds.

Stinging capsules called nematocysts in the cnidarian animal *Hydra* can be discharged by electrical stimulation causing ejection of a stylet adhering to a tubule within 50 ms. Fig. 7.10 shows how *Hydra* was frozen after electrical stimulation in order to study the discharge of nematocysts. After cryofixation the specimens were processed for cryoultramicrotomy and X-ray microanalysis. Fig. 7.11 shows a cryosection of partly discharged nematocysts. This discharge of stinging capsules in *Hydra* could be shown to be accompanied by

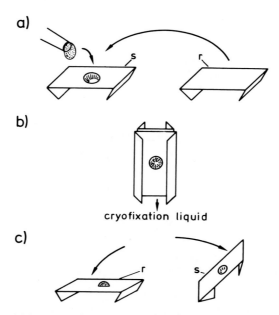

Fig. 7.9.   Rapid freezing of a droplet of cell suspension for cryoultramicrotomy. (a) The droplet is pipetted onto a smoothly polished copper sheet (s) with a central cavity. A flat copper sheet of the same size with a rough surface (r) is placed on the sheet with the droplet. (b) This sandwich arrangement is taken by thin forceps and plunge frozen into liquid propane. (c) In the cold chamber of the cryoultramicrotome the copper sheets are detached from each other by means of forceps. In most cases the droplet remains on the rough copper sheet. This copper sheet is glued on the cryoultramicrotome holder at low temperature by liquid heptane (Steinbrecht & Zierold, 1984). From Zierold (1988).

a slight influx of calcium into the matrix of nematocysts. For detailed results the reader is referred to Zierold, Gerke & Schmitz (1989). By a similarly modified freezing device the exocytosis of trichocysts in paramecia was investigated. Sodium accumulated in intracellular trichochysts was found to be released by exocytosis of these organelles (Schmitz & Zierold, 1989).

   Other reports on biological microanalysis of fast processes concentrate on calcium movements in muscle. Wendt-Gallitelli & Wolburg (1984) and Wendt-Gallitelli & Isenberg (1989) developed special devices based on plunge freezing to follow calcium shifts in heart muscle after electrical stimulation. Somlyo *et al.* (1985) and Ingram *et al.* (1989) showed the calcium depletion of the sarcoplasmic reticulum by X-ray microanalysis of cryosections from muscle fibres which were frozen at defined time periods after electrical stimulation.

   The modifications of existing freezing techniques to arrest ion movements in cells yield a time resolution in the range of 1 ms. Although the real mobility

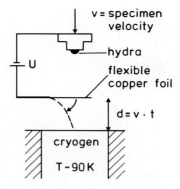

Fig. 7.10. Rapid freezing of biological objects within defined time period (ms range) after electrical stimulation. The biological object, here *Hydra*, is mounted onto a gold freeze-etch specimen holder. The specimen is plunged into liquid cryogen (propane) by pressurised air. Before reaching the cryogen the biological object touches a flexible copper foil. The animal is stimulated by touching the copper foil where an electrical voltage of 20 V is applied. The time period between stimulation and rapid freezing is defined by adjusting the distance between copper foil and cryogen surface. From Zierold *et al.* (1989).

of ions in cells is not known, in a first approximation the mobility of ions in water should be similar. Most electrolyte ions such as Na^+, K^+, Cl^- and Ca^{2+} have a diffusion constant D of about 2×10^{-9} m^2/s at room temperature. The ions of water move slightly faster. The diffusion constants are for H^+: 1×10^{-8} m^2/s and for OH^-: 6×10^{-9} m^2/s. From the above data the diffusion length of electrolyte ions, L within the time period, t, can be estimated according to the formula

$$L = \sqrt{(2 \cdot D \cdot t)}.$$

This calculation leads to $L = 2$ μm within 1 ms, 6 μm within 10 ms and 11 μm within 30 ms. For microanalytical determination of the ion content in homogeneously filled compartments this estimation does not give any reason of concern. However, movement of ions along concentration gradients may proceed too fast to be detected by the described rapid freezing techniques. Thus, spatial resolution in biological microanalysis of ion shifts turns out to be limited by time resolution achievable in cryofixation!

F Conclusion

It was the aim of this chapter to show that none of the different physical approaches of rapid freezing can be considered *a priori* as suitable for

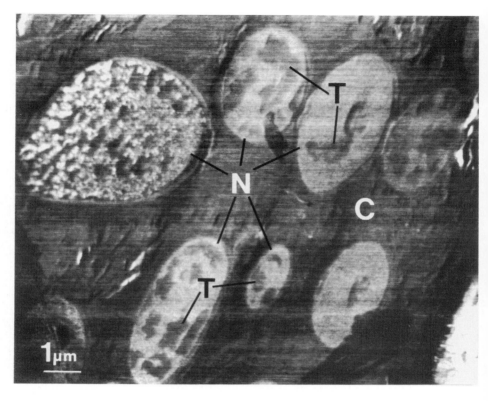

Fig. 7.11. Freeze-dried cryosection of *Hydra* frozen 10 ms after electrical
stimulation showing partly discharged stinging capsules (nematocysts)
with different tubule content indicating different stages of discharge.
C = cytoplasm, N = nematocysts, T = tubule.

cryofixation of cells and tissues for biological electron probe microanalysis.
The choice of a freezing technique has to be made according to the particular
experimental aim in cell biology or physiology. Modifications of existing
techniques are often a prerequisite before reliable microanalytical data related
to functional states or dynamic processes in cells and tissues can be expected.
This was illustrated by selected examples.

Acknowledgement

I would like to thank Mrs S. Dongard for excellent technical assistance and
Mr R. König for the construction of rapid freezing devices.

References

Bald, W.B. (1987). *Quantitative cryofixation*. Bristol, Philadelphia: Adam Hilger.
Bellare, J.R., Davis, H.T., Scriven, L.E. & Talmon, Y. (1988). Controlled environmental vitrification system: an improved sample preparation technique. *Journal of Electron Microscopy Techniques*, **10**, 87–111.
Condron, R.J. & Marshall, A.T. (1990). A comparison of three low temperature techniques of specimen preparation for X-ray microanalysis. *Scanning Microscopy*, **4**, 439–47.
Edelmann, L. (1991). Freeze-substitution and the preservation of diffusible ions. *Journal of Microscopy*, **161**, 217–28.
Elder, H.Y. (1989). Cryofixation. In *Techniques in immunochemistry*, Vol. 4, ed. G.R. Bullock & P. Petrusz, pp. 1–28. London: Academic Press.
Hagler, H.K., Lopez, L.E., Flores, J.S., Lundswick, R. & Buja, L.M. (1983). Standards for quantitative energy dispersive X-ray microanalysis of biological cryosections: validation and application to studies of myocardium. *Journal of Microscopy*, **131**, 221–34.
Hagler, H.K., Morris, A.C. & Buja, L.M. (1989). X-ray microanalysis and free calcium measurements in cultured neonatal rat ventricular myocytes. In *Electron probe microanalysis applications in biology and medicine*, ed. K. Zierold & H.K. Hagler, pp. 181–97. Berlin: Springer-Verlag.
Hall, T.A. & Gupta, B.L. (1982). Quantification for the X-ray microanalysis of cryosections. *Journal of Microscopy*, **126**, 333–45.
Harvey, D.M.R. (1982). Freeze substitution. *Journal of Microscopy*, **127**, 209–21.
Ingram, P., Nassar, R., Le Furgey, A., Davilla, S. & Sommer, J.R. (1989). Quantitative X-ray elemental mapping of dynamic physiological events in skeletal muscle. In *Electron probe microanalysis applications in biology and medicine*, ed. K. Zierold & H.K. Hagler, pp. 251–64. Berlin: Springer-Verlag.
Menco, B.P.M. (1986). A survey of ultra-rapid cryofixation methods with particular emphasis on applications to freeze-fracturing, freeze-etching and freeze-substitution. *Journal of Electron Microscopy Techniques*, **4**, 177–240.
Moor, H. (1987). Theory and practice of high pressure freezing. In *Cryotechniques in biological electron microscopy*, ed. R.A. Steinbrecht & K. Zierold, pp. 175–91. Berlin: Springer-Verlag.
Pavenstädt-Grupp, I., Grupp, C. & Kinne, R.K.H. (1989). Measurement of element content in isolated papillary collecting duct cells by electron probe microanalysis. *Pflügers Archiv, European Journal of Physiology*, **413**, 378–84.
Petzinger, E., Föllmann, W., Acker, H., Hentschel, J., Zierold, K. & Kinne, R. (1988). Primary liver cell cultures grown on gas permeable membrane as source for the collection of primary bile. *In Vitro Cellular and Developmental Biology*, **24**, 491–9.
Petzinger, E. & Frimmer, M. (1988). Comparative investigations on the uptake of phallotoxins, bile acids, bovine lactoperoxidase and horseradish peroxidase into rat hepatocytes in suspension and in cell cultures. *Biochimica et Biophysica Acta*, **937**, 135–44.
Plattner, H. & Bachmann, L. (1982). Cryofixation: A tool in biological ultrastructural research. *International Review of Cytology*, **79**, 237–304.
Robards, A.W. & Sleytr, U.B. (1985). Low temperature methods in biological electron microscopy. In *Practical methods in electron microscopy*, Vol. 10, ed. A.M. Glauert, pp. 1–551. Amsterdam: Elsevier.
Roos, N. & Barnard, T. (1985). A comparison of subcellular element concentrations in frozen-dried, plastic-embedded, dry-cut sections and frozen-dried cryosections. *Ultramicroscopy*, **17**, 335–44.

Schmitz, M. & Zierold, K. (1989). X-ray microanalysis of ion changes during fast processes in cells, as exemplified by trichocyst exocytosis of Paramecium caudatum. In *Electron microscopy of subcellular dynamics*, ed. H. Plattner, pp. 325–39. Boca Raton, USA: CRC Press.

Sitte, H., Edelmann, L. & Neumann, K. (1987). Cryofixation without pretreatment at ambient pressure. In *Cryotechniques in biological electron microscopy*, ed. R.A. Steinbrecht & K. Zierold, pp. 87–113. Berlin: Springer-Verlag.

Somlyo, A.V., McClellan, G., Gonzalez-Serratos, H. & Somlyo, A.P. (1985). Electron probe X-ray microanalysis of post-tetanic Ca^{2+} and Mg^{2+} movements across the sarcoplasmic reticulum *in situ*. *Journal of Biological Chemistry*, **260**, 6801–7.

Somlyo, A.V., Shuman, H. & Somlyo, A.P. (1989). Electron probe X-ray microanalysis of Ca^{2+}, Mg^{2+} and other ions in rapidly frozen cells. *Methods in Enzymology*, **172**, 203–29.

Steinbrecht, R.A. & Zierold, K. (1984). A cryoembedding method for cutting ultrathin cryosections from small frozen specimens. *Journal of Microscopy*, **136**, 69–75.

(ed.) (1987). *Cryotechniques in biological electron microscopy*. Berlin: Springer-Verlag.

Stokes, J.B., Grupp, C. & Kinne, R.K.H. (1987). Purification of rat papillary collecting duct cells: functional and metabolic assessment. *American Journal of Physiology*, **253**, F-251–62.

Wendt-Gallitelli, M.F. & Isenberg, G. (1989). X-ray microanalysis of single cardiac myocytes frozen under voltage-clamp conditions. *American Journal of Physiology*, **256**, H574–83.

Wendt-Gallitelli, M.F. & Wolburg, H. (1984). Rapid freezing, cryosectioning and X-ray microanalysis on cardiac muscle preparations in defined functional states. *Journal of Electron Microscopy Techniques*, **1**, 151–74.

Wroblewski, J. & Roomans, G.M. (1984). X-ray microanalysis of single and cultured cells. *Scanning Electron Microscopy*, **1984/4**, 1875–82.

Wroblewski, J. & Wroblewski, R. (1986). Why low temperature embedding for X-ray microanalytical investigations? A comparison of recently used preparation methods. *Journal of Microscopy*, **142**, 351–62.

Zglinicki, von, T., Bimmler, M. & Purz, H.-J. (1986). Fast cryofixation technique for X-ray microanalysis. *Journal of Microscopy*, **141**, 79–90.

Zierold, K. (1988). Electron probe microanalysis of cryosections from cell suspensions. In *Methods in microbiology, vol. 20, Electron microscopy in microbiology*, ed. F. Mayer, pp. 91–111. London: Academic Press.

(1989). Cryotechniques for biological microanalysis. In *Microbeam Analysis 1989*, ed. P.E. Russel, pp. 109–11, San Francisco: San Francisco Press Inc.

(1990). Low-temperature techniques. In *Biophysical electron microscopy*, ed. P.W. Hawkes & U. Valdrè, pp. 309–46. London: Academic Press.

(1991). Cryofixation methods for ion localization in cells by electron probe microanalysis: a review. *Journal of Microscopy*, **161**, 357–66.

Zierold, K., Gerke, I. & Schmitz, M. (1989). X-ray microanalysis of fast exocytotic processes. In *Electron probe microanalysis applications in biology and medicine*, ed. K. Zierold & H.K. Hagler, pp. 281–92. Berlin: Springer-Verlag.

Zierold, K. & Schäfer, D. (1988). Preparation of cultured and isolated cells for X-ray microanalysis. *Scanning Microscopy*, **2**, 1775–90.

Zierold, K. & Steinbrecht, R.A. (1987). Cryofixation of diffusible elements in cells and tissues for electron probe microanalysis. In *Cryotechniques in biological electron microscopy*, ed. R.A. Steinbrecht & K. Zierold, pp. 272–82. Berlin: Springer-Verlag.

Zierold, K., Tobler, M. & Müller, M. (1991). X-ray microanalysis of high pressure and impact frozen erythrocytes. *Journal of Microscopy*, RP1–2.

8 Radiation damage and low temperature X-ray microanalysis

Thomas von Zglinicki

A Summary

Radiation damage in the electron microscope is reviewed in the context of biological X-ray microanalysis. A phenomenological description of the dependency of radiation induced mass loss on electron dose in frozen–hydrated and frozen–dried biological material is given and some semiquantitative data are presented as a rule of thumb for practical X-ray microanalysis. Quantitative imaging, spatial deconvolution and extrapolation to zero dose are discussed as possibilities to minimise radiation effects in low temperature X-ray microanalysis. The consequences for the measurement of local water fractions by low temperature electron microscopy and X-ray microanalysis are deduced.

B Introduction

X-ray microanalysis can be a very powerful tool in physiological research, because of its ability to track with high resolution the distribution of ions in cells and tissues. This requires low temperature techniques for three reasons. First, there is a beneficial effect of low temperatures in terms of radiation damage. Second, the retention of water during the measurement excludes a number of possible redistribution artefacts. Third, retention of water by cooling is required by most of the techniques to measure local water fractions.

It was clear from the beginning of biological X-ray microanalysis that its potential could be fully exploited only when the distribution of water as well as the distribution of detectable elements could be measured (see, for instance, Gupta, Hall & Maddrell, 1976). However, the need for a *separate* examination of local water fractions was felt more urgent when more and more data arrived showing that monovalent ions might not be simply dissolved in the intracellular water phase (Edelmann, 1984; Cameron, 1985; Kellermayer *et al.*, 1986; von Zglinicki & Bimmler, 1987).

There are three principal approaches to exploit the interaction of electrons with biological matter to measure local water fractions:

1. determination of the oxygen concentration in frozen–hydrated (bulk) samples using a windowless X-ray detector (Marshall, 1980);

2. measurement of the ratio of element concentrations in the frozen–hydrated and frozen–dried state by fully quantitative X-ray microanalysis (Gupta & Hall, 1981);
3. determination of the mass density of sections by any of the electron beam–specimen interactions resulting in a signal linearly related to mass thickness, either in the frozen–hydrated and frozen–dried state or in the frozen–dried state alone (von Zglinicki, 1991a).

Techniques for measuring local water fractions which differ mainly in the approaches used to quantitate the results have been developed. A comparative evaluation of these differences is the subject of a recent review (von Zglinicki, 1991a). However, there are two fundamental problems with any of these techniques, namely, whether the distribution of water as present *in vivo* is artificially altered by the cryopreparation of the sample, and whether the results might be falsified by the measurement itself, i.e. by the interaction of the specimen with the electron beam.

Problems of cryopreparation are reviewed by Zierold (this volume) and have been tackled recently with respect to the measurement of water fractions (von Zglinicki, 1991b), and will not be repeated here. However, it seems appropriate to stress the importance of radiation damage in the electron microscope for low temperature X-ray microanalysis in general and, especially, for water content measurements in biological systems. Problems of radiation damage have been reviewed recently both with respect to high resolution electron microscopy (Zeitler, 1990) and X-ray microanalysis (Echlin, 1991; Lamvik, 1991). The present review will focus on this topic from a rather practical point of view. Some semiquantitative figures will be given as a practical rule of thumb and possibilities to minimise the effects of radiation damage will be considered.

C   Radiation damage: specimen preparation and general operational parameters

For a quantitative analysis by energy dispersive X-ray microanalysis it is necessary to deliver, at the very least, a charge of about 20 nC to the analysed region (Hall & Gupta, 1984). This corresponds to about 10^{11} electrons per analysed area, which has a dimension typically in the order of 1 μm². In other words, the radiation dose on the specimen is as high as if it were close to the centre of a nuclear explosion! This fact was acknowledged very early in the 1930s as the basic problem of biological electron microscopy and analysis, and Stenn & Bahr (1970) concluded in their classic paper that the loss of mainly H, N and O would lead to a carbonisation of any biological specimen during analysis. X-ray microanalysis imposes exposure conditions considerably less favourable to the specimen than electron imaging. Even in high resolution imaging, the dose can be held significantly below 10^3 e/nm² by low

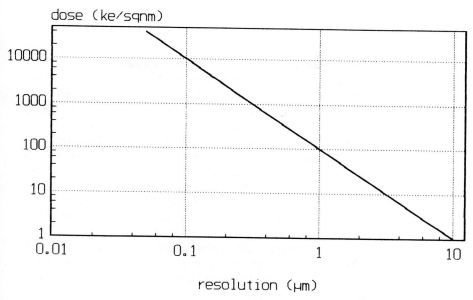

Fig. 8.1. First order approximation of dose (in ke/nm²) vs resolution required for fully quantitative X-ray microanalysis. The graph is computed assuming a charge of 10^{11} e per analytical run (Hall & Gupta, 1984). It might be refined by taking into consideration the energy of the electron beam (Lamvik, 1991) and the relationship between number of interactions and thickness of the sample. This would lead to a slightly better figure for low-resolution work but to even higher doses (about one order of magnitude) for high-resolution X-ray microanalysis.

dose techniques. The dose required for X-ray microanalysis is orders of magnitude higher than this as soon as subcellular resolution becomes the objective (Fig. 8.1).

In contrast to this pessimistic view, many reports have shown the practical possibility of performing meaningful high-resolution biological electron microscopy and X-ray microanalysis. Although our understanding of the basic cascades of events leading to observable radiation damage is still incomplete (Zeitler, 1990), a number of important results have been presented, allowing a phenomenological description of the possibilities and limitations of biological X-ray microanalysis.

Both from a practical and a theoretical point of view it is useful to divide radiation effects into two categories, relating to elemental relocation and mass loss.

1   Relocation artefacts

These involve breakdown of chemical bonds, production of molecular fragments including free radicals, and dispersal or dislocation of these fragments. Free radicals especially will react with other molecules within the specimen, spreading the primary damage out from the irradiation side in a cascade of chemical reactions. However, the mean free path for highly reactive species is rather short, and this type of damage is, therefore, important mainly in high-resolution electron microscopy (Zeitler, 1990). It occurs typically at doses well below 100 e/nm^2 and can be reduced by less than one order of magnitude by cooling to liquid nitrogen temperature. No further significant reduction is possible by cooling below 10 K (International Experimental Study Group, 1986). This type of radiation damage is not amplified by the presence of ice in the frozen sample (Lepault, Booy & Dubochet, 1983).

As long as X-ray microanalysis is not performed at very high resolution, i.e. better than a few nm, dislocation damage is not of interest to X-ray microanalysts. This is not so with the second category of radiation damage, namely mass loss.

2   Mass loss

Mass loss is based on all the effects summarised above. However, in addition it is necessary for the (primary or secondary) molecular fragments produced to reach the surface of the specimen with enough energy to escape. Due to this additional requirement, mass loss as compared to dislocation damage has some distinctive features, which can be summarised as follows.

1. In dry organic material at room temperature the dose, $D(1/e)$, describing the rate of mass loss, is approximately 10^2 to 10^3 e/nm^2, where: $D(1/e)$ is defined as the dose at which all but 37% ($1/e$) of the total damage has occurred (Lamvik, 1991). This is about one order of magnitude higher than the $D(1/e)$ for the fading of diffraction spot intensities (International Experimental Study Group, 1986). Moreover, cooling to liquid nitrogen temperature appears to be effective in reducing mass loss in this type of specimen, often increasing the $D(1/e)$ to values above 10^5 e/nm^2 (see Fig. 8.2). According to the data presented, and in contrast to dislocation damage, there seems to be a possibility that liquid helium cooling might lead to a significant improvement as compared to liquid nitrogen (Fig. 8.2). Mass loss in dry organic material increases logarithmically with the electron dose, at least at room and liquid nitrogen temperature (Egerton, 1982; Cantino *et al.*, 1986; Lamvik, Kopf & Davilla, 1987). This implies a 'single-hit' process as in the case of dislocation damage. At liquid helium temperature, however, the mass–dose relationship in thin collodion films

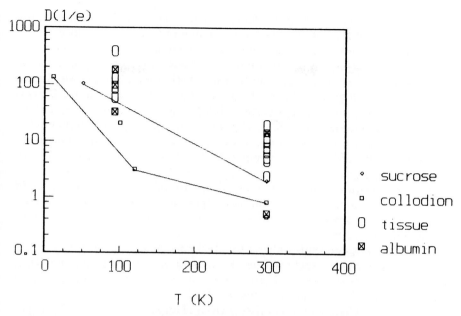

Fig. 8.2.   Values for the *D*(1/e) characteristic for mass loss from some dry organic
materials (in ke/nm²) vs temperature *T* (in K). Data are taken from
Freeman, Leonard & Dubochet (1980), Egerton (1982), Cantino *et al.*
(1986), Lamvik *et al.* (1987) and Lamvik (1991). Data from within one
study are connected by straight lines, if they include measurements made
at liquid helium temperature.

appears to be linear rather than logarithmic (Lamvik *et al.* 1987), which
means that these films are completely etched away after application of high
doses.

2. In dry organic material irradiated at room and liquid nitrogen temperature, a certain fraction of mass remains even after very high doses (up
to about 10^7 e/nm² at liquid nitrogen temperature). This ultimate fraction
of damage appears to be constant for different materials. For example,
Cantino *et al.* (1986) measured dry mass losses of $50 \pm 5\%$ for albumin,
$46 \pm 5\%$ for frozen–dried salivary gland sections and $45 \pm 7\%$ for muscle
homogenate sections irradiated at room temperature with doses around
10^5 e/nm². Wall & Hainfeld (1986) found values around 50% for different
proteins. According to Cantino *et al.* (1986), variations in mass loss are
larger at liquid nitrogen temperatures, if specimens were irradiated with
doses up to 10^7 e/nm². However, there is evidence that at those high doses
mass loss will start to increase again, together with a loss of certain
elements like S or Cl (Cantino *et al.*, 1986). Taken together, these data
suggest that most thin organic films lose about 50% of their mass at

doses below 10^5 e/nm^2. If irradiated further, the remaining mass does not change much for about two orders of magnitude in electron dose.

3. Frozen water and dilute solutions show completely different behaviour (Dubochet *et al.*, 1982). In thin films, mass loss is not dependent on specimen thickness and increases linearly with dose, becoming significant at doses in the order of 10^4 e/nm^2 at liquid nitrogen temperature. There is no big improvement in specimen stability at temperatures down to 25 K. If organic material is diluted in the water, damage starts with 'bubbling' at the ice–organic interphase at a dose up to an order of magnitude lower (Dubochet *et al.*, 1982; Talmon, Adrian & Dubochet, 1986; Zierold, 1988).

   The important point is that, as in the case of collodion films at liquid helium temperature, a thin film of a dilute suspension (a frozen hydrated section of 'normal' wet tissue, say) will be completely etched away at higher doses. Therefore, the proportion of the irradiated specimen that is lost by etching depends only on the thickness of the specimen.

4. Element-specific losses have been shown to occur for: the light elements C, O and N (Egerton, 1982); Cl from microdroplets (LeRoy & Roinel, 1983) and from polyvinylchloride samples (Hall, 1986); S from many organic materials, including different tissues (Rick, Dörge & Thurau, 1982; Hagler & Buja, 1986; Cantino *et al.*, 1986; von Zglinicki & Uhrik, 1988). With all these elements, specific losses occur preferentially at ambient temperature. Liquid nitrogen temperature reduces these losses to insignificant values, at least for doses up to about 10^7 e/nm^2 (Hagler & Buja, 1986; Cantino *et al.*, 1986; von Zglinicki & Uhrik, 1988).

It must be clearly borne in mind that all the data summarised so far have been obtained on, and are strictly valid only for, some model substances under specific experimental conditions. Linearity of mass loss vs dose, and independence of mass loss from specimen thickness, as shown by Dubochet *et al.* (1982) for frozen–hydrated sections, has not been confirmed by Hall (1986) in a study of thick (1 to 2 μm) cryosections of a 100 mM KCl in 20 % dextran solution. Cantino *et al.* (1986) have stressed the importance of carefully controlled experimental conditions such as maintaining a temperature difference between anticontamination blades and the specimen for minimising contamination of cooled samples. It has been found by most authors that deviations between single experiments to determine mass loss were rather large, with differences by a factor of two being normal rather than the exception. Data from different sources are hard to compare, because doses as expressed in e/nm^2 measure the charge delivered to a specimen area, but not the energy deposited onto it. The energy dose depends both on the acceleration of the impinging electrons and on the thickness of the specimens relative to the mean free pathlength of electrons in it. However, even with this constraint some conclusions of practical importance can be drawn from published data.

Calculation of the amount of etching of (originally) frozen–hydrated specimens under the beam is rather easy. Results obtained (Hall, 1986; Zierold, 1988) vary depending on the primary data used, and on the dose taken as necessary for X-ray microanalysis. However, the result under any of the assumptions is that high-resolution X-ray microanalysis (that means resolution better than a few 100 nm) on frozen–hydrated cryosections will hardly ever be possible, because to achieve this goal requires a section thickness considerably smaller than about 1 μm, and those thin sections will be dried away under the beam before any stationary analysis will be completed. In this respect, the data reported by the group of Gupta & Hall, who obtained an analytical resolution in the order of 0.3 to 0.4 μm in 1 μm thick frozen–hydrated sections (Hall & Gupta, 1984), still mark the state of the art.

About 10^{11} electrons per analysed area are necessary to obtain a quantitatively evaluable spectrum (Hall & Gupta, 1984). This necessary electric charge depends on the thickness of the specimen and might well be one order of magnitude larger in thin sections. Therefore mass loss is unavoidable in most X-ray microanalytical examinations. However, the fact that the dry organic mass remains relatively constant over a range of at least two orders of magnitude in electron dose (10^5 to 10^7 e/nm² in specimens at liquid nitrogen temperature) after the initial loss is completed, and the fact that this 'ultimate' mass loss is rather similar for different biological materials (Cantino *et al.*, 1986; Wall & Hainfeld, 1986), explains why quantitative X-ray microanalysis of frozen–dried biological specimens is possible even at high resolution. An important proviso is that standards matching the radiation sensitivity of the biological sample are used and analysed in the same range of radiation doses. In fact, a huge number of studies has obtained physiologically meaningful and thoroughly tested results using electron doses in the order of 10^5 to 10^7 e/nm² (see, for example, Somlyo, Bond & Somlyo, 1985; Hagler & Buja, 1986; von Zglinicki & Bimmler, 1987).

Although the variation in the results obtained at liquid nitrogen temperature might be larger than at room temperature due to increasing variation in the fraction of mass lost (Cantino *et al.*, 1986), cooling is preferable because it effectively blocks element specific losses, especially of S (Rick *et al.*, 1982; Hagler & Buja, 1986; von Zglinicki & Uhrik, 1988). If one accepts this variability of results, the maximum dose tolerable in a dry organic specimen will be in the order of 10^7 e/nm². Above this dose both total mass and the amounts of specific elements, such as S and Cl, start to drop again (Cantino *et al.*, 1986). With this maximum tolerable dose, the best obtainable resolution in stationary spot X-ray microanalysis of dry biological material will be about 0.1 μm (Fig. 8.1). This is quite in accord with the applications cited above, and one order of magnitude better than the value calculated by Roos & Morgan (1990) assuming 10^5 e/nm² as maximum tolerable dose.

It should be stressed that it is of prime importance to have all the water removed from the specimen in order to realise these results. For that reason, Zierold (1988) recommends freeze-drying of cryosections at temperatures as high as 230 K.

D   Further techniques to minimise radiation damage effects during low temperature X-ray microanalysis

Apart from cooling there are various other countermeasures to reduce the effects of radiation damage to biological specimens. The first of these is simple carbon coating of sections. A 20 nm carbon coat was shown to increase the stability of frozen–hydrated sections by a factor of 2 to 4 (Zierold, 1988). Coating also reduces radiation damage effects in dry specimens (Fryer, McNee & Holland, 1984). It is assumed that coatings may trap products of radiolysis (Berriman & Leonard, 1986). Moreover, they increase the conductivity of biological samples which might confer additional beneficial effects (Lamvik, 1991).

A second possibility is the use of spatial averaging techniques in X-ray microanalysis. Spatial averaging comparable to the principles of correlation averaging used in high-resolution electron microscopy (Zeitler, 1990) is, in principle, done in any stationary spot analysis as soon as the results from a number of compartments of the same type are summed up. However, this can be done much more effectively in quantitative imaging (Saubermann & Heymann, 1987; Ingram *et al.*, 1988; Fiori *et al.*, 1988). Due to the exhaustive sampling in quantitative imaging, averaging over the full size even of irregularly shaped compartments is possible. Pixel size and spot size should be equal to obtain the maximum information at a given electron dose.

To give a quantitative example, the charge delivered per pixel might well be lower than 1 nC. With a spot and pixel size of 0.1 μm a dose significantly less than 10^6 e/nm² would be applied. A resolution of even 50 nm has been claimed (Fiori *et al.*, 1988), using a probe current of 4 nA for 200 ms per pixel, which corresponds to a dose of about 2.5×10^6 e/nm². Pixel size might in fact be equal to the analytical resolution if averaging over identical subunits is possible as, for instance, in striated muscle (Cantino & Johnson, 1988).

However, in spite of taking every precaution to reduce the radiation dose applied to the biological sample, irradiation will still result in significant mass loss from the specimen. In fact, due to its exponential decay at lower doses, specimen mass thickness will depend critically on the applied electron dose. Under these conditions it is extremely important that both the specimen and the standard are measured using exactly the same dose, a condition which might not easily be met in practice.

A somewhat contrary approach to spatial averaging is spatial deconvolution. This technique might be used to evaluate the element concentrations in a small compartment embedded within one or more larger ones using a spot size considerably larger than the size of the small compartment and decreasing the electron dose with the second power of the spot diameter (Tormey & Walsh, 1989). All that is needed are: (i) the element concentrations in the large compartment (which can normally be measured separately), and (ii) the volume fractions of the compartments under the measuring spot. If very high resolution is required, as in the case of the analysis of calcium concentrations in the muscle cell sarcolemma, the density distribution in the electron beam may also be required (Tormey & Walsh, 1989). Volume fractions might be measured by morphometry, or compartments like the extracellular space might be marked by some tracer. The statistical accuracy of the spatial deconvolution technique decreases with decreasing gradients between the compartments in question and, especially, with the ratio of compartment size to spot size.

The last possibility is extrapolation to zero dose. This technique has recently been used in a study of the halogen content of extremely radiation sensitive copper phthalocyanides (McColgan, Chapman & Nicholson, 1990). The maximal cumulative dose used in this study was about 6×10^5 e/nm^2 distributed onto 10 spectra from the same specimen area. The basic idea is that if a regression model for the dose dependency is given (which was linear in the cited study, but should be an exponential decay of mass in dry biological samples), the accuracy of the fit, and even of the extrapolation to zero, should still be better than that of a single spectrum. This allows each spectrum to be collected with a lower dose. Once installed, both the acquisition of multiple spectra (with beam blanking in between) and the extrapolation could easily be automated. Thus it would be no more laborious or time-consuming than conventional stationary spot X-ray microanalysis.

E Consequences of radiation damage for water content measurements

Two of the principal methods to measure local water fractions require a fully quantitative X-ray microanalysis in the frozen–hydrated state (approaches 1. and 2. above). Heide (1984) has estimated the dose which will remove a 1 nm layer of ice to be about 3×10^3 e/nm^2. This holds for temperatures low enough to prevent sublimation of ice. Assuming that in a hydrated section at least 1 μm thick the total charge necessary for quantitative X-ray microanalysis will be a few 10^{10} electrons only, one arrives at the most optimistic estimation of the attainable resolution of about 0.3 μm. This corresponds to an applied dose of some 10^5 e/nm^2 which would etch the section to about 10 % of its depth. This is the best resolution ever realised with fully quantitative X-ray microanalysis of hydrated sections (Gupta & Hall, 1979). To my knowledge, this resolution has not hitherto been obtained in any other laboratory. Instead, X-ray microanalysis has been used at lower resolution in bulk frozen–hydrated specimens (Zs.-Nagy, Lustyik & Bertoni-Freddari, 1982; Echlin, Lai & Hayes, 1982; von Zglinicki & Lustyik, 1986), an approach facilitated by use of the background-under-the-peak quantitation method that corrects for X-ray absorption in bulk samples (Boekestein *et al.*, 1984). Etching is insignificant in bulk samples but the resolution is rather low.

Resolution in bulk samples is defined primarily by the penetration depth of electrons. This will be in the order of 2 µm in frozen–hydrated samples but amounts to about 10 µm in frozen–dried specimens at 20 kV accelerating voltage. Water content measurements will be biased as soon as the distribution of ions and/or water is inhomogeneous within the larger volume. Overpenetration of cells by the beam in frozen–dried samples, and the inclusion of extracellular compartments, might be an especially serious source of artefacts.

It can be avoided by the use of semithick (cryostat) sections (Wroblewski, Wroblewski & Roomans, 1987), provided that the retention of full hydration in those sections cut at rather high temperatures can be guaranteed. This requires the careful adjustment of the partial water pressure in the atmosphere of the cryomicrotome (von Zglinicki, 1991b).

The third principal approach, namely the measurement of mass thickness in sections, does not require a fully quantitative X-ray analysis in the frozen–hydrated state. Only background counts have to be measured, and this helps to reduce the radiation dose. The price one has to pay for this advantage is that one has to assume that the impinging electron dose remains constant, and that no shrinkage of sections during freeze-drying occurs, otherwise both effects have to be corrected for. In many cases, the reduction of electron dose is still not sufficient to minimise the etching of thin frozen–hydrated sections to a tolerable amount. Spatial averaging in the frozen–hydrated state has been suggested for those cases (Bulger, Beeuwkes & Saubermann, 1981; Saubermann & Heyman, 1987). Additional problems of low dose, low resolution spatial averaging (for example smoothing of results, and the inherent assumption of constant section thickness) have been discussed earlier (von Zglinicki, 1991a).

The radiation dose can be decreased by orders of magnitude if electron scattering is used to measure the mass thickness of sections instead of X-ray microanalysis. Both bright-field (Zeitler, 1971; Linders, Stols & Stadhouders, 1981; von Zglinicki, Bimmler & Krause, 1987; Strain *et al.*, 1989) and dark-field intensity (Zierold, 1988) are linearly related to mass thickness in thin specimens and might be used. Linearity can be improved by excluding inelastically scattered electrons by energy filtering (Ornberg & Kuijpers, 1989).

In addition to the low doses required, these techniques offer several other advantages:

- linearity is best for ultrathin samples;
- statistics is considerably improved as compared with X-ray microanalysis (Linders *et al.*, 1981);
- measurements in frozen–hydrated and frozen–dried sections can be done in the very same compartment, excluding the influence of section thickness; and
- the techniques lend themselves to simple corrections for beam current drift and shrinkage of sections, especially if photographic recording is used (von Zglinicki, 1991a).

These advantages make the techniques especially well suited for the estimation of water distributions at higher resolution (better than a few 100 nm) in ultrathin cryosections. Problems with these methods arise from two facts. First, shrinkage of ultrathin cryosections during freeze-drying is highly variable, and might be rather significant (up to 30 % in one dimension). The amount of shrinkage in an individual section appears unpredictable, depending probably, among other factors, on the contact between a section and its supporting film. Therefore, an individual shrinkage correction for every section in two perpendicular directions is advised (von Zglinicki, 1991a). Even then, correction factors might be rather large, contributing significantly to the variation of data. Moreover, the method will lead to biased results if the amount of shrinkage is different in different compartments within one section. Empty clefts between compartments as seen in frozen–dried sections might be an indicator of inhomogeneous shrinkage. Those sections should not be used for the measurement of local water fractions. This advice holds even though these clefts seem to appear mainly in rather slowly frozen samples, indicating that they are a consequence of osmotic water shifts during freezing rather than the result of inhomogeneous freeze-drying (von Zglinicki, 1988, 1991b).

A second problem of these electron scattering techniques stems from the fact that partial dehydration of ultrathin sections might take place even under the comparatively low doses involved in overview imaging and recording. The dose required for low magnification (less than $10000 \times$) imaging will be typically in the order of 10^2 e/nm^2 per second. This will leave about 10^2 s for imaging before significant etching of a 0.1 µm section takes place. Low dose searching at very low magnification ($1000 \times$ or less) is necessary, therefore. However, this figures holds only under the assumption that the temperature of the specimen is well below 150 K. The temperature of a cryosection on a grid under vacuum is hardly controlled due to the section's poor contact with the supporting film. The water in the cryosections might easily be removed by freeze-drying due to local rewarming rather than by radiolysis under the beam. Sublimation becomes significant at temperatures above 150 K (Heide, 1984). Sandwiching of sections between two heavily carbon-coated films improves the thermal contact and effectively helps to prevent dehydration during imaging (von Zglinicki, 1991a).

Some techniques for measuring local water fractions have been developed which do not require analysis in the frozen–hydrated state at all (Rick *et al.*, 1982; von Zglinicki *et al.*, 1987; Zierold, 1988). The common feature of these techniques is that they use external data to standardise the mass thickness measurement on frozen–dried specimens in terms of local water content. This circumvents problems of radiation damage in frozen–hydrated specimens completely. Local water fractions can be measured even if no cryostage is available (von Zglinicki & Zierold, 1989). However, additional assumptions are involved, the most important being that section thickness is constant in all the compartments compared. These underlying assumptions have been discussed in detail recently (von Zglinicki, 1991a).

References

Berriman, J. & Leonard, K.R. (1986). Methods for specimen thickness determination in electron microscopy. 2. Changes in thickness with dose. *Ultramicroscopy*, **19**, 349–66.

Boekestein, A., Thiel, F., Stols, A.L.H., Bouw, E. & Stadhouders, A.M. (1984). Surface roughness and the use of peak to background ratios in the X-ray microanalysis of bulk bio-organic samples. *Journal of Microscopy*, **134**, 327–34.

Bulger, R.E., Beeuwkes, R. & Saubermann, A.J. (1981). Application of scanning electron microscopy to X-ray analysis of frozen–hydrated sections. III. Elemental content of cells in the rat renal papillary tip. *Journal of Cell Biology*, **88**, 274–80.

Cameron, I.L. (1985). Electron probe X-ray microanalysis studies on the ionic environment of nuclei and the maintainance of chromatin structure. In *Advances in microscopy*, ed. R.R. Cowden & F.W. Harrison, pp. 223–41. New York: Alan R. Liss.

Cantino, M.E. & Johnson, D.E. (1988). Elemental imaging techniques in studies of striated muscle. In *Microbeam Analysis – 1988*, ed. D.E. Newbury, pp. 427–8. San Francisco: San Francisco Press.

Cantino, M.E., Wilkinson, L.E., Goddard, M.K. & Johnson, D.E. (1986). Beam induced mass loss in high resolution biological microanalysis. *Journal of Microscopy*, **144**, 317–28.

Dubochet, J., Lepault, J., Freeman, R., Berriman, J.A. & Homo, J.C. (1982). Electron microscopy of frozen water and aqueous solutions. *Journal of Microscopy*, **128**, 219–38.

Echlin, P. (1991). Ice crystal damage and radiation effects in relation to microscopy and analysis at low temperatures. *Journal of Microscopy*, **161**, 159–70.

Echlin, P., Lai, C.E. & Hayes, T.L. (1982). Low-temperature X-ray microanalysis of the differentiating vascular tissue in root tips of *Lemna minor L*. *Journal of Microscopy*, **126**, 285–306.

Edelmann, L. (1984). Subcellular distribution of potassium in striated muscle. *Scanning Electron Microscopy*, **1984/II**, 875–88.

Egerton, R.F. (1982). Organic mass loss at 100 K and 300 K. *Journal of Microscopy*, **126**, 95–100.

Fiori, C.E., Leapman, R.D., Swyt, C.R. & Andrews, S.B. (1988). Quantitative X-ray mapping of biological cryosections. *Ultramicroscopy*, **24**, 237–50.

Freeman, R., Leonard, K.R. & Dubochet, J. (1980). The temperature dependence of beam damage to biological samples in the scanning transmission electron microscope. In *Proceedings of the Seventh European Congress on Electron Microscopy*, ed. P. Brederoo & W. de Priester, Leiden, Vol. 2, p. 640.

Fryer, J.R., McNee, C. & Holland, F.M. (1984). The further reduction of radiation damage in the electron microscope. *Ultramicroscopy*, **14**, 357–8.

Gupta, B.L. & Hall, T.A. (1979). Quantitative electron probe X-ray microanalysis of electrolyte elements within epithelial tissue compartments. *Federation Proceedings*, **38**, 144–53.

(1981). The X-ray microanalysis of frozen–hydrated sections in scanning electron microscopy: an evaluation. *Tissue and Cell*, **13**, 623–43.

Gupta, B.L., Hall, T.A. & Maddrell, S.H.P. (1976). Distribution of ions in a fluid-transporting epithelium determined by electron-probe X-ray microanalysis. *Nature*, **264**, 284–6.

Hagler, H.K. & Buja, L.M. (1986). Effect of specimen preparation and section transfer techniques on the preservation of ultrastructure, lipids and elements in cryosections. *Journal of Microscopy*, **141**, 311–18.

Hall, T.A. (1986). Properties of frozen sections relevant to quantitative microanalysis. *Journal of Microscopy*, **141**, 319–28.

Hall, T.A. & Gupta, B.L. (1984). The application of EDXS to the biological sciences. *Journal of Microscopy*, **136**, 193–208.

Heide, H.G. (1984). Observations on ice layers. *Ultramicroscopy*, **14**, 271–8.

Ingram, P., LeFurgey, A., Davilla, S.D., Sommer, J.R., Mandel, L.J., Liebermann, N. & Herlong, J.R. (1988). Quantitative elemental X-ray imaging of biological cryosections. In *Microbeam analysis – 1988*, ed. D.E. Newbury, pp. 433–9. San Francisco: San Francisco Press.

International Experimental Study Group (1986). Cryoprotection in electron microscopy. *Journal of Microscopy*, **141**, 385–91.

Kellermayer, M., Ludany, A., Jobst, K., Szucs, G., Trombitas, K. & Hazlewood, C.F. (1986). Cocompartmentation of proteins and K within the living cell. *Proceedings of the National Academy of Sciences USA*, **83**, 1011–15.

Lamvik, M.K. (1991). Radiation damage in dry and frozen hydrated organic material. *Journal of Microscopy*, **161**, 171–81.

Lamvik, M.K., Kopf, D.A. & Davilla, S.D. (1987). Mass loss rate in collodion is greatly reduced at liquid helium temperature. *Journal of Microscopy*, **148**, 211–18.

Lepault, J., Booy, F.P. & Dubochet, J. (1983). Electron microscopy of frozen biological suspensions. *Journal of Microscopy*, **129**, 89–102.

LeRoy, A.F. & Roinel, N. (1983). Radiation damage to lyophilized mineral solutions during electron probe analysis: quantitative study of chlorine loss as a function of beam current-density and sample mass-thickness. *Journal of Microscopy*, **131**, 97–106.

Linders, P.W.J., Stols, A.L.H. & Stadhouders, A.M. (1981). Quantitative electron probe X-ray microanalysis of thin objects using an independent mass determination method. *Micron*, **12**, 1–5.

Marshall, A.T. (1980). Quantitative X-ray microanalysis of frozen-hydrated bulk biological specimens. *Scanning Electron Microscopy*, **1980/II**, 335–48.

McColgan, P., Chapman, J.N. & Nicholson, W.A.P. (1990). Low-temperature EDX microanalysis of halogenated copper phthalocyanine samples. *Journal of Microscopy*, **160**, 315–25.

Ornberg, L. & Kuijpers, G.A.J. (1989). Application of X-ray microanalysis and electron energy loss spectroscopy to studies of secretory cell biology. In *Electron probe microanalysis*, ed. K. Zierold & H.K. Hagler, pp. 153–68. Berlin: Springer-Verlag.

Rick, R., Dörge, A. & Thurau, K. (1982). Quantitative analysis of electrolytes in frozen dried sections. *Journal of Microscopy*, **125**, 239–47.

Roos, N. & Morgan, A.J. (1990). *Cryopreparation of thin biological specimens for electron microscopy: Methods and applications*. Microscopy Handbooks 21, p. 66. Oxford: Oxford University Press.

Saubermann, A.J. & Heyman, R.V. (1987). Quantitative digital X-ray imaging using frozen hydrated and frozen dried tissue sections. *Journal of Microscopy*, **146**, 169–82.

Somlyo, A.P., Bond, M. & Somlyo, A.V. (1985). Calcium content in mitochondria and endoplasmic reticulum in liver frozen rapidly *in vivo*. *Nature*, **314**, 622–4.

Stenn, K. & Bahr, G.F. (1970). Specimen damage caused by the beam of the transmission electron microscope, a correlative reconsideration. *Journal of Ultrastructural Research*, **31**, 317–39.

Strain, J.J., Kopf, D.A., Le Furgey, A., Ingram, P., Hawkey, L.A. & Davilla, S.D. (1989). Computer-assisted bright-field signal measurement of water content in frozen–hydrated and dehydrated gelatin thin sections: a preliminary report. In *Microbeam analysis – 1989*, ed. P.E. Russell, pp. 97–102. San Francisco: San Francisco Press.

Talmon, Y., Adrian, M. & Dubochet, J. (1986). Electron beam radiation damage to organic inclusions in vitreous, cubic, and hexagonal ice. *Journal of Microscopy*, **141**, 375–84.

Tormey, J.McD & Walsh, L.G. (1989). Strategies for spatial deconvolution of calcium stores in cardiac muscle. In *Microbeam analysis – 1989*, ed. P.E. Russell, pp. 115–17. San Francisco: San Francisco Press.

Wall, J.S. & Hainfeld, J.F. (1986). Mass mapping with the scanning transmission electron microscope. *Annual Review of Biophysics and Biophysical Chemistry*, **15**, 355–76.

Wroblewski, R., Wroblewski, J. & Roomans, G.M. (1987). Low temperature techniques in biomedical microanalysis. *Scanning Microscopy*, **1**, 1225–40.

Zeitler, E. (1971). Trockengewichtsbestimmung von Teilchen mit dem Elektronenmikroskop. In *Methodensammlung der Elektronenmikroskopie*, ed. A. Schimmel & W. Vogell. Stuttgart: Wissenschaftliche Verlagsgesellschaft.

   (1990). Radiation damage in biological electron microscopy. In *Biophysical electron microscopy. Basic concepts and modern techniques*, ed. P.W. Hawkes & U. Valdré, pp. 289–308. London: Academic Press.

von Zglinicki, T. (1988). Intracellular water distribution and ageing as examined by X-ray microanalysis. *Scanning Microscopy*, **2**, 1791–804.

   (1991a). The measurement of water distribution in frozen specimens. *Journal of Microscopy*, **161**, 149–58.

   (1991b). Reliability of intracellular water and ion distributions as measured by X-ray microanalysis. *Scanning Microscopy, suppl. 5* (in press).

von Zglinicki, T. & Bimmler, M. (1987). The intracellular distribution of ions and water in rat liver and heart muscle. *Journal of Microscopy*, **146**, 77–86.

von Zglinicki, T., Bimmler, M. & Krause, W. (1987). Estimation of organelle water fractions from frozen–dried cryosections. *Journal of Microscopy*, **146**, 67–76.

von Zglinicki, T. & Lustyik, G. (1986). Loss of water from heart muscle cells during aging of rats as measured by X-ray microanalysis. *Archives of Gerontology and Geriatrics*, **5**, 283–9.

von Zglinicki, T. & Uhrik, B. (1988). X-ray microanalysis with continuous specimen cooling: is it necessary? *Journal of Microscopy*, **151**, 43–7.

von Zglinicki, T. & Zierold, K. (1989). Elemental concentrations in air-exposed and vacuum-stored cryosections of rat liver cells. *Journal of Microscopy*, **154**, 227–35.

Zierold, K. (1988). X-ray microanalysis of freeze-dried and frozen–hydrated cryosections. *Journal of Electron Microscopical Techniques*, **9**, 65–82.

Zs.-Nagy, I., Lustyik, G. & Bertoni-Freddari, C. (1982). Intracellular water and dry mass content as measured in bulk specimens by energy-dispersive X-ray microanalysis. *Tissue and Cell*, **14**, 47–60.

9 *X-ray microanalysis in histochemistry*

A. T. Sumner

A The suitability of X-ray microanalysis for histochemistry

The great majority of applications of X-ray microanalysis (XRMA) in biology may be regarded as histochemical, in the sense that they are studies of the chemical composition of cells or tissues *in situ*. Conventionally, however, the term histochemistry is used to denote the investigation of the chemistry of cells and tissues *in situ* using specific reactions that produce a distinctive reaction product that can be detected by some microscopical method. For light microscopy, a coloured reaction product is normally used. Although this approach has been extremely successful, not every cellular substance of interest can be identified in this way. In addition, coloured reaction products may diffuse, or dissolve in mountants, and can rarely be quantified satisfactorily. For electron microscopy, reliance is placed on reaction products which are electron dense. As well as the problems of diffusion and quantification noted above for light microscopical histochemistry, it may also be difficult at times to distinguish the reaction product from the normal staining of the section. Alternatively, if the section is left unstained, the histochemical reaction product may be clearly visible, but difficult to localise because of the low contrast of the rest of the material. X-ray microanalysis has the potential to help with all these problems.

Problems of diffusion of histochemical reaction products are likely to be greatest in multi-step reactions, such as those for many enzymes. XRMA can be used to detect intermediate reaction products, at stages when the potential for diffusion is less. Although these intermediate reaction products are not coloured or electron dense, they often contain distinctive elements that can be detected by XRMA. The problem of the reaction product dissolving in the mountant does not occur with XRMA, since of course the specimens are examined dry and unmounted. Quantification of histochemical reaction products is essentially simple with XRMA, since the characteristic X-ray signal is, in fact, proportional to the amount of the substance being analysed. This simple proportionality does not hold for absorbing reaction products, and probably only rarely for fluorescent reaction products. Of course, the stoichiometry of the histochemical reaction also needs to be established for useful quantification, but here again, XRMA can help, since in some cases the substrate for the reaction contains distinctive elements, and the amount of these can be compared with the amount of the reaction product. Indeed, the ease with which XRMA can be used to detect simultaneously two or more substances makes it a very powerful tool in histochemistry, as in its other fields of application.

Finally, XRMA does not require coloured or electron-dense reaction products. In EM histochemistry, this can help to avoid confusion with the ordinary staining of the section. In addition, it should be possible to devise new histochemical procedures specifically for XRMA that require neither a coloured nor an electron-dense reaction product.

Histochemical applications of XRMA have, in fact, been made in all the fields described above, and will be described in more detail later in this chapter. It is, nevertheless, clear that XRMA has not been widely used in histochemical studies, but remains a specialised technique. The probable reasons for this will now be considered briefly.

B   Limitations of X-ray microanalysis in histochemistry

Availability of equipment may be a major reason why XRMA has not been more widely applied to histochemical problems. Whereas any biological or medical department will have many light microscopes (probably including some that can be used for fluorescence), and most will have one or more electron microscopes, equipment for X-ray microanalysis is still not readily available to many biologists. Moreover, the emphasis on measurements of intracellular concentrations of ions has perhaps led to the belief that XRMA is a technique for cell physiologists rather than histochemists. Certainly there are many problems in cell physiology that can only be approached using XRMA (or similar techniques), whereas good histochemical data for solving many problems can be obtained by other methods.

If the histochemist has access to an electron microscope with equipment for XRMA, experience will soon show that this is a very slow technique. Quantification is always a much more time-consuming process than merely taking a quick look to see whether a reaction has worked or not. However, there is no facility with XRMA for taking a quick look; it is always necessary to accumulate slowly an X-ray spectrum. While it is true that the results, once accumulated, are in some sense quantitative, this may not be an important aim of the investigation, and therefore XRMA may not seem an efficient technique. There are also practical advantages in being able to assess at a glance whether a histochemical reaction has worked or not before committing oneself to the more lengthy procedures of measuring the reaction product. These considerations no doubt help to ensure that XRMA is used as a specialised technique in histochemistry, where its particular attributes help to solve specific problems not easily tackled by other methods, rather than becoming a general technique.

It might be supposed that XRMA would have fruitful applications in electron microscope cytochemistry, but here again there are significant limitations. As generally practised, an X-ray microanalytical spatial resolution on thin specimens of biological material in the region of 0.1–0.5 μm can be obtained (Marshall, 1980; Hall & Gupta, 1983, 1984; Sumner, 1983). This is, of course, not significantly better than the resolution limit of light microscopy. Nevertheless, genuine ultrastructural XRMA has been done on cytochemical preparations (Somlyo *et al.*, 1979; Somlyo & Shuman, 1982; Stewart *et al.*, 1980). To obtain improved resolution, an electron source of much higher brightness (such as a field-emission gun) is required. With such a source, sufficient current density can be obtained in a sufficiently small area to provide an adequate X-ray signal. However, a high probe current will probably cause unacceptable radiation damage to the specimen, thereby tending to vitiate any advantages of the improved structural resolution (see von Zglinicki, this volume).

C   Fields of application of X-ray microanalysis in histochemistry

In spite of the limitations described above, XRMA has been used to tackle a wide variety of problems in histochemistry. These are summarised in Table 9.1, and are discussed in more detail in the following sections.

1   Tracers

XRMA has been used to study various tracers, in cell surface labelling, immunocytochemistry, autoradiography, and in certain other applications. Among tracers which can be used for cell surface labelling or immunocytochemistry are iron-containing ferritin and iron dextran, colloidal gold, and the diaminobenzidine reaction product, which can be labelled with silver or gold. Baccetti & Burrini (1977) and Rosenquist & Huff (1986) used concanavalin A (Con A) labelled with iron dextran to detect sites of Con A binding on the surface of cells. In both cases XRMA was used to quantify Con A binding rather than merely to detect it. Ap Gwynn (1981) showed how XRMA could be used to measure the amount of polycationised ferritin bound to negatively charged cell surfaces. Hoyer, Lee & Bucana (1979) showed that colloidal gold particles could be bound to cell surface antigens and detected by XRMA, and they suggested that this procedure could be used to map their distribution. Subsequently, Peschke *et al.* (1986, 1989) have used XRMA to

Table 9.1. *Applications of XRMA in histochemistry*

1.	Tracers
	Cell surface labelling
	Immunocytochemistry
	Autoradiography
	Miscellaneous
2.	Use of colourless or electron-lucent reagents
3.	Validation of cytochemical methods
	Identification of reaction products (including precipitation methods for diffusible ions)
	Study of reaction mechanisms
	Investigation of reaction efficiency
	Stoichiometric studies
4.	Simultaneous localisation of more than one substance: distribution studies
5.	Ultrastructural cytochemistry

quantify the particles bound to cell surfaces, while Eskelinen & Peura (1988) showed that it is possible to analyse individual particles to prove that they are, in fact, composed of gold. Siegesmund, Yorde & Dragen (1979) labelled cells with an immunoperoxidase method; the DAB reaction product was then labelled with gold. It was found that the amount of gold as measured by XRMA correlated well with the amount of antigen in the cells as measured by immunoassay.

XRMA is clearly not a useful method just for the detection of immuno- or cell surface-labels, for which direct observation of electron dense particles by TEM, or of back-scattered electron images by SEM, are no doubt much faster, as well as giving stronger signals. The two fields where XRMA does seem to have useful applications are in the unequivocal identification of the reaction product, and in quantification. Nevertheless, quantification of gold particles can be done quite simply by counting them (Beesley, 1989), and so far no comparative tests seem to have been carried out to see what advantages XRMA might offer for this purpose.

The applications of XRMA to autoradiography have been very similar in concept to those just described for immunolabelling. Thus, Hodges & Muir (1975) proposed the use of XRMA to quantify autoradiographic silver grains, although direct comparison with grain counting to establish the relative merits of the two systems was not reported. However, both they and Ishtani, Miyakawa & Iwamoto (1977) pointed out that XRMA allows unambiguous identification of the silver grains.

Use of a variety of other types of tracers detectable by XRMA has been reported. In one of the earliest such experiments, Clarke, Salsbury & Willoughby (1970) detected the movement along intercellular processes of an iodine-labelled antigen. The experiment could probably be done more easily now, and in living cells, using a fluorescent marker. Walker, Ingram & Shelburne (1984) used the dye erythrosin B, which contains iodine, as a marker for injured cells, and were able to show that such cells had altered electrolyte concentrations when compared with viable cells. Kirkham, Goodman & Chapell (1975) and Hackney & Altman (1983) have applied XRMA to the tracing of neurons injected with cobalt or nickel. Since the tracers could be detected in an electron microscope, it was possible to extend neuroanatomical studies to the ultrastructural level.

Although XRMA is well suited to the study of the concentrations of intracellular ions, it is less appropriate for investigating dynamic changes in such ions. Wroblewski *et al.* (1989) used strontium as an analogue of calcium, and bromide as an analogue of chloride, to study rates of exchange of calcium and chloride in various types of cells, while Krefting *et al.* (1988) used strontium as a tracer for calcium to study transport from blood to the epiphyseal growth plate. Morgan (1981) also used strontium, to study the kinetics of calcium transport from the coelomic fluid of earthworms and the formation of calcite mineral in the calciferous glands. Yet another type of tracer was used by Ingram & Ingram (1985) to measure the water content of tissue. Arguing that the embedding resin replaced the water in the tissue, they used a brominated resin to estimate the water content of the tissue from the bromine X-ray signal.

2 Histochemical procedures with colourless or non-electron-dense reaction products

In principle, the use of XRMA would permit the development of new histochemical techniques which give reaction products that are neither coloured (for light microscopy), nor electron dense (for electron microscopy). In practice, this possibility has scarcely been exploited, perhaps because with such methods there is no possibility of making a quick visual assessment of the reaction. Only two publications dealing with such methods are known to the author. Berry *et al.* (1982) used aluminium as the capture reagent for an acid phosphatase reaction. They regarded the method as being both simpler and more reliable than the Gomori lead precipitation reaction and, moreover, since the reaction product was not electron dense, the structure of the lysosomes was not obscured. Sumner (1984) reported the use of the colourless compound iodophenyl maleimide as a reagent for tissue sulphydryl groups, the iodine atom in this molecule being detectable only by XRMA. However, no further applications of these procedures seem to have been reported.

3   *Validation of histochemical methods*

One of the most fruitful fields of application of XRMA to histochemistry has been in the study of the validity of histochemical procedures. Such studies have ranged from simple quantitative identification of reaction products to quantitative investigations of multi-step procedures for enzymes. There have been a few reports in which XRMA was used simply to identify reaction products. Meisner *et al.* (1973) showed that caesium was present in fixed chromosomes that had been treated with caesium chloride, and suggested that the distribution of caesium determined the subsequent pattern of staining with Giemsa dye, by blocking certain dye-binding sites. Sumner (1978a) showed that Giemsa staining of chromosomes required the incorporation of eosin (identifiable by its bromine atoms) in addition to the binding of thiazine dyes.

More commonly, XRMA has been used to test the specificity of a reaction product for a particular substrate. Gardner & Hall (1969) used XRMA to show that the von Kossa silver technique for calcium is in fact specific for calcium, and does not merely indicate the distribution of phosphate groups associated with the calcium. They showed that there was a quantitative relationship between the amount of silver deposited, and the quantity of calcium present in sections of mineralised bone, although the correlation with phosphorus was much less good. However, the substantial amount of calcium in non-mineralised cartilage was not labelled with silver. Hoefsmit *et al.* (1986) showed that when silver intensification of colloidal gold was carried out, the silver was always deposited on gold particles, and never found without gold, thereby proving the specificity of the method. Danscher, Hansen & Moller-Madsen (1984) also used silver amplification to detect gold, in this case accumulated in kidney lysosomes after treatment with gold-containing anti-arthritic drugs. Again, silver was only found at sites of gold deposition. Whittaker *et al.* (1990) demonstrated the specificity of a silver method for mercury. When silver amplification was used to detect mercury deposits in yeast cells grown in the presence of mercuric chloride, the silver was found in conjunction with mercury and sulphur. Farkas, Szerdahelyi & Kasa (1988) used XRMA to measure zinc in Purkinje cells of rat cerebellum. Although the levels of zinc were too low to be detected directly by XRMA, the zinc could be measured after silver intensification. Studies of model systems showed that there was a linear relationship between the quantity of zinc in the specimen, and the amount of silver deposited.

All the above examples involve silver staining, but Modis *et al.* (1988) used XRMA to study the reaction of a copper-containing dye, Alcian blue, with sulphated glycosaminoglycans in cartilage. They were able to establish the conditions required for optimal specificity of staining, and to show that there was approximately one Alcian blue molecule bound to each sulphate group of chondroitin sulphate. Stockert *et al.* (1991) used XRMA to study the binding of toluidine blue (identified by its sulphur content) to the DNA of chromosomes (identified by its phosphorus content), and were able to measure the stoichiometry of the reaction. Sumner (1986) carried out a comprehensive quantitative study of the binding of the fluorochrome quinacrine to chromosomal DNA, and was able to make deductions about the mode of binding of the dye to DNA, and about the factors influencing its fluorescence.

Before rapid freezing methods were fully established as the only reliable preparative methods for localising diffusible ions in cells and tissues, several methods were devised which depended on precipitation of various ions. Problems with such methods include dislocation and loss of ions during the precipitation process, and such methods are, in general, no longer regarded as acceptable. Nevertheless, during the period when such methods were in vogue, XRMA proved to be a valuable technique for assessing the specificity of these precipitation methods. The pyroantimonate method was originally introduced by Komnick (1962) to immobilise sodium ions, but in fact it was shown that Mg^{2+}, Ca^{2+}, Zn^{2+}, and sometimes K^+, were also precipitated (Simson, Bank & Spicer, 1979; Tisher, Weavers & Cirksena, 1972; van Iren *et al.*, 1979; Wick & Hepler, 1982). Zinc uranyl acetate was shown to precipitate both sodium and potassium ions (Harvey & Kent, 1981), while sulphide precipitated zinc, cadmium and mercury ions (George *et al.*, 1976; de Filippis & Pallaghy, 1975). These studies all showed that the identification of a precipitate was not by itself adequate to identify any specific ion (regardless of the more fundamental considerations of the general reliability of such techniques). In the case of the pyroantimonate precipitation technique, it has been reported that the ratio of calcium to antimony is variable, giving rise to the suggestion that only freely diffusible, and not bound, calcium could react (Chandler & Battersby, 1976; Wick & Hepler, 1982). However, this idea has not been followed up, and in view of the known unreliability of precipitation techniques is unlikely to produce acceptable results.

XRMA has been particularly valuable in studying the validity of several enzyme histochemical methods. Many of these are multi-step methods, with the possibility of inefficient conversion of the primary reaction product to an intermediate product, and thence to the final reaction product. Use of XRMA permits quantitative evaluation of the reaction at each stage, so that it is possible to assess not only the efficiency of conversion of one reaction product to the succeeding one, but also whether the reaction is occurring as expected, or whether it is proceeding in a different way. Some applications of XRMA to enzyme histochemistry are listed in Table 9.2. Earlier studies showed whether or not the reactions were stoichiometric, and also analysed the composition of the reaction products. More recently, measurements of the composition of the reaction products have been used to determine which are genuinely produced by the enzyme being studied (Kyriacou & Garrett, 1980), and as an aid to identify the enzyme itself (Asanuma, 1990).

Table 9.2. *Analysis of histochemical enzyme reaction mechanisms by XRMA*

Enzyme	Observations	References
Acid phosphatase	Different precipitates with different Pb:P ratios, only one of which due to enzyme	Kyriacou & Garrett (1980)
Alkaline phosphatase	Inefficient conversion to final reaction product	Hale (1962)
AMP-PNP-hydrolysing enzyme	Distinguished from ATPase on the basis of the P:Pb ratio of the reaction product	Asanuma (1990)
Muscle ATPase	Inefficient conversion to final reaction product, with diffusion	Engel, Resnick & Martin (1968); Vallyathan & Brody (1977)
Na-K-activated ATPase	Stoichiometric conversion of initial to final reaction product	Beeuwkes & Rosen (1975); Rosen & Beeuwkes (1979)
Esterase	Stoichiometric conversion to intermediate reaction product plus non-specific binding	Vallyathan & Brody (1977)
Esterase (thiolacetic acid method)	Reaction product compared with lead sulphide	Dierkes (1977)
Acetyl-cholinesterase	Nature of final reaction product analysed	Sehgal, Tewari & Malhotra (1981); Tewari, Sehgal & Malhotra (1982)
Succinic dehydrogenase	Final reaction product did not have expected composition	Weavers (1974)

From the foregoing paragraphs it will be clear that XRMA has a valuable, and probably unique, role in studying histochemical reaction mechanisms, and establishing a sound theoretical basis for them. The principal characteristics that make it particularly suitable for such studies are its ability to distinguish easily between different substances, and to quantify them. Unlike techniques such as absorption or fluorescence microspectrophotometry, which have to deal with broad and generally overlapping spectral peaks, the characteristic peaks of an X-ray spectrum are much sharper, and with modern computational techniques can generally be separated quite satisfactorily even when overlapping. Quantification is in principle simple, especially in thin specimens, since the X-ray signal is directly proportional to the amount of the element present (although absorption can, of course, occur). Such a simple proportionality cannot be assumed with fluorescence, or even with light absorption measurements.

4 Simultaneous localisation of more than one substance: distribution studies

Several studies have exploited the ability of XRMA to detect more than one substance simultaneously to make comparative distribution studies. This is a particularly powerful application, which is not so conveniently done using light microscopy (with either absorption or fluorescent staining) because of overlapping spectra, and the ambiguous mixed colours which can arise when two substances coincide.

Goldfischer & Moskal (1966) used XRMA to study lysosomes in the liver cells from patients with Wilson's disease. In this disease, copper is deposited in the cells. Acid phosphatase, used as a marker for lysosomes, was detected using a lead precipitation method. It was found that copper was almost invariably found at the same sites as the lead, indicating that the copper was deposited in the lysosomes. De Bruijn *et al.* (1980) also studied lysosomes, in the Kupffer cells of the liver. The cells were incubated with colloidal gold particles, and once again a lead precipitation method for acid phosphatase was used to identify lysosomes. Three classes of lysosomes could be identified by XRMA: those which contained only lead; those which only contained gold; and those which contained both elements. Bacsy (1982) also found that lysosomes were heterogeneous, this time in adrenocortical cells. Acid phosphatase was detected using a lead precipitation method, and aryl sulphatase using a barium precipitation method. Some lysosomes were found to contain only lead, others only barium, and some contained both elements.

A rather different application was made by Sumner (1981) who studied the distribution of the fluorochrome quinacrine on chromosomes in relation to the DNA distribution. The quinacrine was identified by its chlorine atom, and the DNA by the presence of phosphorus. Although the quinacrine fluorescence varied along the chromosomes, producing the well known Q-banding-patterns (Sumner, 1990), the distribution of the dye itself was relatively uniform, and similar to that of the DNA. However, when the euchromatic and heterochromatic segments of mouse chromosomes were studied (Sumner, 1985) it was found that heterochromatin bound relatively more quinacrine than euchromatin, and also contained more sulphur (presumably from chromosomal proteins).

5 Ultrastructural cytochemistry

As noted earlier, much of the X-ray microanalytical work carried out on biological specimens is done with an analytical resolution comparable with that attainable by light microscopy, although the structural resolution may be better. Nevertheless, with the steady development of brighter electron sources for electron microscopes, more reports of high resolution XRMA are being published, and will no doubt increase in the future, in spite of the problems of radiation damage. It should, however, be noted that radiation damage may be less of a problem in histochemical reactions, where the reaction products may be more stable if they are composed of inorganic, rather than purely organic, compounds.

Somlyo *et al.* (1979) obtained an analytical resolution of at least 40 nm, and probably better than 20 nm when analysing tropomyosin paracrystals whose sulphydryl groups were labelled with mercury. De Bruijn and his colleagues (de Bruijn, 1985; de Bruijn & Cleton-Soeteman, 1985; de Bruijn *et al.*, 1986) have carried out mapping of the distribution of various reaction products in cellular organelles at ultrastructural resolution. Materials studied included lysosomes labelled using a cerium-precipitation method for acid phosphatase, and eosinophil granules labelled with a DAB/platinum procedure for peroxidase.

XRMA has also been used for ultrastructural identification of biogenic amines (Abiko *et al.*, 1983; Hess *et al.*, 1972; Lever *et al.*, 1977; McClung & Wood, 1982; Wood, 1975). After reaction with glutaraldehyde, noradrenaline, dopamine and serotonin bind chromium, but adrenaline does not. After reaction with formaldehyde, only serotonin still binds chromium. These differences permit one to identify vesicles containing different biogenic amines, on the basis of their chromium content after treatment with the different aldehydes, and to correlate the presence of different classes of vesicles with the ultrastructure of the cells.

D Quantification in histochemical X-ray microanalysis

General aspects of quantification in biological XRMA have been amply discussed elsewhere, and this section will be restricted to pointing out some special problems associated with histochemical applications. Before starting any quantitative analysis, it is important to be quite clear what sort of information is required, and what sort of data are useful. For example, it is usual to measure the amounts of endogenous elements in tissues and cells as concentrations (either in terms of wet weight or dry mass). Such information may be inappropriate or unobtainable for histochemical reactions, because the reactions add material to the specimen. The amount of material added to the specimen may, in fact, be very substantial (Sumner, 1978b), and cannot be ignored. It is not possible to estimate the thickness of the specimen (without reaction product) from the continuum X-ray counts at the site where the reaction product is deposited, although it may be possible to get a reasonably good measurement on an adjacent region which lacks reaction product.

Fortunately, there are probably rather few situations in histochemical XRMA where concentration is the desired parameter. It may, for example, be appropriate for enzyme histochemical reactions. However, in many of the examples given in this chapter, where quantification has been used, the most useful measurement has been the ratio between two elements, either in the reaction product itself, or one element in the reaction product and the other in the substrate. Such measurements are quite straightforward (whether given as X-ray counts, or calibrated against standards to give atomic or mass ratios). It must nevertheless be borne in mind that many reaction products contain heavy elements, which implies the possibility of X-ray absorption. This is not necessarily easy to detect or correct for, although no systematic investigation of this problem seems to have been made.

E   Specimen preparation for histochemistry

Specimen preparation for histochemistry in XRMA differs markedly from the large number of applications of XRMA to the study of diffusible ions, in that there is, in general, no requirement for the rapid immobilisation of highly mobile cellular components. As in all histochemical procedures, it is important that the substance to be demonstrated is not lost from the specimen, redistributed or inactivated. The wide variety of substances that can be demonstrated histochemically means that there is no universal method of preparation, but that appropriate procedures have to be selected for the problem under investigation. However, it is worth pointing out that rapid freezing methods without fixation are not appropriate for histochemical studies, since it is highly unlikely that the material will be adequately stabilised to prevent serious damage during the performance of the histochemical reactions.

Another requirement for specimen preparation is that unwanted elements should not be introduced into the tissue. Any added material will, of course, increase the mass of the specimen and thereby affect quantification, and may also absorb other characteristic X-rays. Possibly a greater problem is that many heavy metals used as fixatives or stains (e.g. osmium, lead, uranium) have characteristic X-ray M peaks that overlap the K peaks of the biologically important elements phosphorus and sulphur. Although such overlapping peaks can often be deconvoluted quite successfully by modern computer programmes, it is better to avoid overlaps if possible. In some cases it may be possible, and indeed necessary, to remove endogenous elements from the material before performing the staining reaction. For example, Sumner (1981, 1985, 1986) had to wash chromosome preparations with acetic acid to displace chloride ions before staining with the fluorochrome quinacrine, which was measured by its chlorine content.

Introduction of extraneous elements with the embedding medium is a general problem in biological XRMA, and will not be discussed in detail here. However, it may be noted that resins such as araldite and Spurr's resin contain significant quantities of sulphur and chlorine. On the other hand, methacrylate resins generally seem to be relatively free of such elements. It is, however, advisable to analyse the composition of a resin before using it for analytical purposes. A more serious problem is that resins generally prevent access of the histochemical reagents to the interior of the section. On the other hand, pre-embedding staining of the tissue may also be superficial, because the reagents may be unable to penetrate through the intact tissue. Again, it is not possible to give a general solution to these problems, and the experimenter will have to work out the optimal solution in each case. It is, perhaps, worth noting that certain embedding media are available that not only permit thin sectioning, but can be dissolved away afterwards (Wolosewick, 1980; Capco, Krochmalnic & Penman, 1984; Gorbsky & Borisy, 1986). Although use of such procedures should permit penetration of the reagents throughout the thickness of the sections, they have not, so far as I am aware, been used for histochemical X-ray microanalytical studies.

It is well known that X-ray peaks from grid elements can be detected in spectra obtained from parts of the section well away from grid bars. This is, of course, not a problem peculiar to histochemical studies. What may not be appreciated, however, is that certain grid materials may be unsuitable for histochemical procedures, not merely because their characteristic X-rays overlap with peaks of interest in the specimen, but because they are attacked by certain histochemical reagents. For example, copper grids are not only rather delicate for the repeated handling involved in many histochemical procedures, but they actually dissolve in certain common reagents, such as Tris buffers. Grids made of nickel or titanium are much stronger and more resistant to chemicals than copper grids, and moreover their characteristic X-ray peaks are unlikely to interfere with those of most elements of histochemical interest.

F   Concluding remarks

The widespread use of XRMA in biology has not led to the technique being generally used in histochemical studies, but nevertheless the inherent characteristics of XRMA equipment have led to important applications in this field. The relevant characteristics are the ability to detect and identify multiple elements simultaneously, and the ease of quantification of these elements. Although detection and quantification of more than one substance simultaneously are possible by light microscopy (using either absorption or fluorescence staining), such facilities are not commonly available, whereas they are built into X-ray microanalysers. These characteristics have therefore made XRMA particularly important in two histochemical fields: the validation of reactions, and comparative distribution studies of more than one substance. Regarding the question of validation of reactions, the ability of XRMA to measure, in many cases, the substrate as well as the reaction product, gives the technique a great advantage over other methods, which are unable to detect the presence of the substrate without using the histochemical reaction. The measuring ability of XRMA is vital for establishing the stoichiometry of reactions, and X-ray microanalysis does not suffer the problems of non-linearity between the quantity measured and the actual amount of the substance present, problems which are inherent to fluorescence (due to quenching and energy transfer, for example), and to a lesser extent to absorption staining (which can be affected by absorption shifts on binding, as well as metachromasy).

XRMA is also well adapted to distribution studies of more than one substance. Its primary advantage is the ease with which different spectral peaks can be distinguished and measured, again often a difficult problem by light microscopy, for which overlapping peaks are the rule, and equipment for which is not, in general, designed to measure multiple peaks simultaneously. With the increasing incorporation of mapping facilities in XRMA equipment, this approach will have further advantages over the alternatives. Such systems allow a high degree of automation of the acquisition of X-ray spectra and of the quantification and mapping of the different elements detected (e.g. Foster & Saubermann, 1991), and will no doubt soon make a considerable impact on the field of histochemical mapping as they are doing now for studies of diffusible ions (see the contribution by Wood in this volume).

In the field of ultrastructural histochemistry, there is also an increasing number of applications of XRMA, due to the more widespread availability of high brightness field-emission electron sources, which permit the formation of sufficiently small probes to give genuinely ultrastructural analytical resolution. Combined with mapping facilities, this will make a very powerful tool, and the main limitation is likely to be the reliability of the histochemical procedures being used. As we have seen, however, XRMA is well suited to investigate this aspect of the problem as well.

References

Abiko, Y., Oota, I., Sashida, H., Miyaka, K. & Hashizume, K. (1983). Electron-probe X-ray microanalysis of tissue catecholamines. In *Methods in biogenic amine research*, ed. S. Parvez, T. Nagatsu, I. Nagatsu & H. Parvez, pp. 815–30. Amsterdam: Elsevier.

ap Gwynn, I. (1981). Estimation of surface labelling on whole, critical point dried, cells by microprobe analysis in the scanning electron microscope. *Journal of Microscopy*, **121**, 329–36.

Asanuma, N. (1990). Identification of 5′-adenylylimidodiphosphate-hydrolyzing enzyme activity in rabbit taste bud cells using X-ray microanalysis. *Stain Technology*, **65**, 69–75.

Baccetti, B. & Burrini, A.G. (1977). Detection of concanavalin A receptors by affinity to peroxidase and iron dextran by scanning and transmission electron microscopy and X-ray microanalysis. *Journal of Microscopy*, **109**, 203–9.

Bacsy, E. (1982). Enzymic heterogeneity of adrenocortical lysosomes: an X-ray microanalytical study. *Histochemical Journal*, **14**, 99–112.

Beesley, J.E. (1989). *Colloidal gold: a new perspective for cytochemical marking.* Oxford: Oxford University Press.

Beeuwkes, R. & Rosen, S. (1975). Renal sodium-potassium adenosine triphosphatase optical localization and X-ray microanalysis. *Journal of Histochemistry and Cytochemistry*, **23**, 823–39.

Berry, J.P., Hourdry, J., Sternberg, M. & Galle, P. (1982). Aluminium phosphate visualisation of acid phosphatase activity: a biochemical and X-ray microanalysis study. *Journal of Histochemistry and Cytochemistry*, **30**, 86–90.

Capco, D.G., Krochmalnic, G. & Penman, S. (1984). A new method of preparing embedment-free sections for transmission electron microscopy: applications to the cytoskeletal framework and other three-dimensional networks. *Journal of Cell Biology*, **98**, 1878–85.

Chandler, J.A. & Battersby, S. (1976). X-ray microanalysis of zinc and calcium in ultrathin sections of human sperm cells using the pyroantimonate technique. *Journal of Histochemistry and Cytochemistry*, **24**, 740–8.

Clarke, J.A., Salsbury, A.J. & Willoughby, D.A. (1970). Application of electron probe microanalysis and electron microscopy to the transfer of antigenic material. *Nature*, **227**, 69–71.

Danscher, G., Hansen, H.J. & Moller-Madsen, B. (1984). Energy dispersive X-ray analysis of tissue gold after silver amplification by physical development. *Histochemistry*, **81**, 283–5.

de Bruijn, W.C. (1985). Integration of X-ray microanalysis and morphometry of biological material. *Scanning Electron Microscopy*, **2**, 697–712.

de Bruijn, W.C. & Cleton-Soeteman, M.I. (1985). Application of Chelex standard beads in integrated morphometrical and X-ray microanalysis. *Scanning Electron Microscopy*, **2**, 715–29.

de Bruijn, W.C., Schellens, J.P.M., van Buitenen, J.M.H. & van der Meulen, J. (1980). X-ray microanalysis of colloidal-gold labelled lysosomes in rat liver sinusoidal cells after incubation for acid phosphatase activity. *Histochemistry*, **66**, 137–48.

de Bruijn, W.C., van der Meulen, T., Brederoo, P. & Daems, W.T. (1986). Pt-staining of peroxidatic reaction products at the ultrastructural level. *Histochemistry*, **84**, 492–500.

de Filippis, L.F. & Pallaghy, C.K. (1975). Localization of zinc and mercury in plant cells. *Micron*, **6**, 111–20.

Dierkes, U. (1977). Zur Frage des histochemischen Esterase-nachweises mit der Blei-Thioessigsauremethode beim acellularen Schleimpilz *Physarum confertum*. *Microscopica Acta*, **79**, 23–8.

Engel, W.K., Resnick, J.S. & Martin, E. (1968). The electron probe in enzyme histochemistry. *Journal of Histochemistry and Cytochemistry*, **16**, 273–5.

Eskelinen, S. & Peura, R. (1988). Location and identification of colloidal gold particles on the cell surface with a scanning electron microscope and energy dispersive analyzer. *Scanning Microscopy*, **2**, 1765–74.

Farkas, I., Szerdahelyi, P. & Kasa, P. (1988). An indirect method for quantitation of cellular zinc content of Timm-stained cerebellar samples by energy dispersive X-ray microanalysis. *Histochemistry*, **89**, 493–7.

Foster, M.C. & Saubermann, A.J. (1991). Personal-computer-based system for electron beam X-ray microanalysis of biological samples. *Journal of Microscopy*, **161**, 367–73.

Gardner, D.L. & Hall, T.A. (1969). Electron-microprobe analysis of sites of silver deposition in avian bone stained by the v.Kossa technique. *Journal of Pathology*, **98**, 105–9.

George, S.G., Nott, J.A., Pirie, B.J.S. & Mason, A.Z. (1976). A comparative and quantitative study of cadmium retention in tissues of a marine bivalve during different fixation and embedding procedures. *Proceedings of the Royal Microscopical Society*, **11** (Suppl.), 42.

Goldfischer, S. & Moskal, J. (1966). Electron probe microanalysis of liver in Wilson's disease. *American Journal of Pathology*, **48**, 305–15.

Gorbsky, G. & Borisy, G.G. (1986). Reversible embedment cytochemistry (REC); a versatile method for the ultrastructural analysis and affinity labelling of tissue sections. *Journal of Histochemistry and Cytochemistry*, **34**, 177–88.

Hackney, C.M. & Altman, J.S. (1983). Rubeanic acid and X-ray microanalysis for demonstrating metal ions in filled neurons. In *Functional Neuroanatomy*, ed. N.J. Strausfeld, pp. 96–111. Berlin: Springer.

Hale, A.J. (1962). Identification of cytochemical reaction products by scanning X-ray emission analysis. *Journal of Cell Biology*, **15**, 427–35.

Hall, T.A. & Gupta, B.L. (1983). The localization and assay of chemical elements by microprobe methods. *Quarterly Review of Biophysics*, **3**, 279–339.

(1984). The application of EDXS to the biological sciences. *Journal of Microscopy*, **136**, 193–208.

Harvey, D.M.R. & Kent, B. (1981). Sodium localization in *Suaeda maritima* leaf cells using zinc uranyl acetate precipitation. *Journal of Microscopy*, **121**, 179–83.

Hess, H., Mallasch, M., Marshall, M. & Klingele, H. (1972). Ultrahistochemische Untersuchungen mittels Rontgenmikroanalyse: Nachweis biogener Amine in menschlichen Thrombocyten. *Klinische Wochenschrift*, **50**, 776–82.

Hodges, G.M. & Muir, M.D. (1975). Quantitative evaluation of autoradiographs by X-ray spectroscopy. *Journal of Microscopy*, **104**, 173–8.

Hoefsmit, E.C.M., Korn, C., Blijleven, N. & Ploem, J.S. (1986). Light microscopical detection of single 5 and 20 nm gold particles used for immunolabelling of plasma membrane antigens with silver enhancement and reflection contrast. *Journal of Microscopy*, **143**, 161–9.

Hoyer, L.C., Lee, J.C. & Bucana, C. (1979). Scanning immunoelectron microscopy for the identification and mapping of two or more antigens on cell surfaces. *Scanning Electron Microscopy*, **3**, 629–36.

Ingram, M.J. & Ingram, F.D. (1985). Cell volume regulation studies with the electron microscope. In *Science of biological specimen preparation*, ed. M. Muller, R.P. Becker, A. Boyde & J.J. Wolosewick, pp. 43–9. Chicago: Scanning Electron Microscopy, Inc.

Ishtani, R., Miyakawa, A. & Iwamoto, T. (1977). Elemental analyses of autoradiographic grains by X-ray microanalysis. *Experientia*, **33**, 440–1.

Kirkham, J.B., Goodman, L.J. & Chapell, R.L. (1975). Identification of cobalt in processes of stained neurones using X-ray energy spectra in the electron microscope. *Brain Research*, **85**, 33–7.

Komnick, H. (1962). Elektronen-mikroskopische Lokalisation von Na^+ und Cl^- in Zellen und Geweben. *Protoplasma*, **55**, 414–18.

Krefting, E.-R., Hohling, H.J., Felsmann, M. & Richter, K.-D. (1988). Strontium as a tracer to study the transport of calcium in the epiphyseal growth plate (electronprobe microanalysis). *Histochemistry*, **88**, 321–6.

Kyriacou, K. & Garrett, J.R. (1980). Mitochondrial staining with lead capture techniques in rabbit submandibular glands. *Histochemical Journal*, **12**, 609–14.

Lever, J.D., Santer, R.M., Lu, K.-S. & Presley, R. (1977). Electron probe X-ray microanalysis of small granulated cells in rat sympathetic ganglia after sequential aldehyde and dichromate treatment. *Journal of Histochemistry and Cytochemistry*, **25**, 275–9.

Marshall, A.T. (1980). Principles and instrumentation. In *X-ray microanalysis in biology*, ed. M.A. Hayat, pp. 1–64. Baltimore: University Park Press.

McClung, R.E. & Wood, J. (1982). Analytical electron microscopic evaluation of the effects of paraformaldehyde pretreatment on the reaction of glutaraldehyde with biogenic amines. *Journal of Histochemistry and Cytochemistry*, **30**, 481–6.

Meisner, L.F., Chuprevich, T.W., Johnson, C.B., Inhorn, S.L. & Carter, J.J. (1973). Microanalysis of chromosomes with X-ray energy dispersion. *Lancet*, **2**, 561.

Modis, L., Lustyik, Gy., Adany, R. & Zs.-Nagy, I. (1988). Energy dispersive X-ray microanalysis of sulfated glycosaminoglycans in cartilage matrix stained with alcian blue 8GX. *Basic and applied Histochemistry*, **32**, 415–28.

Morgan, A.J. (1981). A morphological and electron microprobe study of the inorganic composition of the mineralised secretory products of the calciferous gland and chloragogenous tissue of the earthworm, *Lumbricus terrestris* L. The distribution of injected strontium. *Cell and Tissue Research*, **220**, 829–44.

Peschke, T., Gabert, A., Wollweber, L. & Augsten, K. (1986). Effect of several fixatives on the density of concanavalin A binding sites of murine peritoneal macrophages determined by fluorescence microscopy and X-ray microanalysis. *Acta histochemica* (suppl), **33**, 201–6.

Peschke, T., Augsten, K., Sasama, F. & Gabert, A. (1989). Determination of gold-labelled surface receptors on single cells by X-ray microanalysis. *Journal of Microscopy*, **156**, 191–9.

Rosen, S. & Beeuwkes, R. (1979). Renal Na-K-ATPase: quantitative X-ray microanalysis. In *Microbeam analysis in biology*, ed. C.P. Lechene & R.R. Warner, pp. 489–505. New York: Academic Press.

Rosenquist, T.H. & Huff, T.A. (1986). A procedure to measure concanavalin-A binding with atomic spectroscopy and X-ray microanalysis. *Histochemistry*, **84**, 61–5.

Sehgal, S.S., Tewari, J.P. & Malhotra, S.K. (1981). Thiocholine methods for the demonstration of acetylcholinesterase in neuromuscular junctions. *Cytobios*, **30**, 69–82.

Siegesmund, K.A., Yorde, D.A. & Dragen, R. (1979). A quantitative immunoperoxidase procedure employing energy dispersive X-ray analysis. *Journal of Histochemistry and Cytochemistry*, **27**, 1226–30.

Simson, J.A.V., Bank, H.L. & Spicer, S.S. (1979). X-ray microanalysis of pyroantimonate-precipitable cations. *Scanning Electron Microscopy*, **2**, 779–92.

Somlyo, A.P. & Shuman, H. (1982). Electron probe and electron energy loss analysis in biology. *Ultramicroscopy*, **8**, 219–34.

Somlyo, A.P., Somlyo, A.V., Shuman, H. & Stewart, M. (1979). Electron probe analysis of muscle and X-ray mapping of biological specimens with a field emission gun. *Scanning Electron Microscopy*, **2**, 711–22.

Stewart, M., Somlyo, A.P., Somlyo, A.V., Shuman, H., Lindsay, J.A. & Murrell, W.G. (1980). Distribution of calcium and other elements in cryosectioned *Bacillus cereus* T spores, determined by high-resolution scanning electron probe X-ray microanalysis. *Journal of Bacteriology*, **143**, 481–91.

Stockert, J.C., Gosalvez, J., del Castillo, P., Pelling, C. & Mezzanotte, R. (1991).

X-ray microanalysis of toluidine blue stained chromosomes: a quantitative study of the metachromatic reaction of chromatin. *Histochemistry*, **95**, 289–95.

Sumner, A.T. (1978a). Changes in elemental composition of human chromosomes during a G-banding (ASG) and C-banding (BSG) procedure. *Histochemical Journal*, **10**, 201–11.

(1978b). Quantitation in biological X-ray microanalysis, with particular reference to histochemistry. *Journal of Microscopy*, **114**, 19–30.

(1981). The distribution of quinacrine on chromosomes as determined by X-ray microanalysis. I. Q-bands on CHO chromosomes. *Chromosoma*, **82**, 717–34.

(1983). Standards for measuring spatial resolution in biological X-ray microanalysis. *Scanning Electron Microscopy*, **2**, 785–92.

(1984). X-ray microanalysis of protein sulphydryl groups in chromatin. *Scanning Electron Microscopy*, **2**, 897–904.

(1985). The distribution of quinacrine on chromosomes as determined by X-ray microanalysis. II. Comparison of heterochromatic and euchromatic regions of mouse chromosomes. *Chromosoma*, **91**, 145–50.

(1986). Mechanisms of quinacrine binding and fluorescence in nuclei and chromosomes. *Histochemistry*, **84**, 566–74.

(1990). *Chromosome banding*. London: Unwin Hyman.

Tewari, J.P., Sehgal, S.S. & Malhotra, S.K. (1982). Microanalysis of the reaction product in Karnovsky and Roots' histochemical localization of acetylcholinesterase. *Journal of Histochemistry and Cytochemistry*, **30**, 436–40.

Tisher, C.C., Weavers, B.A. & Cirksena, W.J. (1972). X-ray microanalysis of pyroantimonate complexes in rat kidney. *American Journal of Pathology*, **69**, 255–70.

Vallyathan, N.V. & Brody, A.R. (1977). X-ray microanalysis as an adjunct tool in enzyme histochemistry. *Scanning Electron Microscopy*, **2**, 93–102.

van Iren, F., van Essen-Joolen, L., van der Duyn Schouten, P., Boers-van den Sluijs, P. & de Bruijn, W.C. (1979). Sodium and calcium localization in cells and tissues by precipitation with antimonate: a quantitative study. *Histochemistry*, **63**, 273–94.

Walker, S.R., Ingram, P. & Shelburne, J.D. (1984). Energy dispersive X-ray microanalysis of vital dye (erythrosin B) stained cells. *Journal of Microscopy*, **133**, RP3–4.

Weavers, B.A. (1974). An X-ray microanalytical study of the ferricyanide reaction for the electron cytochemical demonstration of succinate dehydrogenase in isolated mitrochondria. *Histochemical Journal*, **5**, 121–31.

Whittaker, S.G., Smith, D.G., Foster, J.R. & Rowland, I.R. (1990). Cytochemical localization of mercury in *Saccharomyces cerevisiae* treated with mercuric chloride. *Journal of Histochemistry and Cytochemistry*, **38**, 823–7.

Wick, S.M. & Hepler, J.K. (1982). Selective localisation of intracellular Ca^{2+} with potassium antimonate. *Journal of Histochemistry and Cytochemistry*, **30**, 1190–204.

Wolosewick, J.J. (1980). The application of polyethylene glycol (PEG) to electron microscopy. *Journal of Cell Biology*, **86**, 675–81.

Wood, J.G. (1975). Analytical electron microscopic identification of cytochemical products in thin sections. *Acta Histochemica*, **52**, 143–51.

Wroblewski, J., Sagstrom, S., Mulders, H. & Roomans, G. M. (1989). Strontium and bromide as tracers in X-ray microanalysis of biological tissue. *Scanning Microscopy*, **3**, 861–4.

10 Sprayed microdroplets: methods and applications

A. J. Morgan and C. Winters

A Introduction

1 Perspective: directly deposited microdroplets

Electron probe X-ray microanalysis (EPXMA) is capable of yielding quantitative data from excited specimen volumes < 1 μm^3. Thus, the analysis of fluid microvolumes in the range 10^{-11} to 10^{-10} litres is comfortably within the scope of this technique. For this reason, EPXMA was advocated at an early stage in its technological evolution for the analysis of nanolitre fluid volumes (Ingram & Hogben, 1967). Subsequently a number of laboratories have independently pursued physiological questions by the analyses of diverse biological fluids. This family of related microdroplet methods is described fully elsewhere in this volume by Roinel and Rouffignac (Chapter 11). However, in order that the sprayed microdroplet technique be seen in a realistic perspective a brief resumé of the more conventional 'direct deposition' microdroplet methods is provided here.

In general, two major variants of the microdroplet method are practised (Fig. 10.1).

1. Microdroplets of known volume. Microdroplets of known volume are dispensed from calibrated micropipettes, mounted on polished solid (bulk) support materials, then dehydrated by freeze-drying to produce the all-important microcrystallites that minimise X-ray absorption. They are then individually analysed by wavelength-dispersive spectrometry (WDS).
2. Microdroplets of unknown volume. Microdroplets of samples and standards of indeterminate volume are dispensed from constant-volume micropipettes, mounted on thin-film supports, then dehydrated by flash evaporation and subsequently analysed by energy-dispersive spectrometry (EDS). A selected list of the combinations of available options adopted by some of the major laboratories in this specialised field is presented in Table 10.1.

It is significant that the laboratories which routinely use bulk supports for their microdroplets usually dehydrate by lyophilisation, and invariably analyse by WDS. Bulk supports cannot be damaged during droplet deposition, neither are they prone to disruption during freeze-drying, but they do generate high background signals. This problem is circumvented by WDS

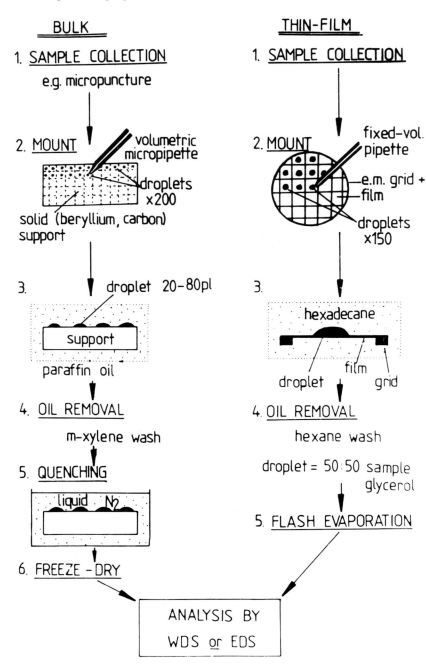

Fig. 10.1.   Schematic diagrams of the two major 'direct deposition' microdroplet preparation procedures for fluid analysis by electron microprobe X-ray microanalysis (see Table 10.1 for references).

Table 10.1. *The methodological options adopted by the major practitioners of the direct deposition microdroplet technique*

Methodology		References
1. Micropipettes	(a) *Calibrated* constriction microcapillaries (known volume), suction-filled	Bonventre, Blouch & Lechene, 1980; Garland, Brown & Henderson, 1978; Ingram & Hogben, 1967; Ingram & Ingram, 1981; Lechene, 1974; Lechene & Warner, 1979; Morel & Roinel, 1969; Roinel, 1975; 1988; Roinel, Meny & Henoc, 1980; Roinel & de Rouffignac, 1982; Van Eekelen *et al.*, 1980
	(b) *Non-calibrated* microcapillaries (constant volume), self-filling	Quinton, 1976; 1978a,b,c; 1979
2. Microdroplet supports	(a) Polished, hydrophilic *solid* (i.e. bulk) blocks of conductive materials, e.g. aluminium, beryllium, carbon, silicon, quartz	The laboratories of Garland, Lechene, Roinel: see above, for references
	(b) Carbon-coated *thin films* of formvar, parloidin or nylon	Rick *et al.*, 1977; Quinton, see above; Van Eekelen *et al.*, 1980
3. Dehydration	(a) *Freeze-drying* after shock freezing	The laboratories of Garland, Lechene, Roinel; Van Eekelen *et al.*, 1980; see above, for references
	(b) *Flash evaporation* (less likely than freeze-drying to damage thin support films), with added solute to retard crystal formation	Quinton, 1976; 1978a,b,c; 1979 (Bostrom *et al.* (1988) have used air-drying of droplets with added dextran)
4. X-ray microanalysis	(a) *Wavelength spectrometry*	Bonventre *et al.*, 1980; Garland *et al.*, 1978; Ingram & Hogben, 1967; Lechene, 1974; Lechene & Warner, 1979; Morel & Roinel, 1969; Roinel, 1975; 1988; Roinel *et al.*, 1980; Roinel & de Rouffignac, 1982; Van Eekelen *et al.*, 1980
	(b) *Energy-dispersive spectrometry*	Rick *et al.*, 1977; Quinton, 1976; 1978a,b,c; 1979; Bostrom *et al.*, 1988

because it has the advantage over EDS of not only superior spectral resolution, but also superior peak-to-background (P/b) ratios (Morgan, 1985). On the other hand, thin film supports reduce the background signal to about 3% of that generated by solid beryllium (Quinton, 1979). Thus, P/b signals are substantially improved with the result that EDS becomes a practical proposition. However, thin films are fragile and prone to fracture during freezing and freeze-drying, and yet the production of fine (< 1 μm) microcrystallites of well dispersed mineral within each mounted microdroplet is an analytical necessity. Quinton (1978a,b,c) elegantly solved these problems by adding glycerol as an anticrystallising agent to the samples, and dehydrating by flash evaporation. Bostrom *et al.* (1988) produced micro-droplets that behave analytically like thin films after air-drying, providing dextran, to a final concentration of 1.5% to 2.5%, is added to retard crystallisation.

Microdroplets mounted on thin films can, of course, be analysed by WDS (Van Eekelen *et al.*, 1980), but the high beam currents that are required with these relatively inefficient X-ray collection systems (Morgan, 1985) can cause film damage, although this was not observed by Van Eekelen *et al.* (1980) even at currents that volatilised constituents of their microdroplets.

2   Sprayed microdroplets: theoretical background

The sprayed microdroplet technique, the subject of this chapter, shares many of the features associated with the technique of EDS analysis of directly deposited microdroplets mounted on thin films. Some of the more important comparisons are noted below.

1. Directly deposited microdroplets may typically have volumes in the range 10 to 150 picolitres (1×10^{-11} l to 1.5×10^{-10} l), with a 'dried diameter' on the supporting surface of 10 to 150 μm. The important point here is that the initial fluid sample need not necessarily have a volume larger than the individual microdroplet taken for analysis. Individual sprayed micro-droplets are usually much smaller than directly deposited droplets, with a 'dried diameter' in the range 0.5 μm to 5 μm (estimated lower volumes of the wet droplet $< 1 \times 10^{-13}$ l), but are derived from fairly large initial sample volumes of several μl or ml.

2. No systematic efforts are usually made to ensure the formation and maintenance of homogeneously dispersed mineral microcrystallites in sprayed microdroplets, although in some variants of the technique the film-coated grid is supported on a hot-plate during spraying to promote rapid air-drying and, thus, to minimise crystal size (Morgan, Davies & Erasmus, 1975; Marshall, 1977). It would be instructive to study the effect of adding crystallisation inhibiting molecules, such as glycerol (Quinton, 1978a,b,c) and dextran (Bostrom *et al.*, 1988), to the initial fluid to

determine their influences on the spraying properties of aqueous media, and on the morphologies of the dried droplets. Rehydration of dried microdroplets by atmospheric moisture can lead to significant, and analytically restrictive, crystal growth (Roinel *et al.*, 1980; Roinel & de Rouffignac, 1982; Roinel, 1988). Sprayed microdroplets should be stored under a dry atmosphere so that they continue to conform to thin specimen criteria and, thus, obviate corrections for absorption, fluorescence and atomic number effects (ZAF).

3. Quantitation is directly achieved in the case of deposited microdroplets from standard curves, obtained by plotting measured X-ray intensities (peak minus background counts) against the known concentrations of a series of standard solutions (Fig. 10.2a). The accuracy of this approach depends on the assumption that the 'unknowns' and 'standards' behave identically in analytical terms, although their compositions may be very different (Roinel *et al.*, 1980). In the cases of microdroplets dispensed from 'calibrated' or 'constant volume' microcapillaries it is essential to maintain constant the accelerating voltage (because it affects the P/b ratio and, thus, analytical sensitivity) and beam current (because it affects the number of measured X-ray counts per unit time).

The population of microdroplets produced by spraying is heterogeneous. The volume of individual droplets is indeterminate, because the diameter of the dried cluster of crystallised mineral is not equal to, and not necessarily directly proportional to, that of the wet microdroplet that yielded it, especially if the supporting film is hydrophobic. Also, the diameter of droplets that are dried relatively slowly may vary with fluid composition.

Quantitation of sprayed microdroplets cannot be achieved directly from count rates versus standards concentration curves (Fig. 10.2a) because, unlike 'deposited microdroplets', neither the volume of individual droplets is known (or measurable), nor is it possible to ensure that the wet volumes (i.e. volume of the projected microdroplet immediately before it impacts the support film and dries) of specimen and standards microdroplets are identical. Morgan *et al.* (1975) and Davies & Morgan (1976) overcame this problem by introducing a known amount of an 'external' reference element, cobalt, into the specimen and standard fluids before spraying. The concentration of elements in the specimen fluids could be determined from curves constructed from the ratio of the measured X-ray intensities of the elements to the measured X-ray intensities of the reference element (e.g. cobalt) against the actual concentrations of the elements in a series of standard solutions (Fig. 10.2b). The inclusion of a fixed concentration of an external reference element (say, cobalt) effectively compensates for differences in wet microdroplet volumes: for a given solution and element x, the larger the droplet the proportionately higher the measured intensity of the cobalt signal, such that $(P-b)_x/(P-b)_{Co}$ remains constant.

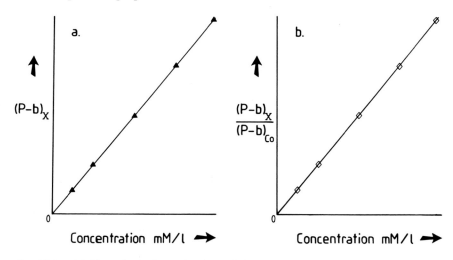

Fig. 10.2. (a) Hypothetical standard graph for the quantitation of the concentration of element *x* in 'directly deposited microdroplets' – see Table 10.1 for references. (b) Hypothetical standard graph for the quantitation of the concentration of element *x* in sprayed microdroplets, where the selected 'external' reference element is cobalt.

Practical consequences of the ratio quantitation procedure are: (i) it is not necessary to maintain constant beam currents for different droplets, because the count rates of all measured elements will be affected equally, providing that the current does not exceed the threshold of volatilisation of any one of the components of the ratio (Morgan & Davies, 1982); (ii) there is no need to correct the measured X-ray intensities of individual elements within the microdroplets for differences in their 'overall relative detection efficiencies' (Morgan *et al.*, 1975) to yield true atomic ratios, providing quantitation is achieved by reference to standard curves (Fig. 10.2b); (iii) it is probable that the standard curves will maintain linearity over a wider concentration range than direct intensity versus concentration curves, providing the measured X-ray energies of the elements of interest and of the 'reference' element are not very different.

Morgan (1983) described a method for calculating element concentrations in sprayed microdroplets, based on the Russ Atomic Ratio Method (Russ, 1974; 1975), that dispenses with the need to construct intensity ratios versus concentration standard curves. For a thin specimen (i.e. assuming negligible X-ray absorption) the key equation is:

$$\frac{C_x}{C_y} = K_{xy} \cdot \frac{(P-b)_x}{(P-b)_y} \qquad (10.1)$$

Table 10.2. *A summary[a] of the Hyatt & Marshall hybrid method for preparing and analysing directly deposited fluid microdroplets*

1. Micropipettes	These were drawn from glass tubes with a microelectrode puller.
2. Droplet mounting	The fluids of interest were held under oil in a small Petri dish. Nylon-filmed nickel finder grids were immersed under the oil in the same Petri dish. A micropipette was held in a micromanipulator and connected to a micrometre syringe. Equal volumes (arbitrarily 'measured' via an eyepiece graticule in a stereomicroscope) of sample solution and 100 mmol/l $Co(NO_3) \cdot 6H_2O$ reference fluid, separated by a small volume of oil, were drawn into the pipette. Droplets containing equal volumes of sample and reference fluids were deposited on the nylon film.
3. Dehydration	Oil was removed from the grid by washing in xylene. Droplets were initially dried in air at room temperature. Subsequently, the droplets were rehydrated deliberately by gently blowing onto chilled grids, and then quenched in liquid propane. The grid was then transferred under liquid N_2 to a freeze-drying unit. (Nylon grids are much more robust than collodion, parlodion and formvar films: they are less easily damaged by micropipettes and solvents; they also withstand freezing and freeze-drying.) The grids were then carbon coated.
4. Analysis	The grids were mounted in an angled carbon support, and analysed by EDS in a scanning electron microscope.

[a] For further practical details see the original paper by Hyatt & Marshall (1985).

where C_x and C_y are the concentrations (mmoles/l) of element 'x' and 'y', respectively, in the given sample; $(P-b)_x$ and $(P-b)_y$ are the measured intensities of 'x' and 'y'; K_{xy} is a proportionality constant that relates the overall relative detection efficiency of the analytical system for 'x' and 'y', and adjusts the ratios of the peak intensities of 'x' and 'y' to yield atomic or molar ratios. K-factors can be determined easily by the analysis of thin standards of known atomic proportions; sprayed microdroplets of isoatomic (i.e. equimolar) solutions of salt mixtures are convenient as such standards (Morgan *et al.*, 1975; Davies & Morgan, 1976; Marshall, 1977; Marshall & Forrest, 1977; Bostrom *et al.*, 1988; Hall, 1989). Once the appropriate K-factors have been determined it can be seen from equation (10.1) that the concentration of element 'x' can be readily determined because the concentration of the reference element 'y' (e.g. cobalt) is known, and $(P-b)_x$ and $(P-b)_y$ are measurable. (Note that this 'standardless method' is identical to the Cliff–Lorimer method for determining weight fraction ratios in thin foils (Cliff & Lorimer, 1975; Goldstein *et al.*, 1977).)

Equation (10.1) indicates that the reference element need not be an introduced or external element. If the concentration of any of the element constituents of the solution can be determined independently, then it can be used as an 'internal' reference. Bostrom *et al.* (1988) successfully used chlorine, measured separately by microcoulometry, as an internal reference in urine samples.

3   *Hybrid techniques: direct deposition with reference elements*

Two publications (Hyatt & Marshall, 1985; Bostrom *et al.*, 1988) describe hybrid techniques that combine direct deposition micropipetting with the incorporation of a reference element (cobalt) for quantitation via equation (10.1). The Hyatt & Marshall (1985) procedure, summarized in Table 10.2, is attractive because it is less demanding than the established direct deposition techniques (see Table 10.1), it does not require standard curves for quantitation, and it does not require the relatively large starting volumes that spraying demands.

B   Practical considerations

Many aspects of the sprayed microdroplet technique were reviewed by Morgan (1983), and a repeat of the exercise is unwarranted. Nevertheless, it is germane for heuristic reasons to highlight certain practical issues.

1   *Spraying procedures*

Sprayed microdroplets were originally dispensed from a commercially available glass nebuliser or microspray (Agar Scientific Ltd., Essex, England; Fig. 10.3a), requiring a minimum of about 1 ml of starting solution to yield a fairly large population of droplets on coated grids, from which a number of roughly equal-sized (∼ 1 μm diameter) droplets could be selected for analysis (Morgan *et al.*, 1975). Davies & Morgan (1976) designed a simple jet sprayer (the 'Mini Spray') that reduced the necessary starting volume to the 2 μl to 10 μl range (Fig. 10.3b). With the 'Mini Spray' it is important to check the quality of the film on each grid individually before spraying, because the yield of microdroplets in the desired size range (diameter about 1 μm to 3 μm) is low. Two further advantages of the 'Mini Spray' over the nebuliser are: (a) it is much faster; and (b) there is no risk of cross-contamination because the microcapillaries from which the solutions are sprayed are disposable (Davies

(a) *Glass nebuliser*

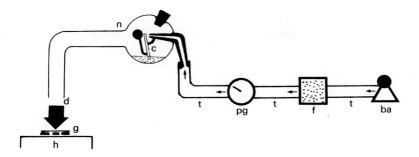

(b) *Mini Spray*

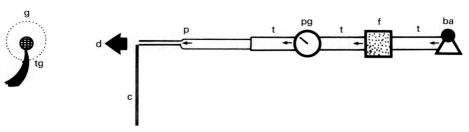

Fig. 10.3. Schematic diagrams of (a) a commercial glass nebuliser; (b) a simple
jet sprayer ('Mini Spray'), which can easily be constructed from
common laboratory supplies and using basic workshop facilities. For
reproducible results with both sprayer systems it is recommended that
the air supply is filtered (or compressed dry nitrogen gas is used as
the propellant), and that the spraying pressure is monitored. In the
'Mini Spray', it may be advantageous to support the grids against
the vertical face of a hot plate to accelerate drying and promote
microcrystallisation. (Note that Marshall (1977) used a modified
standard spray-gun, with compressed CO_2 as a propellant.) Labelling:
ba, bench air-supply (propellant); c, capillary tube; d, microdroplets; f,
filter; g, coated grid; h, hot-plate; n, nebuliser; pg, pressure gauge; t,
pressure tubing; tg, target area within which the grid is held by forceps.

& Morgan, 1976). Spraying pressures can be adjusted to accommodate fluids
of different viscosities; thus, there is no reason why crystal inhibiting
molecules, such as glycerol (Quinton, 1978a,b,c) or dextran (Bostrom *et al.*,
1988) should not be routinely added to the starting fluids.

2 Analytical throughput

It is obvious that any 'external' reference element added for quantitative
purposes must be absent from the original fluid sample (Morgan, 1983). A

frustrating aspect of the sprayed microdroplet technique, as commonly practised, is that only microdroplets from a single starting solution are deposited on a given filmed grid. Consequently, grid changing is a major data acquisition bottle-neck. Since individual sprayed droplets are usually well dispersed with few overlaps over the grid surface it is, in principle, feasible to increase data acquisition rates by adding different reference elements to different sample solutions, and spraying these onto the same grid. Fig. 10.4 upholds this principle for four samples. X-ray dot-mapping (see Wood, this volume) would be useful for rapidly identifying on the basis of composition representatives of the different microdroplet subpopulations prior to their individual analyses. However, in practice the utility of this approach may be questionable, because it may be rather difficult to obtain suitable reference elements that: (i) have sufficiently soluble salts, that are preferably not hygroscopic; (ii) possess X-ray lines that preferably do not overlap with the characteristic signals from the elements to be measured, or with 'extraneous' characteristic X-ray signals from grid and instrumental sources.

Even though the starting volume required for spraying is relatively large (> 2 µl), the EDS analysis of sprayed microdroplets offers certain advantages over more sensitive analytical techniques (e.g. flameless atomic absorption spectrophotometry), notably: it is multi-elemental, and detects both cationic and anionic elements; it does not usually require sample dilution; it does not suffer from signal suppression effects, overlapping characteristic peaks can be deconvoluted with standard software; it is sufficiently sensitive to detect the dominant element constituents in most biological fluids.

3   Precision of the method

The precision of any measurement should be defined statistically, and for the EDS of microdroplets it is a function of several factors, including: accelerating voltage, which determines P/b ratios; beam current and analysis time, factors both of which affect the number of accumulated counts, and thus sensitivity; the precise geometric and physical characteristics of the EDS detector system; the mass density of the sample. Analytical precision, reproducibility, and minimum detectable concentrations have not been evaluated with the degree of rigour in sprayed microdroplets that they have in directly deposited microdroplets (Roinel & de Rouffignac, 1982; Quinton, 1979).

4   Microdroplet stability under electron irradiation

Organic specimens are often susceptible to radiation damage leading to mass loss (Zierold, 1988; Roos & Morgan, 1990; von Zglinicki, this volume). At beam currents that are generally higher than those required for EDS analysis,

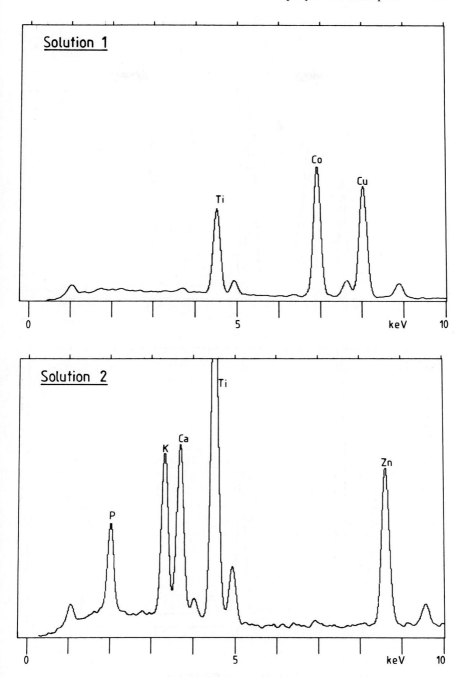

Fig. 10.4. Solutions 1 and 2. For legend see page 163.

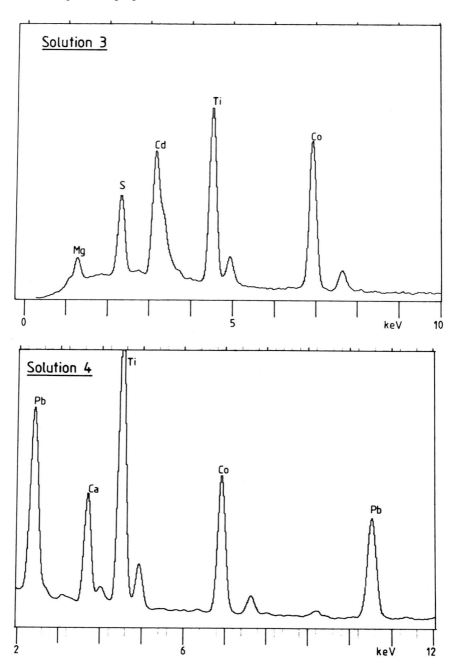

Fig. 10.4. Solutions 3 and 4.   For legend see facing page.

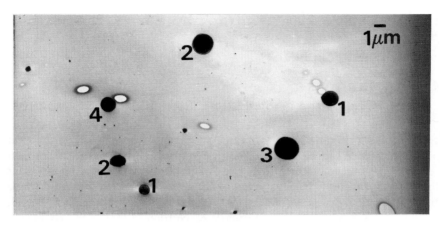

Fig. 10.4. Micrograph and spectra, taken from a region on a filmed titanium grid, containing air-dried sprayed microdroplets from four solutions of different composition. The numbers refer to the solutions from which the individual droplets were derived. The energy spectra were derived from representative microdroplets: solution 1 (Co, Cu); solution 2 (P, K, Ca, Zn); solution 3 (Mg, S, Co, Cd); solution 4 (Ca, Co, Pb). Note that the droplets from the four solutions sprayed from a glass nebuliser are sufficiently well dispersed to enable several different samples containing different 'reference' elements to be deposited on a single grid.

detectable elements, especially halides, can be volatilised from microdroplets (Roinel *et al.*, 1980). Element loss appears to be related to beam current density, and to the mass thickness of the droplets (LeRoy & Roinel, 1983; Roinel & de Rouffignac, 1982; Morgan & Davies, 1982). Morgan & Davies (1982) suggested that element loss is a thermal effect; certainly elements were lost from sprayed microdroplets of mixed-salts solutions, with increasing current densities, in the order of ascending boiling points. Other factors also affect element stability under electron irradiation. First, the presence of an organic matrix such as urea (Roinel, 1975; Morgan & Davies, 1982) promotes stability probably by rendering the specimens more conductive. Not surprisingly, biological fluids are more stable than pure mineral solutions (Roinel, 1988). Second, 'chemical speciation' within the microdroplet is important. For example: chlorine is less stable in the form of $CaCl_2$ than as NaCl (Lechene & Warner, 1979); pH of the biological fluid can affect the stability of Cl in microdroplets derived from it (Quinton, 1979); instability of the S signal has been recorded in solutions of sulphate salts, but not in methionine or cysteine solutions (Roinel *et al.*, 1980). It is encouraging that a number of authors have found no elemental loss during EDS analysis of microdroplets, mounted on thin films, with beam current densities in the range 10^{-13} to 10^{-11} A/μm^2 (Rick *et al.*, 1977; Hyatt & Marshall, 1985; Bostrom *et al.*, 1988).

C   Applications

1   Standards and calibration

For the quantitation of thin inorganic specimens, where X-ray absorption and fluorescence can be assumed to be negligible, the simple ratio approach (equation (10.1)) has been widely used. The K_{xy} factors (or 'Cliff–Lorimer factors') have usually been experimentally determined from thin films made from single phase alloys, ceramics or minerals (Cliff & Lorimer, 1975; Wood, Williams & Goldstein, 1984; Wirmark & Norden, 1987). Sprayed micro-droplets of binary or mixed salt solutions of known composition (and most conveniently isoatomic or equimolar for the elements of interest) have also been used for K_{xy} determinations (Morgan *et al.*, 1975; Graham & Steeds, 1984). The advantages of these sprayed standards are: they are easy to prepare; they allow a wide range of different element combinations to be used, which may not be readily available in alloys and minerals; and they are reliable (K_{xy} factors can be determined with an error of about $\pm 3\%$ with sprayed microdroplets; Graham & Steeds, 1984). In addition, sprayed standards can be very useful for evaluating and optimising certain aspects of the performance of TEM-EDS systems (Marshall, 1977; Marshall & Forrest, 1977). It should also be noted that K_{xy} factors can be computed from established theory (Gauvin & L'Esperance, 1991).

2   The analysis of biological fluids

Surprisingly, the sprayed microdroplet technique appears not to have been used for the direct EDS analysis of biological fluids, although Nott & Mavin (1985) did analyse Na, Mg, P, S and Ca concentrations in samples of the blood of a shrimp, *Crangon crangon*, by treating them as soft tissue samples (see below).

The direct analysis of sprayed biological fluids could easily be achieved in one of two ways, by exploiting analytical principles established and validated by Hyatt & Marshall (1985) and Bostrom *et al.* (1988). First, a measured volume of a cobalt solution of known concentration (e.g. 100 mmol/l $Co(NO_3)_2 \cdot 6H_2O$) could be added as an 'external reference' to a measured volume of the sample; the proportions of the standard and sample would typically be 50:50, but could be adjusted to provide the most acceptable characteristic counts for the elements of interest. After thorough mixing, the Co-spiked sample could then be sprayed. Alternatively, the concentration of a convenient elemental constituent of the fluid sample could be determined by

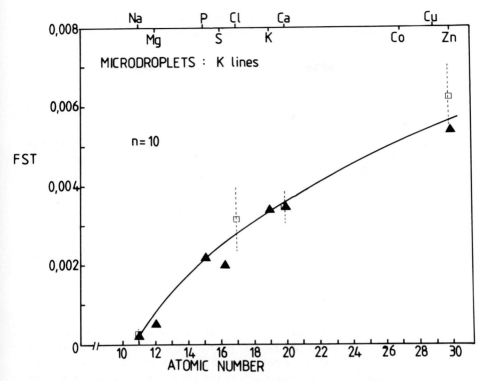

Fig. 10.5. *FST* (K lines) versus atomic number curve (provisional data) for the
lighter elements (A.N. 11 to 30). The FST_x values were determined,
from sprayed microdroplets of isoatomic salt solutions, by the 'ratio
method' using the FST_{Ca} value determined in aminoplastic sections as a
reference (see equation 10.7) (▲——▲). In addition, *FST* values
(mean ± S.D.) for Na, Cl, Ca and Zn (K lines were determined directly
by the analysis of sectioned aminoplastics (*n* = 4); these (□) are
included for comparative purposes. Note that the hand-down '*FST*
ratio' curve falls within the standard deviations of the aminoplastic-
derived values, thus providing a strong indication that the former
method yields valid data.

an independent technique, and used as an 'internal reference' (Bostrom *et al.*,
1988) for quantifying the sprayed microdroplets.

3 In conjunction with the quantitation of thin sections of biological tissues

The most widely used method for quantifying elemental concentrations in
thin biological specimens by EPXMA is the Hall Continuum Normalisation
Method (Hall & Gupta, 1983), which indicates that the concentration of an

element x (C_x) in a thin specimen can be determined from a fairly simple equation, providing a suitable absolute thin standard is available:

$$(C_x)_{\text{specimen}} = (C_x)_{\text{standard}} \cdot \frac{(P_x/W)_{\text{specimen}}}{(P_x/W)_{\text{standard}}} \cdot \frac{G_{\text{specimen}}}{G_{\text{standard}}} \qquad (10.2)$$

where P_x = measured net characteristic counts for element x, W = measured continuum signal, $G = \overline{Z^2/A}$.

In practice, a constant factor (designated the FST_x factor in the Hall & Gupta (1983) and Link Systems Ltd nomenclature) is determined from sectioned standards by applying the following equation:

$$FST_x = \frac{(P_x/W)_{\text{standard}}}{(C_x)_{\text{standard}}} \cdot G_{\text{standard}}. \qquad (10.3)$$

Equation (10.2) can now be simplified by the substitution of equation (10.3):

$$(C_x)_{\text{specimen}} = \frac{(P_x/W)_{\text{specimen}} \cdot G_{\text{specimen}}}{FST_x}. \qquad (10.4)$$

FST values must be determined for each element to be analysed (unless of course, the concentrations of elements in the thin specimen are determined from the measured concentration of one element by the application of equation (10.1)). Warley (1990) reviewed the large number of different thin film standards that have been developed in conjunction with EPXMA quantitation. One type of standard with several attractive properties (Roos & Morgan, 1985) that has been widely used to determining standard factors is the sectioned aqueous aminoplastic (Roos & Barnard, 1984; Morgan & Winters, 1988). Morgan *et al.* (1989) provided provisional data indicating that if a standard factor is carefully determined by the analysis of a concentration series of aminoplastics containing a 'reference element' (e.g. calcium; yielding FST_{Ca}), then the FST values for any other element can be determined from overall relative sensitivity values. Specifically, in a thin film standard contained equimolar concentrations of two elements, x and y, the ratio of their FSTs is:

$$\frac{FST_x}{FST_y} = \frac{[(P_x/W)/C_x] \cdot G}{[(P_y/W)/C_y] \cdot G}. \qquad (10.5)$$

Since in an equimolar standard $C_x = C_y$, equation (10.5) can be simplified to:

$$FST_x = \frac{P_x}{P_y} \cdot FST_y. \qquad (10.6)$$

Table 10.3. *An outline of the major steps in the preparation of microdroplets of homogenised tissues for EDS analysis*

1. Sampling	Option 1:	Tissue samples (50–150 µg wet weight) are placed in individual pre-weighed aluminium 'crucibles' made by pressing a shallow well in the centre of 1 cm × 1 cm sheets. Samples are dried and carefully weighed (Morgan, 1983).
	Option 2:	Tissue samples (any size) are placed in pre-weighed, acid-washed flat-bottomed glass tubes or beakers. Samples are dried to constant-weight and weighed.
2. Digestion or solubilisation	Option 1:	Dried tissue samples in aluminium crucibles (see Sampling; Option 1) are completely ashed at low temperature (< 100 °C) in an oxygen-plasma. The ash is dissolved in a known volume (5 µl for smaller specimens) of 0.5 M HNO_3 containing a known concentration of cobalt reference (e.g. 100 mmol/l Co). The dissolved ash is thoroughly mixed prior to spraying.
	Option 2:	The dried tissue (see Sampling; Option 2) is wet digested in boiling concentrated HNO_3. After the digestion is complete, the acid is completely evaporated and the residue is extracted with a known volume of 0.5 M HNO_3 containing a known concentration of Co. Extraction can be facilitated by heating because evaporation at this stage will not change the measured P_x/P_{Co} ratio values. (Any conventional wet digestion protocol can be used.)
3. Spraying and analysis		The tissue homogenates containing the 'external' Co reference can then be sprayed via a nebuliser, jet-spray, or Mini Spray (see Fig. 10.3) depending on volume. (The 'internal' reference method is also applicable.)

Under given analytical conditions, the ratio of the peak intensities (P_x/P_y) for any pair of elements is an expression of their overall relative sensitivities (K_{xy} values; see equation (10.1)). Thus, equation (10.6) simplifies further to:

$$FST_x = FST_y \cdot K_{xy}. \qquad (10.7)$$

Morgan *et al.* (1989) used FST_{Ca} as FST_y, determined by the analysis of sectioned aminoplastics; $L_x \cdot K_{Ca}$ values for a number of heavy metals were determined by the analysis of sprayed microdroplets of isoatomic solutions by the ratio method (equation (10.1)). The FST_x values (L lines) experimentally determined by this simple and rapid method were statistically comparable with the values determined by the analysis of sectioned standards containing each of the metals (Morgan *et al.*, 1989). Fig. 10.5 indicates that the ratio method may also be used for determining reliable standard factors for the K lines of lighter elements. A number of practical points should be noted concerning this method: (i) any sectioned or otherwise thin standard can be used to determine the reference FST_y value; (ii) it would be advantageous to select a sectioned standard whose matrix has a similar electron-irradiation sensitivity as the tissue specimens (see von Zglinicki, this volume); (iii) the relative stability of the inorganic microdroplets under electron irradiation is not a handicap in this context; (iv) any beam-stable element can be selected to provide the reference FST_y value; (v) the method provides an analytically valid means of determining standard factors, even for the quantitation of cryosectioned tissues, that circumvents the need to section standards containing each of the elements of interest; (vi) having experimentally established sound FST_x versus atomic number curves for a number of selected elements (Fig. 10.5), the FST values for 'missing elements' within the covered atomic number range can be derived from the curve by interpolation.

4   Analysis of digested, solubilised biological tissues

A method, based on the EDS analysis of sprayed microdroplets containing an 'external reference' element (cobalt), was introduced by Davies & Morgan (1976) for the multi-element assay of tissue homogenates. Two variants of the method are outlined in Table 10.3. The key equation for quantitation is:

$$'C_x = \frac{C_{Co} \cdot (P_x/P_{Co}) \cdot K_{xCo} \cdot v}{d}$$

(10.8)

where

$'C_x$ = concentration of element x in the solid sample (mmol/kg dry weight)

C_{Co} = concentration of cobalt in the solubilised tissue homogenate prior to spraying (or, indeed, the concentration of cobalt in individual microdroplets) (mmoles/l),

K_{xCo} = constant, determined for a given EDS system under specified analytical conditions, that relates the 'overall sensitivity in analysis' of element x to that of cobalt (see equation (10.1)),

v = volume of the tissue homogenate prior to spraying (ml),

d = dry weight of the solid sample (g).

The accuracy of the plasma-ashing variant of the tissue homogenate assay method was confirmed by Morgan (1983) for three low atomic number elements (Na, S, Ca) incorporated in a sucrose 'artificial tissue' matrix (Fig. 10.6). Furthermore, Nott & Mavin (1985) showed that their technique of wet

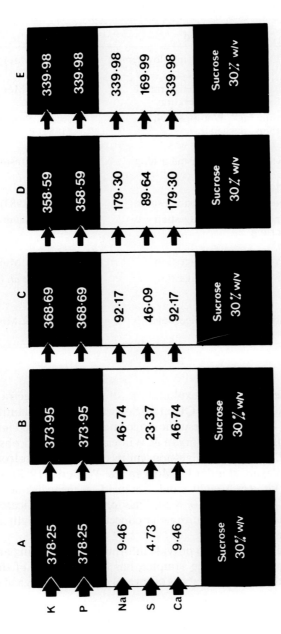

'ARTIFICIAL TISSUES'   mM kg^{-1} dry weight

	A	B	C	D	E
K	378·25	373·95	368·69	358·59	339·98
P	378·25	373·95	368·69	358·59	339·98
Na	9·46	46·74	92·17	179·30	339·98
S	4·73	23·37	46·09	89·64	169·99
Ca	9·46	46·74	92·17	179·30	339·98
	Sucrose 30% w/v	Sucrose 30% w/v	Sucrose 30% w/v	Sucrose 30% w/v	Sucrose 30% w/v

Fig. 10.6.   The composition of five 'artificial tissues', consisting of mixed salt solutions containing a sucrose matrix (Morgan, 1983). The correlation coefficient relating the concentrations of the five 'tissues' measured by EDS of sprayed homogenates to the known concentrations of Na, S and Ca were statistically highly significant: $r_{Na} = 0.9991$; $r_{S} = 0.9996$; $r_{Ca} = 0.9996$. (The method described by Morgan, 1983, for analysing these standards was modified by Nott & Mavin, 1985, for analysing blood samples.)

Table 10.4. *Applications of the sprayed microdroplet method with incorporated 'external standard' for the EDS analysis of homogenised tissues*

Tissue	Method	Reference
Rat heart	O_2-plasma (low temp.) ashing	Davies & Morgan (1976)
Rat heart	High temperature incineration	Davies & Morgan (1976)
Pellets of purified mineral granules from scallop (*Pecten*) kidney	Wet-digestion	George, Pirie & Coombs (1980)
Whole male and female blood flukes (*Schistosoma mansoni*)	O_2-plasma ashing	Shaw & Erasmus (1983)
Female *Schistosoma*	O_2-plasma ashing	Morgan (1983)
Hepatopancreas and blood from the shrimp (*Crangon crangon*)	Wet-digestion	Nott & Mavin (1985)
Whole earthworms	Wet-digestion	Winters & Morgan (1988)
Hepatopancreas of the terrestrial isopod (*Oniscus asellus*)	Wet-digestion	Morgan *et al.* (1990)
Rat cerebellum, kidney and liver	Wet-digestion	Leyshon & Morgan (1991)

digestion, coupled with the exploitation of the EDS-spectra handling capabilities of the Link Systems Quantem Software, yielded quantitative data for Na, Mg, P, S, K and Ca in tissue homogenates that possessed impressively low coefficients of variation. Morgan & Winters (unpublished observations) recently analysed, after HNO_3-digestion, sprayed microdroplets from samples of a Pb-aminoplastic standard (Morgan & Winters, 1988) containing 54.2 mmoles Pb/kg dry weight, and obtained a mean value of 55.7 ± 2.0 mmoles Pb/kg dry weight ($n = 5$). This observation indicates that the method can be used for measuring the concentrations of light and heavy elements in tissue homogenates.

Its simplicity, and capacity to provide information concerning anionic and cationic elements in small tissue samples, has led to the use of the sprayed microdroplet method for the analysis of several different biological specimens (Table 10.4).

Acknowledgements

We would like to thank Miss Alison Richards for typing this paper, and Mr Viv Williams for assistance with the illustrations.

References

Bonventre, J.V., Blouch, K. & Lechene, C.P. (1980). Liquid droplets and isolated cells. In *X-ray microanalysis in biology*, ed. M.A. Hayat, pp. 307–20. Baltimore: University Park Press.

Bostrom, T.E., Gyory, A.Z., Benson, D.C., Horgan, E., Nockolds, C.E., Beck, F.X. & Cockayne, D.J.H. (1988). Energy-dispersive X-ray microanalysis of air-dried microdroplets containing a macromolecular solute. *J. Microsc.*, **149**, 217–32.

Cliff, G. & Lorimer, G.W. (1975). The quantitative analysis of thin specimens. *J. Microsc.*, **103**, 203–7.

Davies, T.W. & Morgan, A.J. (1976). The application of X-ray analysis in the transmission electron microscope (TEAM) to the quantitative bulk analysis of biological micro-samples. *J. Microsc.*, **107**, 47–54.

Garland, H.O., Brown, J.A. & Henderson, I.W. (1978). X-ray analysis applied to the study of renal tubular fluid samples. In *Electron probe microanalysis in biology*, ed. D.A. Erasmus, pp. 212–43. London: Chapman & Hall.

Gauvin, R. & L'Esperance, G. (1991). Determination of the C_{nl} parameter in the Bethe formula for the ionization cross-section by the use of Cliff–Lorimer K_{AB} factors obtained at different accelerating voltages in a TEM. *J. Microsc.*, **163**, 295–306.

George, S.G., Pirie, B.J.S. & Coombs, T.L. (1980). Isolation and elemental analysis of metal-rich granules from the kidney of the scallop, *Pecten maximus* (L.). *J. exp. mar. Biol. Ecol.*, **42**, 143–56.

Goldstein, J.I., Costley, J.L., Lorimer, G.W. & Reed, S.J.B. (1977). Quantitative X-ray analysis in the electron microscope. *Scanning Electron Microscopy*, **1977/I**, 315–24.

Graham, R.J. & Steeds, J.W. (1984). Determination of Cliff–Lorimer K factors by the analysis of crystallized microdroplets. *J. Microsc.*, **133**, 275–80.

Hall, T.A. (1989). Quantitative electron probe X-ray microanalysis in Biology. *Scanning Microscopy*, **3**, 461–6.

Hall, T.A. & Gupta, B.L. (1983). The localization and assay of chemical elements by microprobe methods. *Q. Rev. Biophys.*, **16**, 279–339.

Hyatt, A.D. & Marshall, A.T. (1985). An alternative microdroplet method for quantitative X-ray microanalysis of biological fluids. *Micron and Microscopica Acta*, **16**, 39–44.

Ingram, M.J. & Hogben, C.A. (1967). Electrolyte analysis of biological fluids with the electron microprobe. *Anal. Biochem.*, **18**, 54–7.

Ingram, M.J. & Ingram, F.D. (1981). Simple microforge for picoliter pipet constrictions. In *Microbeam analysis*, ed. R.H. Geirs, pp. 219–21. San Francisco: San Francisco Press.

Lechene, C. (1974). Electron probe microanalysis of picoliter liquid samples. In *Microprobe analysis as applied to cells and tissues*, ed. T. Hall, P. Echlin, R. Kaufman, pp. 351–67. New York: Academic Press.

Lechene, C. & Warner, R. (1979). Electron probe analysis of liquid droplets. In *Microbeam analysis in biology*, ed. C. Lechene & R. Warner, pp. 279–96. New York: Academic Press.

LeRoy, A. & Roinel, N. (1983). Radiation damage to lyophilized mineral solution during electron probe analysis: Quantitative study of chlorine loss as a function of beam current density and sample mass thickness. *J. Microsc.*, **131**, 97–106.

Leyshon, K. & Morgan, A.J. (1991). An integrated study of the morphological and gross-elemental consequences of methyl mercury intoxication in rats, with particular attention on the cerebellum. *Scanning Microsc.*, **5**, 895–904.

Marshall, A.T. (1977). Iso-atomic droplets as models for the investigation of parameters affecting X-ray microanalysis of biological specimens. *Micron*, **8**, 193–200.

Marshall, A.T. & Forrest, Q.G. (1977). X-ray microanalysis in the transmission electron microscope at high accelerating voltages. *Micron*, **8**, 135–8.

Morel, F. & Roinel, N. (1969). Application de la microsonde électronique à l'analyse élémentaire quantiative d'echantillons liquides d'un volume inferieur à 10^{-9} l. *J. Chim. Phys. Phys. Chim.*, **66**, 1084–91.

Morgan, A.J. (1983). The electron microprobe analysis of sprayed microdroplets of solubilized biological tissues: a useful preliminary to localization studies. *Scanning Electron Microsc.*, **1983/II**, 861–72.

(1985). *X-ray microanalysis in electron microscopy for biologists*. Oxford: Oxford University Press, Royal Microscopical Society.

Morgan, A.J. & Davies, T.W. (1982). An electron microprobe study of the influence of beam current density on the stability of detectable elements in mixed-salts (isoatomic) microdroplets. *J. Microsc.*, **125**, 103–16.

Morgan, A.J., Davies, T.W. & Erasmus, D.A. (1975). Analysis of droplets from isoatomic solutions as a means of calibrating a transmission electron analytical microscope (TEAM). *J. Microsc.*, **104**, 271–80.

Morgan, A.J., Gregory, Z.D.E. & Winters, C. (1990). Response of the hepato-pancreatic 'B' cells of a terrestrial isopod, *Oniscus asellus*, to metals accumulated from a contaminated habitat: a morphometric analysis. *Bull. Environ. Contam. Toxicol.*, **44**, 363–8.

Morgan, A.J., Roos, N., Morgan, J.E. & Winters, C. (1989). The subcellular accumulation of toxic heavy metals: qualitative and quantitative X-ray micro-analysis. In *Electron probe microanalysis, applications in biology and medicine*, ed. K. Zierold & H.K. Hagler, pp. 59–72. Berlin: Springer-Verlag.

Morgan, A.J. & Winters, C. (1988). Practical notes on the production of thin aminoplastic standards for quantitative X-ray microanalysis. *Micron and Microscopica Acta*, **19**, 209–12.

Nott, J.A. & Mavin, L.J. (1985). Adaptation of a quantitative programme for the X-ray analysis of solubilized tissue as microdroplets in the transmission electron microscope: application to the moult cycle of the shrimp *Crangon crangon* (L.). *Histochem. J.*, **18**, 507–18.

Quinton, P.M. (1976). Construction of pico-nano-liter self-filling volumetric pipettes. *J. Appl. Physiol.*, **40**, 260–2.

(1978a). Ultramicroanalysis of biological fluids with energy dispersive X-ray spectrometry. *Micron*, **9**, 57–69.

(1978b). SEM-EDS X-ray analysis of fluids. *Scanning Electron Microsc.*, **1978/II**, 391–7.

(1978c). Techniques for microdrop analysis of fluids (sweat, saliva, urine) with an energy-dispersive X-ray spectrometer on a scanning electron microscope. *Am. J. Physiol.*, **234**, F255–9.

(1979). Energy-dispersive X-ray analysis of biological microdroplets. In *Microbeam analysis in biology*, ed. C. Lechene & R. Warner, pp. 327–45. New York: Academic Press.

Rick, R., Horster, M., Dorge, A. & Thurau, K. (1977). Determination of electrolytes in small biological fluid samples using energy-dispersive X-ray microanalysis. *Pflügers Arch.*, **369**, 95–8.

Roinel, N. (1975). Electron microprobe quantitative analysis of lyophilized 10^{-10} l volume samples. *J. Microsc. (Paris)*, **22**, 261–8.

(1988). Quantitative X-ray analysis of biological fluids: the microdroplet technique. *J. Electron Microscopy Technique*, **9**, 45–56.

Roinel, N., Meny, L. & Henoc, J. (1980). Accuracy of electron microprobe analysis of biological fluids: choice of standard solutions, and range of linearity of the calibration curves. *National Bureau of Standards Special Publication*, **533**, 101–30.

Roinel, N. & de Rouffignac, C. (1982). X-ray analysis of biological fluids: contribution of microdroplet technique to biology. *Scanning Electron Microsc.*, **1982/III**, 1155–71.

Roos, N. & Barnard, T. (1984). Aminoplastic standards for quantitative X-ray microanalysis of thin sections of plastic-embedded biological material. *Ultramicroscopy*, **15**, 277–86.

Roos, N. & Morgan, A.J. (1985). Aminoplastic standard for electron probe X-ray microanalysis (EPXMA) of ultrathin frozen-dried cryosections. *J. Microsc.*, **140**, RP3–4.

(1990). *Cryopreparation of thin biological specimens for electron microscopy: methods and applications.* Oxford: Oxford University Press, Royal Microscopical Society.

Russ, J.C. (1974). The direct element ratio model for quantitative analysis of thin sections. In *Microprobe analysis as applied to cells and tissues*, ed. T. Hall, P. Echlin & R. Kaufmann, pp. 269–76. London: Academic Press.

(1975). Evaluation of the direct element ratio calculation method. *J. Microscopie Biol. Cellulaire*, **22**, 283–6.

Shaw, M.K. & Erasumus, D.A. (1983). *Schistosoma mansoni:* changes in elemental composition in relation to the age and sexual status of the worms. *Parasitology*, **86**, 439–53.

van Eekelen, C.A.G., Boekestein, A., Stols, A.L.H. & Stadhouders, A.M. (1980). X-ray microanalysis of picoliter microdroplets: improvement of the method for quantitative X-ray microanalysis of samples of biological fluids. *Micron*, **11**, 137–45.

Warley, A. (1990). Standards for the application of X-ray microanalysis to biological specimens. *J. Microsc.*, **157**, 135–47.

Winters, C. & Morgan, A.J. (1988). Quantitative electron probe X-ray microanalysis of lead-sequestering organelles in earthworms: technical appraisal of air-dried sections and freeze-dried cryosections. *Scanning Microsc.*, **2**, 947–8.

Wirmark, G. & Norden, H. (1987). A study of factors influencing the accuracy of quantitative X-ray analysis of thin foil specimens. *J. Microsc.*, **148**, 167–78.

Wood, J.E., Williams, D.B. & Goldstein, J.I. (1984). Experimental and theoretical determination of K_{AFe} factors for quantitative X-ray microanalysis in the analytical electron microscope. *J. Microsc.*, **133**, 255–74.

Zierold, K. (1988). X-ray microanalysis of freeze-dried and frozen–hydrated cryosections. *J. Electron Microscopy Technique*, **9**, 65–82.

Addendum

Recently (Malone, M., Leigh, R.A. & Tomos, A.D. (1991). Concentrations of vacuolar inorganic ions in individual cells of intact wheat leaf epidermis.

J. Exp. Botany, **42**, 305–9) a microdroplet technique has been applied for the first time to botanical specimens. Samples of undiluted vacuolar sap were collected from single cells with a pressure probe, and microdroplets were dispensed from constant volume micropipettes, mounted on thin films, and freeze-dried for EDS analysis in an SEM.

11 X-ray microanalysis of biological fluids: applications to investigations in renal physiology

Nicole Roinel and Christian de Rouffignac

A Introduction

Determination with an electron probe of the elemental composition of liquid samples of 'small' volumes is not microanalysis *sensu stricto*. Indeed, these volumes are in the range of 10^{-10} to 10^{-11} litre, i.e. still several orders of magnitude larger than the 1 μm^3 volume normally excited in conventional X-ray microanalysis. However, dealing with nanolitre volumes is a common situation for a biologist, who then faces the problem of determining a number of elements in so 'small' a volume. This explains why the technique of X-ray analysis of droplets has been so extensively used over the past two decades. This chapter will not focus on the technical details of sample preparation. Two recent review papers (Quamme, 1988; Roinel, 1988) have abundantly dealt with this problem. After a brief overview of the technique (sample preparation and characteristics, sensitivity, accuracy and precision), we will show how efficient the use of the technique has been in renal physiology, especially *in vivo* for the study of the handling of magnesium by the kidney and *in vitro* for studies involving the microperfusion of isolated tubules.

Although the droplet technique has so far mainly been used in investigations of the renal physiology of mammals, it has also been applied to the study of the excretory function of insects, and to the study of other physiological functions in man (Ferrary *et al.*, 1988), mammals and various animals (see bibliographic reviews in Roinel & Rouffignac, 1982, and more recently Quamme, 1988).

B The microdroplet technique

1 Principle

The principle of the method is the following: identical volumes of biological fluids and standard solutions of appropriate composition are deposited on a holder and desiccated to crystals thin enough to be transparent to both primary electrons and emitted X-rays. Each dried sample is fully irradiated by

(a)

(b)

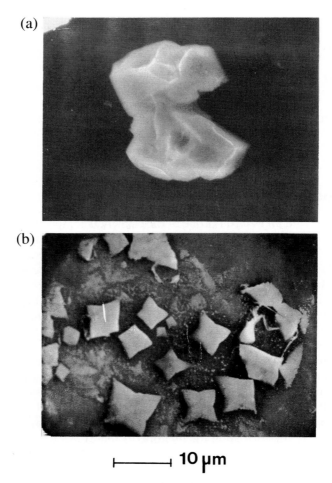

├───────┤ **10 μm**

Fig. 11.1. a and b.   For legend see facing page.

a defocused electron beam of adequate surface area and energy, to make it capable of emitting X-rays characteristic of all the contained elements. If the intensity of the characteristic X-rays of an element is proportional to its concentration inside the sample, an intensity/concentration relationship can be established from calibration curves constructed by irradiating standard solutions of known composition for this element (Morel & Roinel, 1969). Any element with an atomic number above fluorine can be analysed from a single sample.

(c)

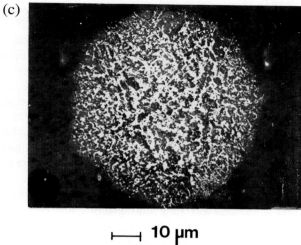

⊢——⊣ **10 µm**

Fig. 11.1.   Scanning electron microscope images of dried droplets of 0.1 nl
solutions of 150 mmol·l⁻¹ NaCl. (a) Droplet just dehydrated in the
open air; (b) same procedure as (c) but sample too thick for correct
analysis, leading to underestimation of the concentration (this sample
must be rehydrated, shock-frozen and lyophilised); (c) droplet shock-
frozen and lyophilised after complete rehydration.

2   Sample preparation

Sample preparation comprises two steps, which can be separated in time and
place (Roinel, 1975): (1) the depositing on a support of droplets with
reproducible volumes; (2) the obtaining of sub-micrometer size crystals.

Supports must give as low a continuum as possible. They can be either
plastic films, or massive holders. Plastic films are fragile to handle but
necessary for energy-dispersive analysis (EDS). A massive holder is not fragile
and can work for wavelength-dispersive analysis (WDS); pure beryllium has
so far proven to be the best choice. Reproducible volumes are easily obtained
from constriction micropipettes or constant bore pipettes (Quinton, 1976).
Both types of pipettes have to be siliconised in order to become hydrophobic.

All the fluids are manipulated under paraffin oil, in order to avoid
spontaneous evaporation of such small aqueous volumes. This oil must be
saturated with water for obvious reasons. The surface of the microdroplet
support is covered with oil, as also are the tiny droplets of biological fluid
deposited on the holder. When all the droplets (biological fluids and standard
solutions) have been deposited, the paraffin oil on the holder is dissolved in a
bath of chloroform, or any adequate solvent, and liquid droplets are allowed
to dry in the open air.

Sub-micrometer size crystal preparations of the salts contained in the biological fluids and the standard solutions are then obtained as follows. The dried droplets are fully rehydrated by cooling the support to 4 °C (by means of a Peltier plate or, more simply, by a brass cup cooled with ice) and letting atmospheric water preferentially condense on them. The support is then quickly frozen (by contact with dry ice, or by plunging in hexadecane cooled by liquid nitrogen), and lyophilised (one-stage vacuum of 1 Pa is sufficient). After lyophilisation, which is immediate, the support is allowed to warm up at room temperature. This procedure reproducibly gives dried samples composed of tiny crystals (Fig. 11.1). There is no need for carbon coating prior to analysis. Since the preparation is immediately placed under vacuum inside the electron microscope, there is no danger of rehydration of the dried crystallised droplets.

Interestingly, this procedure of rehydration and shock-freezing can be carried out quite separately (in time and place) from the deposition of droplets. Another advantage of this two-step procedure is that the step of rehydration and rapid freezing (with lyophilisation) may be performed again if crystals of adequate size have not been obtained (either because of incomplete rehydration, or a support still cold when the vacuum in the lyophiliser is broken; see Fig. 11.1(b)). Variations of this preparative procedure have been proposed by Quinton (1978) and others (see review in Roinel & Rouffignac, 1982). It should be emphasised that the accuracy of the results depends upon the lack of absorption of X-rays and electrons inside all dried samples, and that careful examination must be carried out on the structural appearance of samples to determine how thin they are.

3   Electron probe analysis procedure

The procedure described here is used for WDS analysis. The electron beam is defocused to a surface area sufficient to cover the largest sample. Electron energy and intensity are chosen, according to the mean sample mass thickness and the nature of the element, to ensure stability of the counting rate. For a mass thickness of about 0.05 mg·cm^{-2} (corresponding to a volume of 0.1 nl and 60 µm diameter), an 11 keV electron energy is adequate for full excitation of the sample. The background is measured at the Bragg angle at random places on the support to ensure good counting statistics (fifteen measurements with a statistical test). Each solution whose composition is to be determined is deposited as five droplets of identical volume, with a statistic test to discard any wrong value (dust, or large crystals). The need for X-ray calibration for each series of samples requires standard solutions to be present upon each support, as five replicates for each solution. This frequently leads to several hundred droplets having to be analysed for a single experiment. When low concentrations are to be measured, counting times are long. Automation procedures able to monitor all the operations, although not strictly necessary, are really convenient, so far as routine procedures can be monitored. At

Saclay, the WDS analysis of seven elements in 350 to 400 droplets represents total counting times of 17 h; this is carried out routinely by a computer.

4   Results

(i) LINEARITY OF THE CALIBRATION LINES

Characteristic signals proportional to the element concentration in the deposited volumes are routinely obtained for volumes smaller than 0.2 nl and concentrations up to 200 mmol·l^{-1} of NaCl, with mixtures of other elements up to 20 mmol·l^{-1} in a NaCl matrix. Factors affecting the slope of the calibration lines have been described at length (Morel & Roinel, 1969); they are the beam intensity, the beam cross-section, the electron energy, and the pipette volume. There is a direct proportionality of the X-ray intensities with the element's concentration, without the need for correction factors.

(ii) MINIMUM DETECTABLE CONCENTRATION

The literature reports a variety of formulae for calculating the minimum detectable concentrations (MDCs). The values quoted are calculated from one, two or three times the standard deviation of the background, converted into the element's concentration. More rigorous but sophisticated formulae do not give results significantly different from that mentioned above. They all predict MDCs in the 0.05 mmol·l^{-1} range, using beryllium supports and WDS analysis. More interesting is the experimental and routine confirmation of such values (Roinel & Rouffignac, 1982), for the elements Mg, P and Ca in the presence of a matrix of 150 mmol·l^{-1} NaCl and 100 s analysis time (Fig. 11.2). There is little prospect that this figure will be significantly improved with improved technology of wavelength-dispersive spectrometers.

EDS analysis gives poorer sensitivity, in terms of the ratio of characteristic to continuum signals, than WDS analysis, due to the different technology of the X-ray detector (Quinton, 1978). This ratio is independent of a better collection of photons. The continuum can be lowered by using plastic films as holders (Van Eekelen *et al.*, 1980), but those films are very fragile and difficult to use routinely.

(iii) MAXIMUM DETECTABLE CONCENTRATION

It is interesting to determine this value as many biological fluids are not iso-osmotic with mammalian plasma, for whose range of concentrations all calibration curves have so far been determined. The knowledge of the highest measurable concentration is important for hyperosmotic fluids. Do they have to be diluted or not?

The proportionality of the measured characteristic signal to the concentration is limited only by electron and X-ray absorption in dried samples of homogeneous thickness. When thickness increases, calibration curves

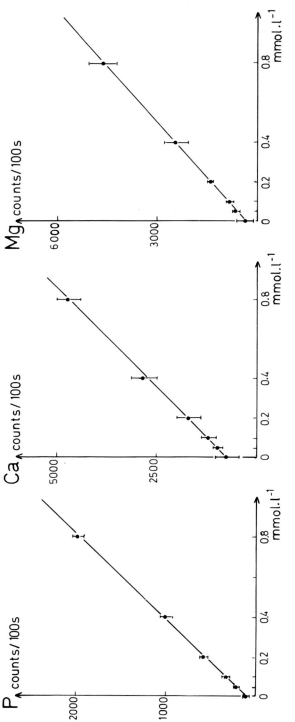

Fig. 11.2. Calibration lines for P, Ca and Mg in the concentration range 0 to 1 mmol·l⁻¹, in the presence of 150 mmol·l⁻¹ NaCl. Mean of five replicates + standard deviation. Electron energy, 13 keV, beam current, 300 nA, electron beam dimensions, 80 × 80 µm; sample volume, about 0.1 nl.

exhibit a saturation effect, leading to underestimation of the concentration. The useful electron range increases with the electron energy, and can be modulated through this parameter. In contrast to the absorption of electrons, X-ray absorption depends only on the overall absorption coefficient of each dried sample (depending on its composition) for the emitted photon and the X-ray take-off angle of the analysis system. Curves have been experimentally and theoretically established for six elements – Na, Cl, K, Ca, P and Mg – in samples composed mainly of NaCl, analysed at several electron energies, and two take-off angles of 18° and 40° (reviewed by Roinel, 1988). It was noted that, even for the lowest electron energy of 12 keV, at 40° take-off angle, linearity of the calibration lines is still obtained for Na and Mg, elements yielding the least energetic X-rays, in samples of 0.1 mg·cm^{-2} mass thickness. This last figure should be compared with the mass thickness of less than 0.05 mg·cm^{-2}, which is easily achieved for original liquid volumes of 0.1 to 0.2 nl.

(iv) PITFALLS

The pitfalls described here do not result from the presence of large crystals in dried deposits. They arise from general properties of electrons and X-rays, and from the special conditions of manipulation under oil of very small volumes of fluids. These pitfalls may lead to systematic errors and inaccuracy of the determinations.

(a) Stability of the counting rates Mineral and organic solutions are subjected to radiation damage which manifests itself as the loss of characteristic signals with time. The signal of any element, especially of halides, can decrease with time, depending on the beam current density, the sample mass thickness, and the overall composition of the solution. Solutions of hydrated salts are particularly sensitive to beam damage, and have to be avoided as standards. However, conditions of irradiation exist where characteristic signals are stable; as reported by Leroy & Roinel (1983) for beam current densities of 10^{-3} to 1.5 × 10^{-2} A·cm^{-2} with samples of KCl and NaCl of mass thicknesses ranging from 0.05 to 0.75 mg·cm^{-2}. Radiation damage represents a real risk of systematic error in the determination of elemental concentrations. Consequently, the stability of signals of all the elements in any kind of solution, biological and standard, has to be checked systematically by WDS analysis.

Such radiation effects have not been reported for X-ray analysis by EDS because the beam current density is a thousand times lower than in WDS analysis. As to the effect of electron energy, it has not really been observed for analyses performed in the short range of 13 to 18 keV.

(b) Chemical shift The degree of oxidation of an element may affect its X photon energy, depending on the oxidation level and the element's atomic

number (Faessler & Goehring, 1952). In the case of sulphur, for example, an increase in its K energy level of 1.2 eV with the degree of oxidation reduces the wavelength by 0.03 nm. This results in a significant variation of the corresponding Bragg position in WDS. These effects have been investigated at length for the various elements we are routinely measuring (Roinel, Meny & Hénoc, 1980). It was concluded that the composition of the standard solutions must also match the degree of oxidation of the element in the analysed biological fluid.

(c) Storage of samples under oil   Volumes in the nanolitre range always present a large surface-to-volume ratio, which may induce surface phenomena when stored for long periods under oil. The paraffin oil is water-saturated, and measurements of the radioactivity of nanolitre droplets stored under oil for several hours never showed any sign of water pumping by the oil. In contrast to this, significant decreases with time of the concentration of calcium, and to a lesser extent of magnesium, have been reproducibly observed and studied by Quamme's group (Muhlert, Julita & Quamme, 1982; Quamme & Dirks, 1986a; Quamme, 1988), the loss being the highest at high sample pH. At Saclay, we observed such calcium loss, but only for plasma fluids, and after an initial 2 h period of concentration stability. At any rate, the stability with time of the composition of any liquid stored under oil has to be checked.

C   Applications in renal physiology: renal handling of electrolytes, with special references to magnesium

In renal physiology, the technique of droplet analysis described above has allowed the first simultaneous determination of the respective concentrations of most of the elements in the microsamples collected by micropuncture from kidney structures (Morel, Roinel & Le Grimellec, 1969). Moreover, this is one of the few techniques available for measuring the Mg concentration in such samples, and it is very convenient for assessing the Ca and P concentrations. In this respect, it is significant that most of the studies devoted to the simultaneous tubular handling of Ca, P and Mg in the kidney come from laboratories that have used an electron probe analyser.

Magnesium plays an important part in the life of organisms by its participation in biological events such as, intracellular enzyme chemistry (as a cofactor), bone mineralisation, and complex formation with extracellular proteins (Lasserre & Durlach, 1991). The factors which are involved in the maintenance of Mg balance are not fully understood. It is clear, however, that the kidney plays an important part in these homeostatic mechanisms. The kidneys adapt the excretion of Mg, through hormonal and non-hormonal controls, to the variations in its intake in the same way as they do for other constituents of the body fluids. During magnesium deprivation, the kidneys reabsorb most of the filtered load and the Mg delivery into urine drops to very low levels, whereas during high dietary magnesium intake, the excretion of Mg is increased to maintain the concentration in body fluids.

Several reviews have been devoted to the renal handling of magnesium (Quamme & Dirks, 1980; 1986b; Quamme, 1989; Rouffignac, 1990; 1991). The two last reviews have given new information relative to the sites and the regulation of Mg transport. The present chapter will be limited to the regulation of the transport of Mg and other related electrolytes in the mammalian kidney, studied both with *in vitro* techniques (applied to tubular segments obtained from microdissected kidneys) and with an *in vivo* model (the 'hormone-deprived' model) adapted to this particular investigation.

1 In vitro *studies of the effects of ADH, PTH, calcitonin, glucagon and isoproterenol on the transport of electrolytes and Mg in the thick ascending limb*

Here, we shall focus the illustration of the use of electron probe analysis on new data obtained *in vitro*, by associating this technique to the microperfusion technique of isolated tubules (Fig. 11.3). A detailed bibliography of these data will be found in Rouffignac *et al.* (1991b). Microperfusion of isolated nephron segments has been successfully used by many laboratories involved in renal physiology research. In our laboratory, this technique was used to disclose the physiological effects of hormones on the movement of electrolytes in the thick ascending limb of the mouse. The mouse was chosen because its thick ascending limb is responsive to several hormones, including antidiuretic hormone (ADH), parathyroid hormone (PTH), calcitonin, glucagon and the beta-agonist isoproterenol.

Table 11.1 shows adenylate cyclase activities measured in the presence of these hormones in the three species commonly studied either with the *in vitro* microperfusion technique (mouse and rabbit) or by micropuncture *in vivo* (rat) (Morel, 1981; Morel, Imbert-Teboul & Chabardès, 1981; 1987). In the rat and mouse, all five ligands stimulate the cyclase system in the cortical thick ascending limb (cTAL), whereas for the medullary thick ascending limb (mTAL) the two species show certain differences. The latter segment is responsive to ADH and glucagon, but not to PTH, in both species, but is unresponsive to calcitonin in the mouse and to isoproterenol in the rat. In the rabbit, the cyclase responsiveness to the hormones is quantitatively different from those of the rat and mouse but the sensitivities to ADH and PTH show some similarities. Unlike the thick ascending limb of the rat and mouse, that of the rabbit is unresponsive to isoproterenol (Table 11.1).

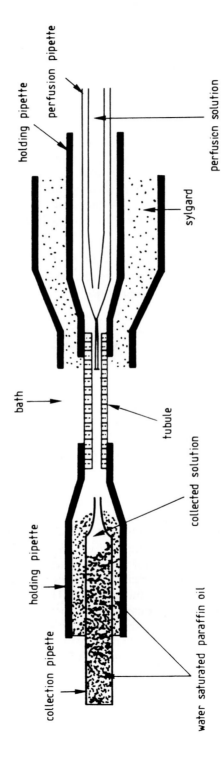

perfusion side

collection side

holding pipette

perfusion pipette

perfusion solution

sylgard

bath

tubule

holding pipette

collection pipette

collected solution

water saturated paraffin oil

Fig. 11.3. Microperfusion of isolated renal tubules. The tubule is maintained in a bath at 37 °C. The perfusion solution is delivered at a known flow rate through a microperfusion pipette situated at one end, and is collected under water-saturated paraffin oil at the other end of the tubule. For a given element, the concentration differences between the perfused and collected solutions, as determined by electron probe analysis, combined with the determined flow rate, make it possible to determine the transport properties of the tubule for this element.

Table 11.1. *Hormonal effects on cyclase responsiveness and electrolyte transport in isolated segments of rabbit, mouse and rat cortical and medullary thick ascending limbs*

Adenylate cyclase	Cortical thick ascending limb			Medullary thick ascending limb		
	Rabbit	Mouse	Rat	Rabbit	Mouse	Rat
A. cyclase basal	**0.30**	**1.10**	**0.80**	**0.40**	**0.70**	**0.80**
ADH	1.00	26.70	7.00	3.70	36.90	27.40
Glucagon	n.t.	44.30	23.80	n.t.	37.90	34.60
PTH	4.10	36.00	30.00	N.S.	N.S.	N.S.
Calcitonin	2.80	18.40	24.20	12.60	N.S.	6.30
Isoproterenol	N.S.	9.90	6.00	N.S.	3.90	N.S.

A. cyclase: adenylate cyclase activity ($fmol \cdot mm^{-1} \cdot min^{-1}$) obtained in presence of indicated hormones. Gives the activity due to the hormone (stimulated minus basal activity).
N.S. = not statistically significant; n.t. = not tested.
From Morel, 1981; Morel *et al.*, 1981, 1987.

(i) EFFECTS ON SODIUM CHLORIDE TRANSPORT

(a) **Medullary thick ascending limb (mTAL)**　In the mouse mTAL, ADH always increases the transepithelial Na^+ net fluxes (JNa^+)† (Fig. 11.4, top), and also stimulates JCl^- (data not shown). Glucagon exerts an effect very similar to that of ADH on this nephron segment. These data confirm previous observations reported in this species. Isoproterenol significantly increases JNa^+ and JCl^- (Table 11.1). On the other hand, human calcitonin (HCT) has no effect on the JNa^+ (Fig. 11.4, top) and JCl^- fluxes, as is to be expected since there is no calcitonin-sensitive adenylate cyclase in the mTAL segment of the mouse.

Thus the effects elicited by ADH in the mouse appear to be similar to those reported in the rat, in which the hormone and cAMP analogues increase the chloride transport. In the rabbit, ADH failed to elicit any change in Cl^- transport. It is therefore likely that the different species show different physiological responses to hormone stimulation of the thick ascending limb.

(b) **Cortical thick ascending limb (cTAL)**　In the mouse cTAL, ADH, glucagon, PTH, calcitonin and isoproterenol reversibly increase JNa^+ (Fig.

---

† The transepithelial net flux of an element JX ($pmol \cdot min^{-1} \cdot mm^{-1}$) is calculated from the difference in the concentrations of the element in the perfused and the collected solutions, as determined by electron probe analysis; it is the product of the element concentration difference, in $mmol \cdot l^{-1}$, and the perfusion flow rate in $nl \cdot min^{-1}$, divided by the length of the perfused tubule, in mm.

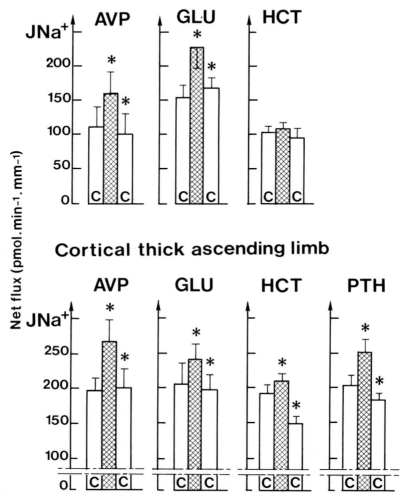

Fig. 11.4.   *In vitro* microperfusion studies. Effects of peptide hormones on transepithelial Na⁺ net flux (JNa⁺) in single cortical (bottom) and medullary (top) portions of mouse thick ascending limbs. The tubular segments were microperfused in the absence (C) and presence of either ADH, glucagon, calcitonin or PTH, added to the bath. The tubular fluid was collected for 30 min before (C) and during (hatched column) addition of the hormone to the bath and after removal (C) of the hormone. Each period comprised three collection periods of 10 min. A 20 min equilibration period was allowed before each 30 min period. The perfusion rates ranged between 0.8 and 4 nl·min⁻¹. The number of tubules varied from 7 to 22 in each series. The concentrations of the

11.4, bottom, and Table 11.1) and JCl^- (data not shown). In these experiments, the increases with HCT were less marked than with the other hormones probably because JNa^+ and JCl^- declined between the pre- and post-experimental periods (Fig. 11.4, bottom).

In the rat, no *in vitro* microperfusion experiments have as yet, to our knowledge, been performed to explore the effects of peptide hormones on NaCl transport in the cTAL. In the rabbit, the available data indicate that PTH, calcitonin and cAMP analogues do not alter the net reabsorption fluxes of NaCl. This absence of effect of both hormones in the rabbit compared with the marked effects of the same hormones in the mouse confirms that species differences do clearly exist in this respect.

(ii) EFFECTS ON MAGNESIUM AND CALCIUM TRANSPORT

PTH, HCT, glucagon, ADH and isoproterenol stimulate Mg^{2+} and Ca^{2+} transport in the mouse cTAL (Fig. 11.5). Propranolol (10^{-6} M in the bath) inhibits the isoproterenol-mediated increase of Mg^{2+} and Ca^{2+} and NaCl transport. There are as yet no data available from studies on the rat. That PTH and cAMP analogues stimulate the transport of these electrolytes in the cTAL had already been well established in the rabbit. It was found that calcitonin stimulates Ca^{2+} transport in the rabbit mTAL but not in the cTAL, whereas the reverse occurs in the mouse, as shown in Table 11.1 (Morel, 1981; Morel *et al.*, 1981, 1987). It seems, therefore, that Ca^{2+} transport can be stimulated by peptide hormones in both cTAL and mTAL segments in the rabbit, whereas in the mouse, Ca^{2+} transport is hormonally stimulated only in the cortical part of the TAL. The observation that calcitonin does not stimulate Ca^{2+} reabsorption in the rabbit cTAL is well explained by the absence of an adenylate cyclase system sensitive to calcitonin in this segment. Thus there is good agreement between the effects of PTH and calcitonin on divalent ion transport in the mouse and rabbit. In addition, it is relevant to note at this stage that ADH and glucagon, in addition to PTH and calcitonin, may also stimulate Ca^{2+} and Mg^{2+} transport in the mouse cTAL.

In the mouse mTAL, all the experiments carried out to date at either Saclay, Bichat Hospital or Freiburg University revealed that Mg^{2+} and Ca^{2+} transport were very limited in this nephron segment. These findings are at variance with those from another study that reported large Ca^{2+} fluxes (*c.* 5 $pmol \cdot min^{-1} \cdot mm^{-1}$) in the mTAL of mice of the same strain as ours (Friedman, 1988). This is surprising, since in this latter study the Ca^{2+} concentration in the bath was twice as high (2 $mmol \cdot l^{-1}$) as ours and it is

hormones were: ADH as arginine vasopressin AVP, 10^{-10} $mol \cdot l^{-1}$, glucagon as synthetic porcine glucagon, 1.2×10^{-8} $mol \cdot l^{-1}$, calcitonin as synthetic human calcitonin, 3×10^{-8} $mol \cdot l^{-1}$ or cibacalcin RC 47175 Ba, PTH as bovine PTH, 1–34 fragments, 10^{-8} $mol \cdot l^{-1}$, Sigma. The asterisks (*) indicate values significantly different from the preceding period. (From Rouffignac *et al.*, 1991b.)

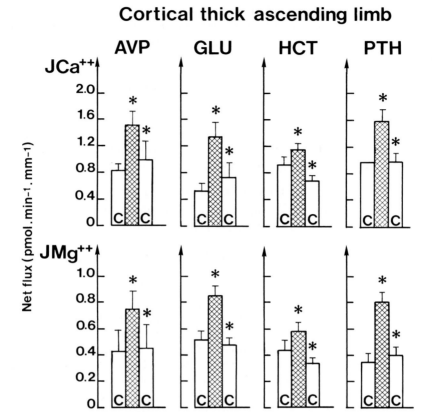

Fig. 11.5. In vitro microperfusion studies. Effects of peptide hormones on transepithelial Ca^{2+} and Mg^{2+} net fluxes (JCa^{2+} and JMg^{2+}) in mouse cortical thick ascending limbs. The tubular segments were perfused in the absence (C) and presence of either ADH, glucagon, calcitonin or PTH, added to the bath (same concentrations as for Fig. 4). The tubular fluid was collected for 30 min before (C), during addition of the hormone to the bath (hatched column) and after removal of the hormone (C). Each period comprised 3 collections periods of 10 min. A 20 min equilibration period was allowed before each 30 min period. The perfusion rates were between 0.8 and 4 $nl \cdot min^{-1}$. The number of tubules varied from 7 to 22 in each series. The asterisks (*) indicate values significantly different from the preceding period. (From Rouffignac *et al.*, 1991b.)

known, at least in the rabbit, that an increase of Ca^{2+} at the basolateral side tends to decrease Ca^{2+} reabsorption in cTAL segments perfused *in vitro*. It is not easy to explain these discrepancies. It must be pointed out, however, that it is impossible to evaluate in Friedman's study either the sensitivity or the accuracy of the method used to determine the Ca^{2+} concentrations in the microsamples, or the concentration differences of Ca^{2+} between samples of collected and perfused fluid because these parameters (or the collection rates permitting their calculation) were not given. By electron probe analysis, a $0.05\ mmol \cdot l^{-1}$ variation of the Ca^{2+} concentration is easily measurable. Thus, in our experiments, net fluxes as low as $0.2\ pmol \cdot min^{-1} \cdot mm^{-1}$ would have been significantly detected.

2   In vivo *studies of the hormonal control of the transport of* Mg^{2+} *and other related electrolytes*

The nephrons are target sites for many hormones including steroids and hormones stimulating the protein kinase A (cAMP), the protein kinase C (Ca, diacylglycerol), or the tyrosine kinase receptor complex (insulin), etc., transduction pathways. Apart from those hormones acting via the cAMP transduction pathway, little is known, at the nephron level, about the effects of the hormones acting through other pathways on Mg transport.

(i) PROXIMAL TUBULE

The transport of Mg in this nephron segment is very limited, and neither PTH, glucagon, calcitonin nor ADH, modified this situation. In parathyroid-ectomised rats Mg concentration along the proximal tubule rose as a function of water removal, as in the normal rats. A more detailed bibliography may be found in Rouffignac (1991) and Rouffignac *et al.* (1991a, 1993).

(ii) LOOP OF HENLE

In vivo, the effects of four hormones (ADH, PTH, calcitonin and glucagon) were tested by micropuncture in the rat using the **hormone-deprived model.** A detailed bibliography can be found in Rouffignac, Elalouf & Roinel (1987) and in Rouffignac *et al.* (1993). This hormone-deprived model was elaborated from the discovery that in this species, as we have seen, several hormones act on the same target cells in several nephron segments. Moreover, all of these hormones act on the same adenylate cyclase pool in these cells, and thus generate the intracellular cAMP responsible for the final biological effect (Morel *et al.*, 1982), as described for the thick ascending limb in the preceding section. Thus, to explore *in vivo* the effect of one of these hormones, it is necessary to remove from the blood all the other hormones which, due to the similarity of their physiological effects, could mask the effect of the hormone under investigation. Brattleboro D.I. rats, lacking circulating ADH, were acutely thyroparathyroidectomised to suppress circulating PTH and cal-citonin and infused with somatostatin to inhibit glucagon secretion. Such rats were called 'hormone-deprived'. Thus, most of the peptide hormones acting on the thick ascending limb were removed.

In the rat, the distribution along the thick ascending limb of the sensitivity of the adenylate cyclase system to these hormones (ADH, glucagon, PTH, calcitonin and isoproterenol) is very similar to that of the mouse, except that in the rat the mTAL is sensitive to calcitonin and insensitive to isoproterenol (Table 11.1). All four hormones (ADH, glucagon, PTH and calcitonin), when administered individually to such hormone-deprived rats, increased K, Mg and Ca as well as (except PTH) NaCl transport in the loop of Henle. These findings confirm and extend previous data showing that PTH in the hamster and calcitonin in the rat stimulate Mg and Ca transport in the loop of Henle. In the rat, in previous experiments, the stimulatory effect of PTH on Mg reabsorption in the loop was observed only when Mg reabsorption was reduced by either hypomagnesemia or furosemide. At any rate, it is clear that the effects elicited *in vitro* by the hormones on the thick ascending limb are fully reflected *in vivo* in the loop of Henle. In the presence of the other hormones the effects still persist *in vivo*, but their detection is less easy.

In the kidney, the thick ascending limb crosses two regions, the medulla and the cortex, which are anatomically distinct and functionally separated (Rouffignac, 1990). The fate of electrolytes delivered in these two regions will therefore depend on the local organisation of the tubular and vascular structures surrounding the thick ascending limbs. In the rat, at the tip of the long loops (i.e. at the end of their descending thin limbs), the Mg (and Ca) deliveries remained unchanged by either glucagon or PTH. This is to be expected, if in this species the hormone-mediated increase in Ca and Mg reabsorptions occurs exclusively in the cortex, as the *in vivo* data obtained in the mouse suggest. Calcitonin significantly reduced Mg and Ca deliveries at the hairpin turn and ADH showed a similar tendency. It was found that both hormones, in contrast to PTH and glucagon, reduced the fluid delivered at this site by water removal from the descending thin limb. We suppose that such water removal resulted in increased Ca and Mg concentrations in the tubular fluid, allowing these divalent cations to diffuse out of the lumen before reaching the hairpin turn, probably at the same site along the descending limb as that at which NaCl is also escaping, i.e. in the inner medulla.

(iii) DISTAL TUBULE

A detailed bibliography will be found in Rouffignac (1991), and Rouffignac *et al.* (1991a, 1993). The adenylate cyclase system of the rat distal tubules is also sensitive to many hormones including ADH, PTH, calcitonin and glucagon. To determine the effects of these four hormones on the distal tubule, late and early accessible distal convolutions of the same superficial nephrons were punctured either in hormone-deprived rats or in such rats receiving one of the missing hormones. PTH, calcitonin and glucagon, but not ADH, elicited a significant stimulation of Mg transport in the distal tubule. The cellular mechanisms of Mg transport in the distal tubule are totally unknown, for the simple reason that the portion of the distal tubule accessible to micropuncture comprises at least three cell populations showing great differences in morphology, ultrastructural organisation and cyclase responsiveness. Which one of these populations is responsible for Mg transport is unknown.

(iv) HORMONAL CONTROL ON THE RENAL HANDLING OF Mg

A detailed bibliography will be found in Rouffignac (1991) and Rouffignac *et al.* (1991a, 1993). The first information on the hormonal control of Mg transport in the kidney was obtained *in vivo* from clearance and micropuncture studies. The first hormone tested was PTH. This hormone was shown to reduce Mg excretion by enhancing the reabsorption of this ion within the kidney, in addition to the effect of calcitonin. Calcitonin was observed to reduce Mg excretion in Mg-loaded rats: in such rats, salmon calcitonin (SCT, 60 mU · min^{-1} · rat^{-1}) significantly reduced the Mg fractional excretion from 24.7 % during control periods to 15.3 % during SCT administration ($p <$ 0.001) in spite of a slight increase in filtered Mg during SCT infusions (S. Corraza & C. de Rouffignac, unpublished observations). From the observations reported for the loop of Henle and the distal tubule, it is clear that these effects resulted from a direct action of PTH and calcitonin at these sites.

However, it is apparent from the data obtained in hormone-deprived rats that not only PTH and calcitonin, but also ADH and glucagon, may decrease the urinary excretion of divalent cations. This reduction was dose-dependent: for ADH it was observed in a range evaluated to be between 1 and 5 pg · ml^{-1} and for HCT, as measured by radio-immunoassay, in the range 0.5 to 5.0 ng · ml^{-1}. Both titres are in the physiological ranges of the two hormones. During maximal stimulation of the endogenous secretion process, the concentration of circulating hormones ranges between 20 to 50 pg · ml^{-1} for ADH and 5 to 10 ng · ml^{-1} for glucagon. In addition, in these rats the effects of doses resulting in a maximal reduction of Mg excretion were additive when ADH and glucagon, or ADH and HCT (but not HCT and glucagon), were used in combination. Since the decrease in Mg excretion in final urine was demonstrated to be essentially accounted for by hormone-mediated stimulation of Mg reabsorption in the loop of Henle, these observations indicate that one hormone can exert its effect on the thick ascending limb, even in the presence of the other hormones. This was verified indirectly by micropuncture, in either hormone-deprived intact Brattleboro rats or in Wistar rats. Thus, it was possible to conclude that the ion reabsorption in the loop of Henle was still increased by one peptide hormone when the others acting on the same target cells of the thick ascending limb were present. The effects, however, were less marked than in hormone-deprived rats during testing of individual hormones. At the whole kidney level, the stimulation effects of either glucagon or dDAVP on Ca and/or Mg reabsorption were also observed in intact rats, i.e. in the presence of the other peptide hormones. Thus, all these observations indicate that, in addition to PTH and calcitonin, glucagon and ADH could potentially participate in Mg and Ca homeostasis.

Before accepting this hypothesis, it will be necessary to discover the physiological or physiopathological circumstances in which these hormones are called into play in homeostatic control.

As far as steroids are concerned, little is known about their effects on Mg excretion. It appears, however, that aldosterone may affect magnesium metabolism. In both normal and adrenalectomised rats, it increased excretion of magnesium in the urine. A similar observation was reported in humans. The site of inhibition of Mg transport by aldosterone is unknown. More generally, chronic administration of gluco- and mineralo-corticoids leads to increased urinary excretion of Mg (and Ca) in both animals and humans. Whether these inhibitions result from the direct effects of steroids on the tubular epithelia or, more likely, reflect steroid-associated alterations of extrarenal factors, such as extracellular fluid volume expansion, remains to be determined.

Acknowledgements

We would like to thank all our friends and colleagues who contributed to the work presented here, and particularly F. Morel, D. Chabardès, and M. Imbert-Teboul (Collège de France, Paris), C. Amiel and C. Bailly (Inserm U251, Paris), L. Bankir (Inserm U90, Paris), R. Greger, R. Nitschke and P. Wangemann (Freiburg University, Germany) and B. Corman, A. Di Stefano, J.-M. Elalouf, B. Mandon and M. Wittner at Saclay. This technique was introduced and developed at Saclay by F. Morel and M. Roinel in 1967.

References

Faessler, A. von, & Goehring, M. (1952). Röntgenspektrum und Bindungzustand. Die K Fluoreszenzstrahlung des Schwefels. *Naturwissenschaten*, **39**, 169–77.
Ferrary, E., Tran Ba Huy, P., Roinel, N., Bernard, C. & Amiel, C. (1988). Calcium and the inner ear fluids. *Acta otolaryngologica* (suppl.), **460**, 13–17.
Friedman, P.A. (1988). Basal and hormone activated calcium absorption in mouse renal thick ascending limbs. *American Journal of Physiology*, **254**, F62–70.
Lasserre, B. & Durlach, J. (1991). *Magnesium, a relevant ion*. London: John Libbey.
Leroy, A. & Roinel, N. (1983). Radiation damage to lyophilized mineral solutions during electron probe analysis: quantitative study of chlorine loss as a function of beam current density and sample mass thickness. *Journal of Microscopy*, **131**, 97–106.
Morel, F. (1981). Sites of hormone action in the mammalian nephron. *American Journal of Physiology*, **240**, F159–64.
Morel, F., Chabardès, D., Imbert-Teboul, M., Le Bouffant, F., Hus-Citharel, A. & Montégut, M. (1982). Multiple hormonal control of adenylate-cyclase in distal segments of the rat kidney. *Kidney International*, **21**, S55–62.
Morel, F., Imbert-Teboul, M. & Chabardès, D. (1987). Receptors to vasopressin and other hormones in the mammalian kidney. *Kidney International*, **31**, 512–20.

(1981). Distribution of hormone-dependent adenylate cyclase in the nephron and its physiological significance. *Annual Review of Physiology*, **43**, 569–81.

Morel, F. & Roinel, N. (1969). Application de la microsonde électronique à l'analyse élémentaire quantitative d'échantillons biologiques liquides d'un volume inférieur à 10^{-9} l. *Journal de Chimie Physique et de Physico-chimie Biologique*, **66**, 1084–91.

Morel, F., Roinel, N. & Le Grimellec, C. (1969). Electron probe analysis of tubular fluid composition. *Nephron*, **6**, 350–64.

Muhlert, M., Julita, M. & Quamme, G.A. (1982). Disappearance of calcium and other electrolytes from microvolume samples. *American Journal of Physiology*, **242**, 202–6.

Quamme, G.A. (1988). X-ray analysis of biological fluids: an update. In *Scanning electron microscopy*, Vol. 2, no. 4, ed. O. Johari, pp. 2195–205. Chicago AMF O'Hare: SEM Inc.

(1989). Control of magnesium transport in the thick ascending limbs. *American Journal of Physiology*, **256**, F197–210.

Quamme, G.A. & Dirks, J. (1980). Magnesium transport in the nephron. *American Journal of Physiology*, **239**, F393–401.

(1986a). Micropuncture techniques. *Kidney International*, **30**, 152–65.

(1986b). The physiology of renal magnesium handling. *Renal Physiology*, **9**, 257–69.

Quinton, P. (1976). Construction of picoliter-nanoliter self-filling volumetric pipettes. *Journal of Applied Physiology*, **40**, 260–2.

(1978). Ultramicro-analysis of biological fluids with energy-dispersive X-ray spectrometry. *Micron*, **9**, 57–69.

Roinel, N. (1975). Electron microprobe quantitative analysis of lyophilized 10^{-10} l volume samples. *Journal de Microscopie*, **22**, 261–8.

(1988). Quantitative X-ray analysis of biological fluids: the microdroplet technique. *Journal of Electron Microscopy Technique*, **9**, 45–56.

Roinel, N., Meny, L. & Hénoc, J. (1980). Accuracy of electron microprobe analysis of biological fluids. Choice of standard solutions and range of linearity of the calibration curves. In *NBS special report no. 533*, ed. K.F.J. Heinrich, pp. 101–30. Washington D.C.: National Bureau of Standards.

Roinel, N. & Rouffignac, C. de (1982). X-ray analysis of biological fluids: contribution of microdroplet technique to biology. In *Scanning electron microscopy*, Vol. 3, ed. O. Johari, pp. 1155–71. Chicago AMF O'Hare: SEM Inc.

Rouffignac, C. de (1990). The urinary concentrating mechanism. In *Urinary concentrating mechanisms*, ed. R. Kinne, E. Kinne-Saffran & K. Beyenbach, pp. 31–102, Basel: Karger.

(1991). Regulation of magnesium excretion. In *Disorders of bone and mineral disease*, ed. F. Coe & M. Favus. New York: Raven Press.

Rouffignac, C. de, Di Stefano, A., Wittner, M., Chabane-Sari, D. & Elalouf, J.-M. (1991a). The renal handling of magnesium. Influences of parathyroid hormone, calcitonin, antidiuretic hormone and glucagon. In *Magnesium, a relevant ion*, ed. B. Lasserre & J. Durlach, pp. 145–67. London: John Libbey.

Rouffignac, C. de, Di Stefano, A., Wittner, M., Roinel, N. & Elalouf, J.-M. (1991b). Consequences of the differential effects of ADH and other peptide hormones on the thick ascending limb of the mammalian kidney. *American Journal of Physiology*, **260**, R1023–35.

Rouffignac, C. de, Elalouf, J.-M. & Roinel, N. (1987). Physiological control of the urinary concentrating mechanisms by peptide hormones. *Kidney International*, **31**, 611–20.

Rouffignac, C. de, Roinel, N. & Elalouf, J.-M. (1993). Comparative effects of peptide hormones on water and electrolyte transport along the proximal and distal tubules of the mammalian nephron. In *New insights into vertebrate kidney function*, ed. J.A. Brown, R.J. Balment & J.C. Rankin. Cambridge: Cambridge University Press (in press).

Van Eekelen, C., Boekestein, A., Stols, A. & Stadhouders, A. (1980). X-ray microanalysis of picoliter microdroplets: improvement of the method for quantitative X-ray analysis of samples of biological fluids. *Micron*, **11**, 137–45.

SECTION D
APPLICATIONS OF X-RAY MICROANALYSIS
IN BIOLOGY

X-ray microanalysis has become increasingly used to study the elemental composition of biological specimens and has provided valuable information on the identity and occurrence of diffusible and bound ions, inorganic deposits and elements that are structural components of macromolecules. The attractions of X-ray microanalysis to the potential investigator in terms of spatial resolution and range of elements that can be detected have already been mentioned in Section A. The analytical sensitivity in terms of minimal detectable mass (10^{-19} g) also presents considerable opportunity to the analyst, though the value for sensitivity on a mass fraction basis (normally no better than 0.1 % with energy dispersive systems) poses major limitations.

The diversity of applications can be considered both in terms of general areas of study and type of specimen examined.

A Applications of X-ray microanalysis to different subject areas of biology

X-ray microanalysis has been used in a variety of disciplines within biology, as exemplified by the range of chapters in this section of the book. These areas include microbiology, plant biology, animal biology (vertebrate and invertebrate tissues, cultured animal cells), medicine and environmental biology. A computer survey of the relative numbers of papers in these different areas (Table D1) indicates a clear preponderance in the spheres of animal biology and medicine.

The role of X-ray microanalysis in human physiology and medical research has recently been reviewed by Shelburne et al. (1989), and is important in the study of aspects such as the activity of Ca in cellular processes, ion transport in epithelia and the possible importance of cations in mitogenesis and oncogenesis. X-ray microanalysis has particular potential in the study of dieases that may be related to changes in elemental composition (cystic fibrosis, Alzheimer's Disease) and is an important tool in human pathology for identifying the chemical nature of particles in diseases such as silicosis and asbestosis (Churg, 1989). The study of biomaterials presents a new and exciting application for X-ray microanalysis in medicine, where the response of the human body to implants (biodegradation, bone-bonding) can be studied at the cellular level.

Table D1. *Relative numbers of X-ray microanalytical publications in different areas of biology*

Microbiology	17	Bacteria	10
		Fungi	4
		Protozoa	2
Plant Biology	10	Higher plants	7
		Algae	3
Animal Biology	41	Vertebrates	21
		Invertebrates	14
		Cultured cells	6
Medicine	18		
Environmental biology (Toxicology)	11		
Technique-orientated (range of material)	3		

Numbers of papers in different areas are expressed as a percentage of the total sample of 71 papers abstracted in the computer output from Cambridge Life Sciences, 1988–90. Each paper was assigned to a particular area, based on its major theme.

X-ray microanalysis is also becoming increasingly used in environmental studies, where the presence of elements that relate to nutrient status (e.g. phosphorus) and pollution (e.g. heavy metals) has been monitored in a wide range of biota, including invertebrates, bacteria (Booth, Sigee & Bellinger, 1987) and phytoplankton (Clay, Sigee & Bellinger, 1991).

B Applications of X-ray microanalysis to different types of biological specimen

The importance of specimen preparation in the interpretation of X-ray microanalytical data has been considered in Section B, and related problems such as loss or displacement of soluble ions, radiation damage and beam over-penetration are becoming increasingly realised. What is perhaps less well appreciated is that X-ray microanalysis can be used with a considerable diversity of specimens. Although many users of this technique might automatically consider hydrated cryosections to be the specimen of choice, X-ray microanalysis can also be applied to other types of preparation, some of which are described in Section B. These include whole cell preparations, sections of chemically treated cells, tissue homogenates, fluid microdroplets and isolated macromolecules. The related use of X-ray analysis to study inorganic deposits produced during histochemical procedures represents a further area of application for this technique (Section B).

References

Booth, K.N., Sigee, D.C. & Bellinger, E. (1987). Studies on the occurrence and elemental composition of bacteria in freshwater phytoplankton. *Scanning Microscopy*, **1**, 2033–42.

Churg, A. (1989). Quantitative methods for analysis of disease induced by asbestos and other mineral particles using the transmission electron microscope. In *Microprobe analysis in Medicine*, ed. P. Ingram, J.D. Shelburne & V.L. Roggli, pp. 79–95. New York: Hemisphere.

Clay, S., Sigee, D.C. & Bellinger, E. (1991). X-ray microanalytical studies of freshwater biota: changes in the elemental composition of *Anabaena spiroides* during blooms of 1988 and 1989. *Scanning Microscopy*, **5**, 207–17.

Shelburne, J.D., Tucker, J.A., Roggli, V. L. & Ingram, P. (1989). Overview of applications in medicine. In *Microprobe analysis in Medicine*, ed. P. Ingram, J.D. Shelburne & V.L. Roggli, pp. 55–77. New York: Hemisphere.

12 Electron probe X-ray microanalysis of bacterial cells: general applications and specific studies on plant pathogenic bacteria

D.C. Sigee and N. Hodson

A Introduction

1 General applications of the technique to bacterial cells

Electron probe X-ray microanalytical studies have been carried out on a wide range of bacterial cells, using a variety of preparation techniques. These studies have involved many different applications and have relevance to areas of environmental microbiology, studies on cultured cells, human pathology, plant pathology, metal toxicity and biochemical research. Some of these studies are summarised in Table 12.1, which emphasises the broad contribution that this technique has made to bacteriology.

The main purpose of this chapter is to explore the potential uses of electron probe X-ray microanalysis in bacterial research, and to consider the different experimental approaches that can be used with these cells. The chapter will concentrate particularly on bacterial pathogens of higher plants, since this has been an area of special interest in this laboratory.

2 Elemental composition of plant pathogenic bacteria

Plant pathogenic bacteria are facultative parasites, able to grow and multiply both outside and inside the plant – where they may cause disease. The elemental composition of these organisms is of considerable interest in relation to a number of key aspects of their existence (Sigee, Hodson & El-Masry, 1989). Changes in the concentration of specific cations, in particular, have been implicated in various functional processes – including hydrolysis of bacterial toxins (Levi & Durbin, 1986), interactions between microorganisms at the plant surface (De Weger *et al.*, 1988) bacterial changes during disease (El-Masry & Sigee, 1989), chemical control measures (Sekizawa & Waka-bayashi, 1990) and *in vitro* effects of heavy metals (Sigee & Al-Rabaee, 1986; Hodson & Sigee, 1991).

Table 12.1. *Variety of X-ray microanalysis studies on bacterial cells*

Bacterial type/ species	Preparation and investigation	Reference
(1) ENVIRONMENTAL STUDIES		
Marine fouling bacteria	Determination of biomass and elemental composition by SEM XRMA of whole cells	Heldal, Norland & Tumyr (1985)
Leptospira spp.	Whole cell XRMA of halophilic strains	Hovind-Hougen *et al.* (1981)
Freshwater bacteria	Elemental composition of bacteria in plankton	Booth, Sigee & Bellinger (1987)
(2) CULTURED CELLS		
Bacillus spp.	Detection of Ca in sections of spores	Scherrer & Gerhardt (1972)
Escherichia coli	Analysis and mapping of freeze-dried cryosections	Chang, Shuman & Somlyo (1986)
Escherichia coli	STEM XRMA of freeze-dried cryosections	Zierold (1988)
Pseudomonas syringae	Use of a range of preparation techniques	Sigee, El-Masry & Al-Rabaee (1985)
(3) HUMAN PATHOLOGY		
Bacteria in urinary tract	Role of bacteria in gall stone formation	Kaufman *et al.* (1989)
(4) PLANT PATHOLOGY		
Pseudomonas syringae	Bacterial changes during leaf infection	El-Masry & Sigee (1989)
(5) METAL TOXICITY		
Methanosarcina	Fixed, resin sections of heavy metal-treated cells	Scherrer & Bochem (1983)
Pseudomonas syringae	Effects of Ni toxicity – XRMA of whole cells	Sigee & Al-Rabaee (1986)
Erwinia amylovora	Effects of Cu toxicity – XRMA of whole cells	Hodson & Sigee (1991)
(6) BIOCHEMICAL STUDIES		
Pseudomonas syringae	Analysis of bound ions in extracted DNA	Sigee & El-Masry (1987)

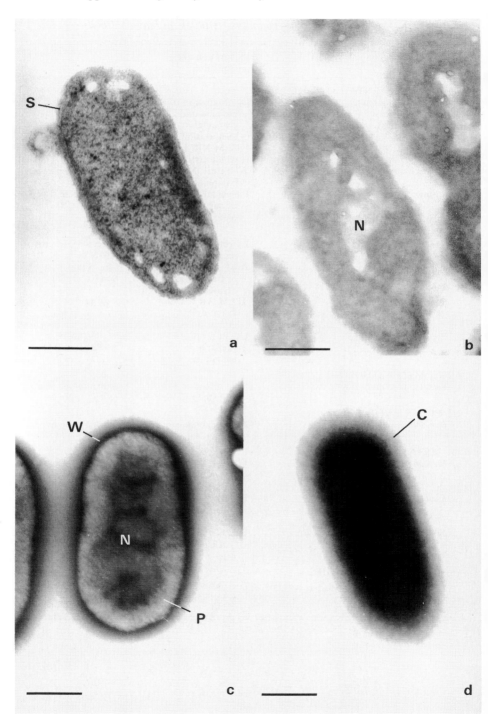

Fig. 12.1.   For legend see facing page.

B   Experimental use of X-ray microanalysis with plant pathogenic bacteria

The application of X-ray microanalysis to the study of the elemental composition of these bacteria will be considered from three main aspects.

1. Specimen preparation and elemental localisation in bacterial cells.
2. Changes in the elemental composition of bacteria during laboratory culture in sterile medium (*in vitro* culture), and the use of *in vitro* model systems to study chemical toxicity and strain differences.
3. Changes in the elemental composition of bacteria during their growth in infected plant tissue.

1   Specimen preparation and elemental localisation in bacterial cells

X-ray microanalysis can be carried out on three major types of preparation (Sigee *et al.*, 1985; Sigee, 1988) – cryosections, whole cell preparations and extracted cell constituents.

(i) CRYOSECTIONS

Although cryosections of bacterial cells can be readily obtained from pellets of cultured samples (Chang *et al.*, 1986; Sigee *et al.*, 1985; Zierold, 1988) the fine-structural quality of the preparations is limited (compare Figs 12.1a and 12.1b) by the fact that fixatives cannot be used to stabilise cell structure where X-ray analysis of diffusible elements is being carried out. Even the use of

---

Fig. 12.1.   Air-dried and cryopreparations of plant pathogenic bacteria for transmission electron microscopy. (a) Freeze-dried cryosection of *Pseudomonas syringae*, fixed in formaldehyde/glutaraldehyde and cryoprotected in 2.3 M sucrose. The cell shows good ultrastructural preservation, with clear surface membranes (S) and dense ribosomal protoplasm containing a dispersed nucleoid. Bar scale = 0.5 μm. (b) Freeze-dried cryosection of chemically unfixed cell of *Pseudomonas syringae*. In this stationary phase cell, the nucleoid (N) appears as a discrete central electron-transparent structure. Bar scale = 0.5 μm. (c) Frozen cell of *Erwinia amylovora*, maintained at −185 °C on a cold stage. The cell is surrounded by a halo of frozen liquid (W) and contains a central electron-dense nucleoid (N) surrounded by transparent ribosomal protoplasm (P). Bar scale = 0.5 μm. (d) Air-dried cell of *Pseudomonas syringae*, showing an outer capsule (C). The main body of the bacterium is uniformly electron-dense, with no internal differentiation. Bar scale = 0.25 μm.

Table 12.2. *Comparison of hydrated and dehydrated bacterial cells*

Element	Hydrated[a]	Air-dried[b]
P	458 ± 84	665 ± 110
S	38 ± 13	56 ± 19
Cl	104 ± 34	144 ± 39
K	373 ± 82	379 ± 41
Ca	27 ± 10	25 ± 10

Elemental concentrations are expressed as μmoles/g total mass within the probe area, and are the mean value (with a given element) for 20 cells selected at random. None of the mean levels for hydrated and air-dried cells, respectively, are statistically different at the 95 % probability level.

[a] Cells plunge-frozen in liquid nitrogen slush and analysed on a cold stage at −185 °C.

[b] Cells air-dried and analysed at room temperature.

cryoprotectants should be avoided to minimise elemental disturbance. The use of cryosections is also limited by the small mass of material in the probe area – resulting in the generation of a low X-ray signal, with correspondingly small elemental peaks (Sigee *et al.*, 1985; Hodson & Sigee, 1991). Small peaks reduce the statistical confidence that can be attached to quantitative data.

X-ray emission spectra obtained from cryosections can, however, provide useful qualitative information on elemental localisation within major areas such as nucleoid, peripheral cytoplasm, cell wall complex and the poly-saccharide capsule. In cryosections of *Erwinia amylovora*, for example, Hodson & Sigee (1991) showed clear peaks of P and Cl in the protoplasm and cell wall complex plus detectable levels of K and Ca in the protoplasm and cell wall respectively. The lack of any detectable elements in the external polysaccharide suggests that this region has much lower ion levels compared with the main body of the cell, and that the elements detected in whole cell preparations (see later) are largely internal (rather than cell surface) constituents.

(ii) WHOLE CELL PREPARATIONS

The small size of bacterial cells (normally 1–2 μm diameter) means that even at low accelerating voltages there is little electron scattering, X-ray absorption or fluorescence under the electron probe. Individual cells thus present suitable specimens for transmission X-ray microanalysis, where they qualify for thin film quantitation (Hall, 1971), but are not suitable for scanning electron microscope (bulk) analysis where over-penetration of the probe will generate X-ray emission from the underlying support matrix.

Whole cell preparations of cultured cells can be obtained by collecting the sample by centrifugation, briefly washing the cells in deionised water then allowing the cells to sediment onto an electron microscope grid for 2–5 min. The grid is then drained of excess liquid and either plunge-frozen to examine the cells in the frozen (fully-hydrated) state on a cold stage, or air-dried for analysis as a dehydrated preparation.

Although hydrated and dehydrated whole cells have different appearances by transmission microscopy (Fig. 12.1c,d), their elemental composition is broadly similar, suggesting that no substantial change in the elemental composition was occurring during the air-drying process. Comparison of elemental mass fractions (Table 12.2) does indicate higher mass fractions of the major elements P, S, Cl and K in air-dried cells (consistent with the loss of water matrix), but this difference was not significant at the 95 % probability level.

A typical X-ray emission spectrum taken from an air-dried bacterial cell is shown in Fig. 12.2. Cultured cells of a range of plant pathogenic bacteria – including *Pseudomonas syringae* pv. *tabaci*, *P.s.* pv. *phaseolicola*, *Erwinia amylovora* and *Erwinia herbicola* – showed closely similar X-ray emission spectra, with major peaks of P, S, Cl and K being routinely present, plus occasional peaks of Ca and transition metals. The presence of an Na peak varied with bacterial strain and culture medium.

Although whole cell preparations give useful information on overall elemental levels in bacterial cells, they give no information on elemental occurrence in relation to fine structural detail. Information on elemental localisation within the bacterial cell can be obtained from three sources – cryosections (see previously), extracted cell constituents and elemental correlations.

(iii) EXTRACTED CELL CONSTITUENTS

The elemental composition of major macromolecules within the cell can be determined by extraction, deposition on an electron microscope grid and direct analysis. This approach has so far been carried out on bacterial DNA and outer membrane proteins.

(a) *Analysis of bacterial DNA*   The extraction and X-ray microanalysis of bacterial DNA has been reported by Sigee & El-Masry (1987). DNA isolation followed the procedure of Sato & Miura (1963) and involved bulk cell lysis, phenol separation of chromatin, ethanol precipitation of crude DNA, RNA and protein removal, with final precipitation in ethanol. The DNA fraction was dissolved in sterile distilled water, deposited and air-dried on a hydrophilic EM grid, then carbon-coated prior to analysis.

The DNA molecules appeared as discrete, highly branched structures when viewed by transmission microscopy – possibly corresponding to individual bacterial nucleoids. They appeared to be highly stable under the electron

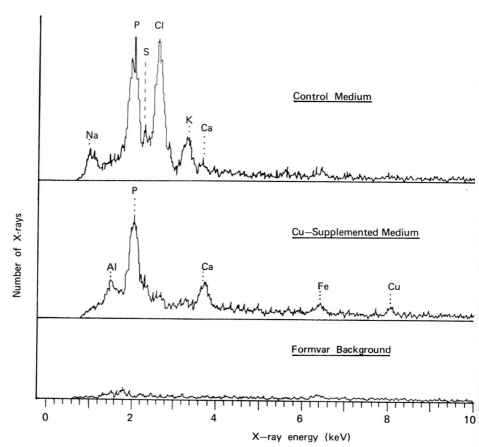

Fig. 12.2.   X-ray emission spectra from air-dried cells of *Erwinia amylovora*, cultured in standard nutrient broth (control medium) and nutrient broth supplemented with 10^{-3} M Cu. No extraneous elements are detectable in the top spectrum (to which the formvar background relates) but the bottom spectrum has an extraneous Al peak. The cells have been cultured for 9 h and show marked differences in elemental composition between the two types of culture (see text).

beam at beam currents of up to 60 nA, and were analysed over a livetime of 200 s. X-ray emission spectra showed high levels of P (major structural component), plus the major monovalent and divalent cations K and Ca. Peaks of transition metals Mn, Fe, Ni, Cu and Zn were also routinely present, plus occasional peaks of Mg, Cr and Co. No S peak was present, in accordance with the high degree of purity (no contaminating proteins) of the preparations. In these samples, approximately 30 monovalent cations and 20 divalent cations were present for every 100 atoms of P (or per 100 nucleotides). Although the use of such drastic extraction procedures would seriously affect the presence of diffusible cations, bound elements would be expected to be retained with minimal alteration. The range of divalent cations detected in these DNA samples is in agreement with earlier biochemical studies involving spectrophotometric analysis of bulk DNA samples from other sources (Skrinska *et al.*, 1978).

(b) *Analysis of extracted proteins* X-ray analysis has considerable potential for determining the elemental composition of protein micro-samples, particularly where the macromolecule has covalently bound metal atoms. Using already purified biochemicals, Shuman, Somlyo & Somlyo (1976) and El-Masry & Sigee (1986) have analysed a variety of metalloprotein films, and Shuman & Somlyo (1976) have carried out high resolution analysis on single ferritin molecules on an electron microscope grid.

The full potential of this approach would be realised where the analysed proteins are selected from a range of molecules that have been extracted from the test organism and separated by a technique such as polyacrylamide gel electrophoresis (PAGE). To test the feasibility of this approach, various metalloproteins were run on duplicate polyacrylamide gels and the band identified on one track by staining. A corresponding region from the unstained track was then removed, the band cut out, and the protein eluted into a small volume of distilled water. The protein was then precipitated by ammonium sulphate, deposited onto an EM grid, and ethanol dehydrated prior to X-ray analysis. Fig. 12.3 shows the appearance of a precipitate of Conalbumin that has been recovered from a polyacrylamide gel, with the X-ray emission spectrum showing clear retention of Fe by the protein.

A similar approach has subsequently been used for analysis of surface membrane proteins of *Erwinia amylovora* that have been extracted using the techniques of Ames (1974). Examination of a range of deposited proteins from both normal and Cu-treated cells have so far failed to reveal any metal associations, with spectra showing varying levels of sulphur, but no cation peaks.

(iv) ELEMENTAL CORRELATIONS

The variability between cells within bacterial populations is such that it is normally necessary to collect at least 15 spectra from separate cells to obtain significant mean concentration values. Although this may seem to be a drawback in comparison to bulk analytical methods (where a single determination will give a reliable mean value), it does provide useful information on the degree of variability within the population and also allows for the estimation of elemental correlations.

Information about the related occurrence of particular elements within the cell can be determined from their correlations within individual spectra.

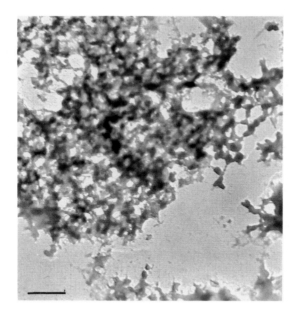

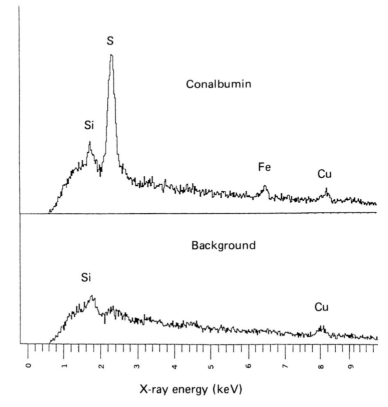

Fig. 12.3. For legend see facing page.

Table 12.3. *Elemental correlations with phosphorus in various samples of plant pathogenic bacteria*

Sample	Significant[a] positive correlations	Significant negative correlations	No significant correlations
Isolated DNA[b]	K, Ca, Zn	None	Mn, Fe, Ni, Cu
P.s. pv. *tabaci*[c]	K (Ca)	None	S, Fe
E. amylovora[d]			
1. Control	K (Na, Cl, S)	(Ca)	
2. Cu-treated	Na, K, Ca (Cl, S)	(Cl)	

[a] Significant – correlation coefficient greater than 0.5.
       X – routinely significant over 5–6 separate samples.
      (X) – only occasionally significant over 5–6 samples.
[b] Sigee & El-Masry (1987). [c] El-Masry & Sigee (1989). [d] Hodson & Sigee (1991).

The direct association between two particular elements will result in a positive correlation, while a negative correlation might arise where two chemical species are competing for the same site or compartment within the cell. Elemental correlations become particularly useful where one of the elements acts as a reference point for a particular macromolecule. As an example of this, P is present largely in nucleic acids within the cell, and is distributed uniformly along the polynucleotide molecule. The correlation of elements with P in various types of specimen is shown in Table 12.3.

The great majority of significant correlations with P (and between other elements) are positive – indicating that the data are providing information on elemental association rather than competition.

In the case of isolated DNA, all of the major cations are correlated with P – in line with both monovalent and divalent cations being associated with the electronegative phosphate groups. Monovalent (K^+) cations do not show any correlations with divalent cations. Each divalent cation, however, shows correlation with at least one other divalent cation, suggesting that monovalent and divalent cations are occupying different binding sites on the macromolecule (Sigee & El-Masry, 1987).

Fig. 12.3.   Test extraction of metalloprotein from polyacrylamide gel for X-ray microanalysis. Micrograph: ethanol-dehydrated precipitate of Conalbumin on electron microscope grid. The metalloprotein has been extracted from an excised band in a non-denaturing polyacrylamide gel. Bar scale = 2 μm. Spectra: X-ray emission spectra from Conalbumin and a region of clear formvar (background), showing significant peaks of S and Fe derived from the metalloprotein, plus extraneous Si and Cu peaks.

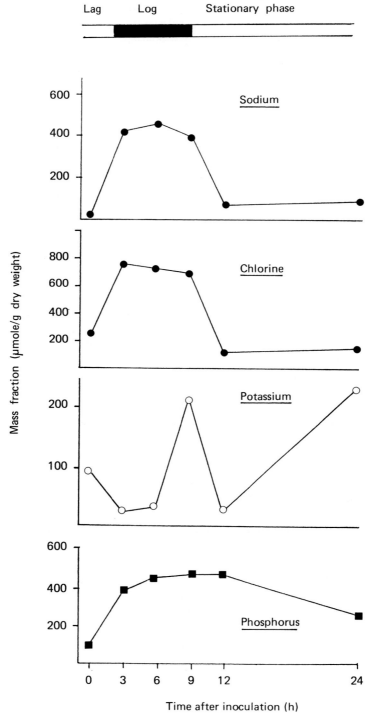

Fig. 12.4.   For legend see facing page.

Whole cell preparations generally showed a high degree of correlation between P and K, probably reflecting electrostatic associations between the major anions (PO_4^-) and cations (K^+) within the cell. Results obtained from normal (control) cultures of *Erwinia amylovora*, for example, revealed routine correlation between these elements over a 24 h period. Addition of toxic levels of Cu to the culture medium completely altered the balance of ionic associations, with Na and Ca (but not K) now showing increased correlation with P (Hodson & Sigee, 1991).

Correlations between elements do not simply reflect chance occurrences of the most commonly occurring cell constituents, but appear to represent elemental associations that are biologically relevant. This is indicated by the specific differences in monovalent and divalent cation asociations seen in the DNA sample, and also in the *Erwinia amylovora* whole cell preparations, where two of the most commonly occurring elements, K and Cl, showed only one positively significant correlation out of 16 samples.

2   X-ray microanalysis of bacteria cultured in vitro

Bacterial samples for X-ray microanalysis may be readily obtained from nutrient broth cultures or by resuspending bacteria from agar surfaces, so both liquid and solid *in vitro* cultures are potential sources of material.

With liquid (batch) cultures, cells can be harvested along a time course within the culture period, and the elemental content of cells related to growth phase. This has now been carried out with various phytopathogenic bacteria, including *Pseudomonas syringae* pv. *tabaci* (El-Masry, 1989) and *Erwinia amylovora* (Hodson & Sigee, 1991). These studies show clear and statistically significant changes in the elemental composition of bacterial cells during the culture period, as illustrated in Fig. 12.4 for *Erwinia amylovora*. In this particular example, the cells entered log phase very rapidly after inoculation into fresh nutrient broth. With a rapid rise in the cellular levels of Na and Cl and a fall in the level of K. The second half of log phase is marked by a reversal in the levels of Na and K, with a further fall in the mass fraction of K as the cells enter stationary phase.

Although the precise elemental changes that occur during batch culture are varied (probably depending on the type and physiological state of the bacterium and the physico-chemical state of the culture medium), some general features are emerging. The fall and rise of K during log phase, for example, has now been observed with a number of liquid cultures and may be a fundamental aspect of bacterial growth under these conditions.

Fig. 12.4.   Changes in elemental mass fractions during *in vitro* (nutrient broth) culture of *Erwinia amylovora*. Each mass fraction value is derived as the mean from 20 spectra. (Redrawn from Hodson & Sigee, 1991.)

In addition to providing information on elemental composition, X-ray microanalysis can also be used to monitor changes in bacterial dry mass (dehydrated preparations), since the background-corrected continuum level is directly proportional to the total mass of material in the probe area under standard analytical conditions. The mean dry mass of both *Pseudomonas syringae* (El-Masry & Sigee, 1989) and *Erwinia amylovora* (Hodson & Sigee, 1991) cells show significant decline during log phase, in line with the active cell division and smaller cell size at this stage of the growth cycle.

(i) *IN VITRO* CULTURES AS EXPERIMENTAL MODEL SYSTEMS

Although *in vitro* systems are experimentally convenient, their resemblance to the natural environment – inside or outside the plant – may seem remote. In spite of this, they do have the potential to provide useful models for certain environmental situations, including the effect of chemical control agents (e.g. Cu-containing compounds) and correlations between elemental composition and ice-nucleation activity.

(a) Copper toxicity in **Erwinia amylovora**   *Erwinia amylovora* is an important pathogen of apples and pears, where it infects the plant via the flowers and where it can be controlled by spraying with Cu-containing compounds. The major site of bacterial growth in the flower is the moist stigmatic surface, and the Cu^{2+} ions are thus acting in a wet, high nutrient environment.

As a model system, the effect of Cu^{2+} ions was investigated by growing bacteria in nutrient broth at various levels of added Cu (Hodson & Sigee, 1991). The results showed that inhibition of population growth occurred at Cu levels of 10^{-3} M and above, with marked changes in the element composition of the cells (Fig. 12.2). X-ray analysis of whole-cell preparations showed that non-toxic (10^{-4} M) levels of Cu had no significant effects on the elemental composition of cells (i.e. similar to control cultures with no added Cu). At 10^{-3} M Cu the following effects were observed.

1. There was no effect on changes in the mean level of Na during early log phase, but the subsequent rise of K (Fig. 12.4) was completely abolished, presumably due to inhibition of either K^+ uptake or retention.
2. Internal levels of Cu^{2+} and Ca^{2+} increased, becoming detectable at 9 h after inoculation and showing a rapid rise to the end of the experiment.

The effect of Cu^{2+} ions on the entry or loss of K^+ is consistent with a direct toxic effect at the cell surface, and is in line with other experiments which have shown leakage of K^+ from bacterial cells to be induced by a range of chemical agents, including the anaesthetics, chlorhexidine, phenol, and phenethyl alcohol (for references, see Hodson & Sigee, 1991).

Entry of Cu into bacterial cells does not become detectable until long after the inhibition of population growth has become apparent, suggesting that

Table 12.4. *Mean elemental mass fractions in ice-nucleating (INA⁺) and non-nucleating (INA⁻) phylloplane[a] bacteria*

Element	P	Cl	K	Ca
INA⁺	1080 ($\pm$217)	20 ($\pm$18)	373 ($\pm$117)	46 ($\pm$34)
INA⁻	952 ($\pm$144)	37 ($\pm$37)	436 ($\pm$130)	55 ($\pm$41)

[a] Isolated from leaf surfaces.
Concentrations are expressed as μmoles/g, with 95 % confidence limits, and are derived from four separate strains: 20 spectra were analysed for each strain.

high internal levels are not a prime factor in toxicity. Cu influx may arise either due to passive entry through a damaged (leaky) surface membrane, or by increased rate of entry through a specific transport system. Mokhele, Tang & Clark (1987) have demonstrated the presence of a specific outer membrane Cu-binding protein (NosA) which is implicated in Cu transport in *Pseudomonas stutzeri*, and a similar situation may occur in *Erwinia amylovora*.

(b) Elemental composition of bacterial strains in relation to ice-nucleation activity Aerial surfaces of plants have large numbers of permanently associated epiphytic bacteria (both pathogens and saprophytes), some of which have the ability to act as active centres of ice nucleation, causing premature frost injury as the temperature falls below 0 °C (Lindow, Arny & Upper, 1982). Although the major factor that determines ice-nucleation activity is the production of specific nucleation proteins, other factors may also be important, such as the ionic composition of the cells. Ice-nucleation active cells might have lower internal ion concentrations (particularly K⁺, for example) resulting in a lower osmotic pressure – which would also promote freezing activity.

To investigate this possibility, ice-nucleation active (INA⁺) and non-active (INA⁻) strains of *Pseudomonas syringae* were grown *in vitro* under conditions that promoted maximal expression of the nucleation activity, and sampled for X-ray analysis. Preliminary studies, comparing four INA⁺ and four INA⁻ strains, indicate that no significant differences in mean elemental composition occur between the two phenotypes (Table 12.4), suggesting that ionic composition does not affect or determine ice-nucleation activity.

3 Elemental composition of bacteria in infected plant tissue

Pathogenic bacteria have a marked effect on the elemental composition of plant cells in infected tissue (Lyon & Wood, 1976). This has been shown by conductivity measurements of the experimental medium around infected leaf disks, demonstrating that these bacteria cause electrolyte leakage and major loss of cations. The possibility of a reciprocal effect on the bacterial cells, with plant metabolites or specific anti-microbial substances (phytoalexins) causing elemental changes in the pathogen, has received little attention.

X-ray microanalysis is uniquely suited to analyse the elemental composition of bacteria in plant tissue, and may involve either the direct analysis of bacterial cells in cryosection or the analysis of bacterial cells that have been extracted and deposited on electron microscope grids.

Studies by El-Masry & Sigee (1989) on elemental changes in *Pseudomonas syringae* pv. *tabaci* during wildfire disease adopted the second approach because of a number of problems associated with the use of cryosections. The experimental procedure involved inoculation of pathogenic bacteria into leaves of tobacco plants, followed by regular sampling of the infiltrated regions over a 96 h time course. Individual samples of infected leaf tissue were rapidly macerated in distilled water, leaf debris removed by filtration, and the bacteria deposited on an electron microscope grid by sedimentation from suspension. The results obtained were of interest both in comparison with the *in vitro* situation and in relation to the host/pathogen interaction.

(i) COMPARISON WITH BACTERIA CULTURED *IN VITRO*

The results obtained showed that the elemental composition of bacteria *in planta* was closely similar to that for cells cultured *in vitro*. The only exception to this was Ca, which typically occurred at levels of 150–200 μmol/g *in planta*, compared with practically undetectable levels in cells grown *in vitro*. The higher level of bacterial Ca that occurred in infected tissue may relate to the high levels of Ca (as calcium pectate) that are present within intercellular spaces.

Some bacterial changes during the time course of the experiment could be related to the exponential growth phase of the bacterial cells within the tissue, which occurred 6–48 h after inoculation. These included a significant increase in the level of P, and a fall in the mean continuum level – similar to the changes observed *in vitro*.

(ii) HOST/PATHOGEN INTERACTIONS

Although some elemental changes occurred during disease development (largely in relation to bacterial phases of growth and senescence), the overall composition of bacterial cells was relatively stable during most of the sampling period. This indicates that there is no period of prolonged cation leakage equivalent to that seen in the plant cell. Only at the end of the sampling period (96 h sample), when bacteria were showing clear ultra-structural signs of degeneration, did the X-ray emission data reveal a sharp fall in the level of K, plus other elemental changes.

The concentration data also showed that there were no significant changes in major bacterial cations during the time of electrolyte leakage from plant cells, so that any uptake of these cations by bacteria did not lead to detectable changes in the elemental composition of the cells.

C General conclusions and future prospects

Transmission X-ray microanalysis has provided useful general information on the elemental composition of plant pathogenic bacteria growing *in vitro* and *in planta*. The use of this technique, however, has so far been limited to a relatively small number of bacterial species and pathovars, and future work will broaden the data base to include other pathogens and a greater range of plant diseases.

Future work will also extend those areas of study outlined in this report – including elemental changes during different types of plant/pathogen interaction, improved techniques for analysing extracted cell constituents and new approaches to *in vitro* experimentation and the use of model systems.

One aspect of bacterial plant pathology that has received particular attention in recent years is the activity and role of bacteria on the plant surface. X-ray microanalysis has considerable potential for examining the chemical composition of pathogenic bacterial cells at this site, where elemental levels may relate to such factors as taxonomic identity, degree of hydration, cell viability, naturally occurring microbial competition and the specific effects of chemical and biological control measures. The difficulty with direct analysis of bacteria on plant surfaces is that transmission microscopy is inappropriate and scanning microanalysis is not feasible due to beam over-penetration with such a small object, so new experimental approaches will need to be developed to overcome these problems. One technique that may have future use in this respect is secondary ion mass spectrometry (SIMS), which in the microprobe mode can provide chemical information that is localised to a few μm in depth (the size of a bacterial cell) at the sample surface (Thellier, Ripoll & Berry, 1991).

References

Ames, G.F. (1974). Resolution of bacterial proteins by polyacrylamide gel electro-phoresis on slabs. *J. Biol. Chem.*, **249**, 634–44.

Booth, K., Sigee, D.C. & Bellinger, E. (1987). Studies on the occurrence and elemental composition of bacteria in freshwater plankton. *Scanning Microscopy*, **1**, 2033–42.

Chang, C.F., Shuman, H. & Somlyo, A.P. (1986). Electron probe analysis, X-ray mapping and electron energy loss spectroscopy of calcium, magnesium and monovalent ions in log-phase and in dividing *Escherichia coli* B cells. *J. Bacteriol.*, **167**, 935–9.

De Weger, L.A., van Avendonk, J.C., Recourt, K., van der Hofstad, G., Weisbeck, P.J. & Lugtenberg, B. (1988). Siderophore-mediated uptake of Fe^{3+} by the plant growth-stimulating *Pseudomonas putida* strain WC3358 and by other rhizosphere microorganisms. *J. Bacteriol.*, **171**, 4693–8.

El-Masry, M.H. (1989). Electron microscope studies on intact and lysed cells of the phytopathogenic bacterium *Pseudomonas syringae* pv. *tabaci*. Ph.D. Thesis, University of Manchester.

El-Masry, M.H. & Sigee, D.C. (1986). The use of metalloproteins as standards for X-ray microanalysis of biological material. *J. Biochem. Biophys. Meth.*, **13**, 305–14.

(1989). Electron probe X-ray microanalysis of *Pseudomonas syringae* pv. tabaci isolated from inoculated tobacco leaves. *Physiol. Mol. Plant Pathol.*, **34**, 557–73.

Hall, T.A. (1971). The microprobe assay of chemical elements. pp. 157–275. In *Physical techniques in biological research*, ed. G. Oster, New York: Academic Press.

Heildal, M., Norland, S. & Tumyr, O. (1985). Determination of biomass, elemental composition and sizes of marine fouling bacteria by use of scanning transmission electron microscopy and X-ray microanalysis. *J. Ultrastruct. Res.*, **91**, 247.

Hodson, N. & Sigee, D.C. (1991). Copper toxicity in *Erwinia amylovora*: an X-ray microanalytical study. *Scanning Microscopy*, **5**, 427–38.

Hovind-Hougen, K., Cinco, M., Roomans, G.M. & Birch-Anderson, A. (1981). Electron microscopy and X-ray microanalysis of a halophilic leptospire. *Arch. Microbiol.*, **130**, 339–43.

Kaufman, H.S., Magnuson, T.H., Lillehoe, K.D., Frasca, P. & Pitt, H.A. (1989). The role of bacteria in gall bladder and common duct stone formation. *Ann. Surg.*, **209**, 584–91.

Levi, C. & Durbin, R.D. (1986). The isolation and properties of a tabtoxin-hydrolysing aminopeptidase from the periplasm of *Pseudomonas syringae* pv. *tabaci*. *Physiol. Mol. Plant Pathol.*, **28**, 345–52.

Lindow, S.E., Arny, D.C. & Upper, C.D. (1982). Bacterial ice-nucleation: A factor in frost injury to plants. *Plant Physiol.*, **70**, 1084–9.

Lyon, F. & Wood, R.K.S. (1976). The hypersensitive reaction and other responses of bean leaves to bacteria. *Ann. Bot.*, **40**, 479–91.

Mokhele, K., Tang, Y.J. & Clark, M.A. (1987). A *Pseudomonas stutzeri* outer membrane protein inserts copper into N_2O reductase. *J. Bacteriol.*, **169**, 5721–6.

Sato, H. & Miura, K.I. (1963). Preparation of transforming deoxyribonucleic acid by phenol treatment. *Biochim. Biophys. Acta.*, **72**, 619–29.

Scherrer, R. & Bochem, H.P. (1983). Energy dispersive X-ray microanalysis of the methanogen *Methanosarcina barkei* 'Fusaro' grown in methanol and in the presence of heavy metals. *Current Microbiol.*, **9**, 187–94.

Scherrer, R. & Gerhardt, P. (1972). Location of Ca within *Bacillus* spores by electron probe microanalysis. *J. Bact.*, **112**, 559–68.

Sekizawa, Y. & Wakabayashi, K. (1990). Bactericides. In *Methods in phytobacteriology*, ed. Z. Klement, K. Rudolph & D.C. Sands, pp. 319–26. Budapest: Akademiai Kiado.

Shuman, H. & Somlyo, A.P. (1976). Electron probe analysis of single ferritin molecules. *Proc. Nat. Acad. Sci. USA*, **73**, 1193–5.

Shuman, H., Somlyo, A.V. & Somlyo, A.P. (1976). Quantitative electron probe microanalysis of biological thin sections: methods and validity. *Ultramicroscopy*, **1**, 317–39.

Sigee, D. C. (1988). Preparation of biological samples for transmission electron microscopy: a review of alternative procedures to the use of sectioned material. *Scanning Microscopy*, **2**, 925–35.

Sigee, D.C. & Al-Rabaee, R.H. (1986). Nickel toxicity in *Pseudomonas tabaci*: single cell and bulk sample analysis of bacteria cultured at high cation levels. *Protoplasma*, **130**, 171–85.

Sigee, D.C. & El-Masry, M.H. (1987). Electron probe X-ray microanalysis of bacterial DNA. *J. Biochem. Biophys. Meth.*, **15**, 215–28.

Sigee, D.C., El-Masry, M.H. & Al-Rabaee, R.H. (1985). The electron microscope detection and X-ray quantitation of cations in bacterial cells. *Scanning Electron Microscopy*, **1985/III**, 1151–63.

Sigee, D.C., Hodson, N. & El-Masry, M.H. (1989). X-ray microanalytical determination of the ionic composition of plant pathogenic bacteria. *Proc. 7th. Int. Conf. Plant Pathogenic Bacteria*, pp. 509–14. Budapest: Akademiai Kiado.

Skrinska, V.A., Messineo, L., Towns, R.L.R. & Pearson, K.H. (1978). Transition metals in calf thymus deoxyribonucleoprotein. *Experientia*, **34**, 15–17.

Thellier, M., Ripoll, C. & Berry, J.P. (1991). Biological applications of secondary ion mass spectrometry. *Microscopy & Analysis*, **23**, 13–15.

Zierold, K. (1988). Electron probe microanalysis of cryosections from cell suspensions. *Meth. Microbiol.*, **20**, 91–111.

13 Ion localisation in plant cells using the combined techniques of freeze-substitution and X-ray microanalysis

M.A. Hajibagheri and T.J. Flowers

A X-ray microanalysis of plant cells

Ions play vital roles in the water relations of plant and animal cells and in the regulation of cellular metabolism. An important component of the function of these elements is often a change in subcellular distribution. Investigation of the subcellular distribution of elements such as sodium, potassium, calcium, magnesium and chlorine, which are highly mobile in the aqueous phase, is a most challenging problem in biology. In order to study the distribution and transport of elements within cells a technique with high sensitivity and high resolution is required. The chosen method must not, in itself, alter the native elemental distribution during sample preparation or during measurement. Measurement of the (inorganic) elemental content of compartments (cell walls, cytoplasm and vacuole) within plant cells can be approached by a number of different methods, including the analysis of efflux of radioactive ions, analysis of tissues with different compartmental volume fractions, the use of microelectrodes specific to particular ions, by NMR in some instances and by X-ray microanalysis of individual cells. Of these, X-ray microanalysis offers more than any of the other methods in terms of both resolution and sensitivity. The use of microelectrodes is restricted in plant biology because of the physical strength of the walls of many cell types, and the difficulties of visualising the electrode tip in intact tissues in other than surface cells. As a technique, X-ray microanalysis also has the potential to resolve concentration differences within the cytoplasmic phase (e.g. between different organelles). X-ray microanalysis looks at a specimen 'fixed' at the instant of sampling; it can thus provide the temporal resolution to allow experimental manipulation of cell ion relations, as opposed to simple description of them. Since 1980, X-ray microanalysis has been used in plants in a number of investigations to study the quantity of ions in individual cell types and within different cell compartments, without the need for cell fractionation (e.g. Harvey et al., 1981; Storey, Pitman & Stelzer, 1983; Echlin & Taylor, 1986; Van Steveninck, Van Steveninck & Läuchli, 1982; Hajibagheri, Harvey & Flowers, 1987).

Microanalysis is performed by determining the energy of X-rays emitted from tissues bombarded by a beam of electrons produced in a conventional electron microscope: the spectrum of an element is unique because of the discrete energy changes in its orbiting electrons. Three methods of tissue preparation have been used prior to analysis – cryomicrotomy, cryofracturing and freeze-substitution – all of which rely on rapid freezing of the fresh tissue to immobilise cellular solutes. Subsequently, the material may be analysed in its frozen state or embedded in resin. In the former method the tissue is cryomicrotomed or fractured to provide a clean surface (termed 'bulk-frozen'). In both cases the specimen remains frozen in the electron beam during analysis, and freezing followed by sectioning or fracturing are the only preparative steps involved.

B   Preparation of freeze-substituted tissue

Freeze-substitution depends upon the replacement of the frozen aqueous phase of a specimen with organic solvent, in discrete molecular events. It is a relatively slow process and, in order to prevent structural damage due to thawing or the growth of crystals, should be carried out at the lowest possible temperature with respect to the freezing point of the organic solvent in use. In outline, frozen tissue ($-170\ °C$) is transferred into acetone or ether at low temperature ($-40\ °C$) in the presence of freshly activated molecular sieve (added to maintain the anhydrous state of the solvent) in sealed vials in a desiccator. This is followed by gradual infiltration with a mixture of the substitution solvent and vinylcyclohexene dioxide (Harvey *et al.*, 1981; Table 13.1) at $-40\ °C$ to $-20\ °C$ and then, by stages, into Spurr's low viscosity epoxy resin mixture (Spurr, 1969). By this time there are no water molecules remaining in the tissue and 'water soluble' ions are deemed to be immobilised. During infiltration with the embedding medium the tissue is allowed gradually to reach room temperature and polymerisation is carried out in the usual manner (e.g. 1 day at 60 °C). Thin sections (about 100 nm) are presently cut dry with a glass knife (the most difficult stage of the whole procedure) for analysis.

The particular value of X-ray microanalysis of freeze-substituted material lies in its ability to provide quantitative estimates of subcellular ion concentrations, which may be repeated within a cell and between different cells to provide sufficient data for statistical analysis: no other technique can, at present, provide this information so directly, provide it in single cells at a single time, or allow repeat measurements within the same cell. Although the X-ray microanalysis of freeze-substituted material has been in use for about 19 years (Pallaghy, 1973), this use has been very limited. The majority of the work that has been done has concerned the distribution of ions in plants subjected to salinity because this was a technique used at the University of Sussex some 12 years ago to investigate this particular problem. The technique

Table 13.1. *Standard freeze-substitution scheme of X-ray microanalysis*

Freeze	1 mm³ segments of tissue in 8% methylcyclohexane in 2-methylbutane cooled to −170 °C by liquid nitrogen ↓
Substitute	in anhydrous acetone, −72 °C, 1 day, followed by −40 °C, 2 days in the presence of activated molecular sieve in sealed vials in desiccator ↓
Infiltrate	50% acetone, 50% ERL 4206 (vinylcyclohexene dioxide), −40 °C, 4 h ↓
	100% ERL 4206, −40 °C, 16 h ↓
	100% ERL 4206, −40 °C to −15 °C, 3 h ↓
(vials uncapped only in dry atmosphere)	50% ERL 4206, 50% complete embedding medium (ERL 4206, 10 g; dibutylphthalate, 12 g; nonenylsuccinic anhydride, 26 g; DY 064, 1.6 g), −15 °C, 6 h ↓
	25% ERL 4206, 75% complete embedding medium, −15 °C to −4 °C, 16 h ↓
	100% complete embedding medium, −4 °C to +20 °C, 3 h ↓
	100% complete embedding medium, 20 °C, 1 day ↓
	encapsulate, polymerise, 60 °C, 1 day

Harvey, Hall & Flowers (1976).

has given valuable information concerning the response of plants to such a situation. We believe, however, that the technique has a much wider potential than has been realised to date. There are numerous other questions in biology that can be approached with identical, existing methodology.

In recent years, freeze-substituted dry-sections of plant material prepared at the University of Sussex have been analysed using microscopes at the National Physical Laboratory at Teddington (EMMA-4 Link Systems energy dispersive analytical microscope and Jeol JEM-2000 FX electron microscope with AN 10,000 X-ray microanalysis, Link system), to provide measurements of elemental concentrations within plant cells (Harvey *et al.*, 1981; Hajibagheri *et al.*, 1987; Hajibagheri & Flowers, 1989; Flowers *et al.*, 1990). This has been possible because realistic calibration standards have been made by incorporating inorganic ions into the resins by using cyclic ethers (Harvey,

Table 13.2. *The balance of the four major osmotic components in the majority of* Suaeda maritima *leaf cells*

Component	Cytoplasm	Vacuole
glycinebetaine (mol m^{-3})	830	—
Na (mol m^{-3})	109	565
K (mol m^{-3})	16	24
Cl (mol m^{-3})	21	388
sum (mol m^{-3})	976	977

Reproduced from Harvey *et al.* (1981) with permission of the publisher, Springer-Verlag.

Table 13.3. *X-ray microanalysis of ion distribution within the major subcellular compartments of* Suaeda maritima *root cortical cells* (*10–20 mm behind the root tip*) *grown in the presence of 200 mol m^{-3} NaCl*

Cell compartment	Ion concentration (mol m^{-3}) of analysed volume		
	Na	K	Cl
Cytoplasm	118 (112–125)	55 (40–60)	90 (86–95)
Vacuole	432 (410–460)	35 (20–48)	445 (430–465)
Cell wall	95 (92–105)	83 (78–110)	120 (111–130)

Reproduced from Hajibagheri & Flowers (1989) with permission of the publisher, Springer-Verlag.

Flowers & Kent, 1984). The value of this technique is perhaps best appreciated by mention of some particular achievements. These are described briefly together with other applications.

1   Ion localisation within the halophyte Suaeda maritima

Work on the growth and ion relations of *Suaeda* has shown that this halophyte grows fast at high salinities and that a necessary feature of its salt tolerance is compartmentation of NaCl within the cell (Flowers, Troke & Yeo, 1977; Wyn Jones, Brady & Speirs, 1979; Greenway & Munns, 1983).

Table 13.4. *Elemental compositions* (*mol m⁻³*) *of the root cortical cells* (*10–20 mm from the tip*) *of two varieties of* Zea mays *grown for 15 days, in culture solution alone* (*control*) *or with 100 mol m⁻³ NaCl*

Variety	Treatment	Element	Cytoplasm	Vacuole	Cell wall
LG_{11}	Control	Na	58	8	36
		K	140	58	151
		Cl	50	32	60
	+ NaCl	Na	138	229	180
		K	90	55	59
		Cl	151	98	106
Protador	Control	Na	69	6	28
		K	162	60	178
		Cl	61	65	68
	+ NaCl	Na	92	178	171
		K	120	52	95
		Cl	128	93	130

Hajibagheri *et al.* (1987).

The compartmental model of the halophyte cell was first verified using X-ray microanalysis by Harvey *et al.* (1981) and now receives general acceptance: there is a consistency between biochemical observations, compartmental analysis, and direct measurements by analytical X-ray microanalysis (Flowers, 1972; Yeo & Flowers, 1980; Harvey *et al.*, 1981). The great majority of ions involved in osmotic adjustment are restricted in the leaf cell vacuoles, the osmotic potential of the cytoplasm being adjusted with glycinebetaine (Table 13.2).

We have also recently studied the ion content of compartments within cortical cells of mature roots of *Suaeda maritima*, grown in 200 mol m⁻³ NaCl, by X-ray microanalysis of freeze-substituted thin sections (see Hajibagheri & Flowers, 1989). Sodium and chloride were found in the vacuoles at about four times the concentration in the cytoplasm or cell walls, whereas K was more concentrated in the cell walls and cytoplasm than in the vacuoles (Table 13.3). The vacuolar Na concentration was 12 times higher than that of K. The Na concentration of cell walls of cortical cells was about 95 mol m⁻³ of analysed volume. The cytoplasmic K concentration within the mature cortical cells was estimated to be 55 mol m⁻³ of analysed volume, confirming the earlier conclusion that the cytoplasmic K concentration of mature cells of halophytes is lower than is commonly deduced for glycophytes.

2   Ion distribution in crop plants

The relationship between the degree of salt-tolerance and the distribution of sodium, potassium and chloride within main subcellular compartment of crop plants can also be determined by X-ray microanalysis. The distribution of ions, particularly those of Na, K and Cl, was determined within the subcellular compartments of root cells using X-ray microanalysis of freeze-substituted tissue (Hajibagheri *et al.*, 1987). Salinity induced a greater increase (about 1.7 times) in cytoplasmic Na concentration in the salt-sensitive maize variety (LG_{11}) than in resistant varieties (e.g. Protador). The mean K : Na ratio in the cytoplasm of the root cortical cells in Protador grown for 15 d in saline conditions (100 mol m^{-3} NaCl) was twice that found for LG_{11} (Table 13.4). Our X-ray microanalysis data indicated that varieties which can keep higher potassium and lower sodium and chloride in the cytoplasm with sodium and chloride mostly localised in vacuoles can survive better under saline conditions.

Ion accumulation in the cell walls of rice plants growing under saline conditions has recently been measured by X-ray microanalysis (Flowers, Hajibagheri & Yeo, 1991). The X-ray microanalysis data proved that an important factor in salt damage in rice is dehydration due to the extracellular accumulation of salt as suggested in the Oertli hypothesis (Table 13.5).

X-ray microanalysis has given valuable information concerning salt-tolerance in the halophytic wild rice *Portersia coarctata*. These plants accumulated sodium and chloride ions in the leaves, but maintained a Na : K ratio as low as 0.7 when grown in 25 % seawater where the Na : K ratio was 34 (Flowers *et al.*, 1990). The results of X-ray microanalysis of the contents of the hairs (presumably salt-glands) on the leaves is powerful evidence in support of the contention (Bal & Dutt, 1986) that the secreted salt comes from the hairs. The Na : K ratio in the hairs of plants raised on NaCl was similar to that of secreted salt and very much higher than that in the mesophyll cells, namely 7.3 in the hairs and as opposed to 0.9 in the mesophyll, for plants growing in 100 mol m^{-3} NaCl (Table 13.6). The salt secreted by the hairs is an important factor in the salt-balance of the leaves of this species.

3   Ion compartmentation in halophytic microorganisms

Dunaliella is a unicellular green alga, species of which grow in a wide range of external salinities, from 0.1 to 5 kmol m^{-3} NaCl (Ben-Amotz & Avron, 1978). The strategies adopted by *Dunaliella* in response to salinity have been well documented: a major factor is an adjustment of the internal concentration of the compatible solute glycerol (Gimmler & Moller, 1981; Gilmour, Hipkins & Boney, 1984). However, the role of inorganic ions in balancing the external osmotic potential has been a source of controversy. For example, Avron (1986) found that the intracellular Na concentrations increased from 20 to

Table 13.5. *Sodium plus potassium concentrations (mM) in cytoplasm and cell walls of leaves of the cultivars Amber and IR 2153-3-6-5-2 of rice grown in the presence of sodium chloride (50 mol m^{-3})*

	Tissue sodium (mmol g^{-1} dry weight)	Cell wall	Cytoplasm
IR2153			
	0.18	86	263
	0.30	194	244
	0.49	140	247
	0.53	271	241
	0.90	383	236
	0.91	523	351
	2.00	394	243
	2.21	511	347
Amber			
	0.19	43	251
	0.21	94	270
	0.30	54	269
	0.45	169	251
	0.57	246	321
	0.81	474	314
	0.92	734	361
	1.97	626	460
	2.23	840	673

Flowers *et al.* (1991).
The values in the table are concentrations expressed on the basis of water contents of 70% and 35% for the cytoplasm and cell walls, respectively. The values for the cell walls were calculated after subtracting the sodium + potassium concentration in walls of non-salinised plants and so represent 'soluble Na + K'.

100 mol m^{-3} when the external salinity was increased from 0.5 to 4.0 kmol m^{-3}, while in contrast, Gimmler & Schirling (1978) found intracellular Na concentration in excess of 590 mol m^{-3} in *Dunaliella* cells grown in 1.5 kmol m^{-3} NaCl. We have measured the Na content of compartments within *Dunaliella* cells by X-ray microanalysis to provide a definitive answer to the question of its internal concentration (Hajibagheri *et al.*, 1986). The internal concentration of Na$^+$ in *D. parva* cells adapted to 1.5 mol m^{-3} NaCl was 105 mol m^{-3} (Table 13.7). This result is in reasonable agreement with the low internal Na measurement reported by Katz & Avron, 1985 (20 mol m^{-3}); Pick, Karni & Avron, 1986 (25 mol m^{-3}); and in good agreement with the work of Bental, Degani & Avron, 1988, in *Dunaliella*, using ^{23}Na-NMR spectroscopy (90 mol m^{-3}).

Table 13.6. *The ion concentrations within the vacuoles of cells from leaves of* Porteresia coarctata

Culture conditions (NaCl, mol m^{-3})	Element	Elemental concentrations in various cell types (mmol l^{-1})		
		Epidermal	Mesophyll	Hair
100	Na	110 ± 42	79 ± 26	255 ± 39
	K	67 ± 30	88 ± 18	35 ± 20
	Cl	100 ± 23	123 ± 37	203 ± 55
200	Na	168 ± 55	158 ± 39	323 ± 96
	K	70 ± 43	50 ± 22	58 ± 19
	Cl	219 ± 40	172 ± 55	280 ± 78

Flowers *et al.* (1990).

Table 13.7. *Published measurements of internal concentration of Na in* Dunaliella *cells grown in media containing 1.5–2.0 kmol m^{-3} NaCl*

Methods	External NaCl (kmol m^{-3})	Internal Na concentration (mol m^{-3})	Reference
Rapid washing isotonic glycine	1.64	7	Ehrenfeld & Cousin (1982)
Li	1.50	20	Katz & Avron (1985)
Cation exchange columns	2.0	25	Pick *et al.* (1986)
X-ray microanalysis	1.5	105	Hajibagheri *et al.* (1986)
^{23}Na-NMR spectroscopy	1.5	90	Bental *et al.* (1988)
Blue Dextran	1.5	623	Ginzburg (1978)
Silicone oil centrifugation	1.5	590–800	Gimmler & Schirling (1978)
^{14}C-Dextran	1.6	292	Ehrenfeld & Cousin (1982)
Blue Dextran	1.5	770	Zmiri and Ginzburg (1983)
Blue Dextran	1.5	500	Ginzburg & Richman (1985)

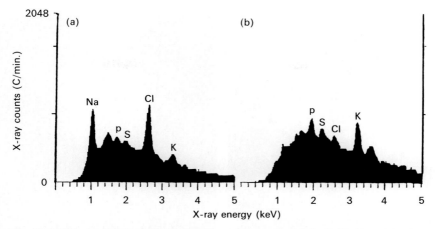

Fig. 13.1. X-ray spectra from leaf epidermal vacuoles of the halophyte *Suaeda maritima* which has been grown in culture solution with the addition of 340 mol m^{-3} NaCl. (a) Guard cell vacuole of an open stoma; (b) guard cell of a closed stoma.

X-ray microanalysis was also performed on hyphae of the filamentous marine fungus *Dendryphiella salina* growing at different salinities to give sodium, potassium and chloride concentrations in the cytoplasm, vacuole and cell wall. Clipson, Hajibagheri & Jennings (1990) found that the compartmentation in *D. salina* is rather different from that of many halophilic cells: that the vacuole has little significance in generating the overall osmotic potential within the protoplasm.

C Frozen–hydrated bulk tissue

X-ray microanalysis of frozen–hydrated bulk tissue samples have frequently been used to examine the regulation of ion uptake and transport in plants grown under saline conditions (Yeo, Läuchli & Kramer, 1977; Stelzer & Läuchli, 1978; Stelzer, 1981; Storey *et al.*, 1983; Harvey *et al.*, 1985; Stelzer, Kuo & Koyro, 1988. For a general review of this method, see Echlin & Taylor, 1986. We describe briefly here an example of application of frozen–hydrated vacuolated plant tissue. Recently we attempted to investigate the relative importance of potassium and sodium in stomatal opening in *S. maritima* (Hajibagheri *et al.*, in preparation), since there has been a suggestion that Na can replace K in this respect in halophytes (Eshel, Waisel & Ramati, 1974) and the ion relations of the shoots of *S. maritima* are clearly dominated by sodium (Yeo & Flowers, 1980). X-ray microanalysis of stomatal function in *S. maritima* suggests that opening and closing results from fluxes of sodium, rather than potassium (Figs 13.1 and 13.2, Table 13.8).

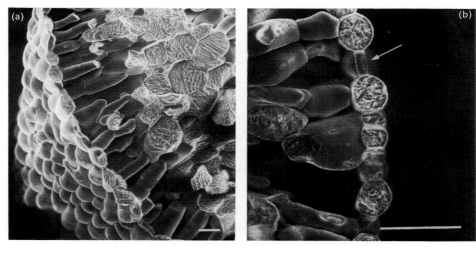

Fig. 13.2. Scanning electron micrographs of transversely fractured leaves of the halophyte *Suaeda maritima* in the frozen–hydrated state. (a) Survey showing no fractured stomata. The arrow in (b) points at a transversely fractured stoma (the fractured across the intercellular space between the guard cells). Bars = 10 μm.

Table 13.8. *Sodium and potassium concentrations in guard cells and neighbouring (nearest) epidermal cells of leaves on the halophyte* Suaeda maritima *that had been grown in sodium chloride (200 mol m^{-3}) for 30–35 days*

Cell type	Number of cells analysed	Ion concentration and range (mol m^{-3}) when	
		stomata open	stomata closed
Sodium			
Guard	6	197 (164–237)	94 (54–127)
Epidermal	10	127 (79–174)	173 (115–218)
Potassium			
Guard	6	50 (26–77)	22 (15–30)
Epidermal	10	44 (31–72)	33 (22–54)

Whether the stomata were predominantly open or closed (plants in the dark) was judged prior to freezing and X-ray microanalysis. The results are derived from energy dispersive X-ray analyses on frozen–hydrated vacuoles of the leaf cells (Fig. 13.2; data of M.A. Hajibagheri and R. Stelzer).

Table 13.9. *The internal Na and K concentrations of* Dunaliella parva *grown in 0.4 kmol m^{-3} NaCl*

Methods	Na	K
1. Spurr resin	105	110
2. Nanoplast MUV116	122	140
3. Molecular Distillation Dryer	78	74

The internal Na and K calculated on the estimates of vacuole and cytoplasmic volumes as percentages of cell volume.
Hajibagheri & Flowers (1991).

D   Conclusions and the prospects for future research using X-ray microanalysis

There are applications in which the rapidity of sample preparation by the frozen–hydrated bulk technique outweighs the limited resolution. However, there are many cases where quantification at high resolution is required. The answer is to use a transmission microscope in which the beam passes through a section whose thickness can be determined. Thin sections can, however, only be prepared from resin embedded plant material and usefully analysed provided care has been taken to preserve the mineral content of cells: this can be achieved by freeze-substitution. The freeze-substitution procedure, together with dry sectioning, has been shown to be a satisfactory method of cell preservation in that there is little re-location of ionic solutes during preparation procedure. We have evaluated freeze-substitution against two other methods of estimating ion contents (efflux analysis and analysis of serial sections; Hajibagheri *et al.*, 1988) and found, for higher plant tissues, good agreement between all three methods for cytoplasmic potassium.

A major concern in the preparation of biological soft tissues for X-ray microanalysis is the redistribution of ions during the preparative stages. Currently, redistribution is restricted during dehydration by low temperatures, but the temperature is raised during infiltration of the resin and polymerisation. We have recently compared the measurement of sodium, potassium and chloride contents in a salt-tolerant unicellular alga, *Dunaliella parva*, following either freeze-substitution (using two different resins) or molecular distillation drying (see Hajibagheri & Flowers, 1992, for details). All three procedures gave similar results: after freeze-substitution, ion contents were marginally (but not significantly) higher following embedding in Nanoplast MUV 116 resin than in Spurr resin (Table 13.9). Since the Nanoplast can be polymerised at low temperatures, it has advantages over the Spurr resin.

At present it is necessary to section freeze-substituted material using a dry knife, since ions are lost rapidly from a section 100 nm thick when sections are cut on to water – presumably a function of water uptake through pores in the epoxy resin. Dry sectioning is at present the limiting factor in the processing of material for analysis: not only is it slow, but the quality of section produced hinders visualisation of structure. A search will be needed for other resins that might display improved retention in presence of aqueous solvents, for alternative combinations of resins and solvents to be used during sectioning.

References

Avron, M. (1986). The osmotic components of halotolerant algae. *Trends in Biochem. Sci.*, **11**, 5–6.
Bal, R.R. & Dutt, S.K. (1986). Mechanism of salt tolerance in wild rice (*Porteresia coarctata*). *Plant and Soil*, **92**, 399–404.
Ben-Amotz, A. & Avron, M. (1978). On the mechanisms of osmoregulation in *Dunaliella*. In *Energetics and structure of halophilic microorganisms*, ed. S.R. Caplan & M. Ginzburg, pp. 529–41. Amsterdam: Elsevier/North Holland.
Bental, M., Degani, H. & Avron, M. (1988). 23Na-NMR studies of the intracellular sodium ion concentration in the halotolerant alga *Dunaliella salina*. *Plant Physiol.*, **87**, 913–17.
Clipson, N.J.W., Hajibagheri, M.A. & Jennings, D.H. (1990). Ion compartmentation in the marine fungus *Dendryphiella salina* in response to salinity: X-ray microanalysis. *J. Exp. Bot.*, **41**, 199–202.
Echlin, P. & Taylor, S. (1986). The preparation and X-ray microanalysis of bulk frozen hydrated vacuolate plant tissue. *J. Microsc.*, **141**, 329–48.
Ehrenfeld, J. & Cousin, J.L. (1982). Ionic regulation of the unicellular green alga *Dunaliella tertiolecta*. *Journal of Membrane Biol.*, **70**, 47–57.
Eshel, A., Waisel, Y. & Ramati, A. (1974). The role of sodium in stomatal movements of a halophyte: a study by X-ray microanalysis. In *7th International Colloquium on Plant Analysis and Fertilizer Problems*, ed. J. Wehrmann, Vol. 3, pp. 1–9. Hannover: German Society for Plant Nutrition.
Flowers, T.J. (1972). Salt tolerance in *Suaeda maritima* (L) Dum. The effect of sodium chloride on growth, respiration, and soluble enzymes in comparative study with *Pisum sativum* L. *J. Exp. Bot.*, **23**, 310–21.
Flowers, T.J., Flowers, S.A., Hajibagheri, M.A. & Yeo, A.R. (1990). Salt tolerance in the halophytic rice, *Porteresia coarctata* Tateoka. *New Phytol.*, **114**, 675–84.
Flowers, T.J., Hajibagheri, M.A. & Yeo, A.R. (1991). Ion accumulation in the cells walls of rice plants growing under saline condition; evidence for the Oertli hypothesis. *Plant Cell Environ*, **14**, 319–25.
Flowers, T.J., Troke, P.F. & Yeo, A.R. (1977). The mechanism of salt tolerance in halophytes. *Ann. Rev. Plant Physiol.*, **28**, 89–121.
Gilmour, D.J., Hipkins, M.F. & Boney, A.D. (1984). The effect of osmotic and ionic stress on the primary processes of photosynthesis in *Dunaliella tertiolecta*. *J. Exp. Bot.*, **35**, 18–27.
Gimmler, H. & Moller, E.M. (1981). Salinity-dependent regulation of starch and glycerol metabolism in *Dunaliella parva*. *Plant Cell Environ.*, **4**, 367–75.
Gimmler, H. & Schirling, R. (1978). Cation permeability of the plasmalemma of the halotolerant alga *Dunaliella parva*. II. Cation content and glycerol concentration of the cells as dependent upon external NaCl concentration. *Zeitschrift fur Pflanzenphysiol.*, **87**, 435–44.

Ginzburg, B.Z. (1978). Regulation of cell volume and osmotic pressure in *Dunaliella*. In *Energetics and structure of halophilic microorganisms*, ed. S.R. Caplan & M. Ginzburg, pp. 543–60. Amsterdam: Elsevier/North Holland.

Ginzburg, M. & Richman, L. (1985). Permeability of whole *Dunaliella* cells to glucose. *J. Exp. Bot.*, **36**, 1959–68.

Greenway, H. & Munns, R. (1983). Interaction between growth, uptake of Cl⁻ and Na⁺, and water relations of plants in saline environments. II. Highly vacuolated cells. *Plant Cell Environ.*, **6**, 575–89.

Hajibagheri, M.A. & Flowers, T.J. (1989). X-ray microanalysis of ion distribution within root cortical cells of the halophyte *Suaeda maritima* (L.) Dum. *Planta*, **177**, 131–4.

(1992). The use of freeze-substitution and molecular distillation drying in the preparation of *Dunaliella parva* for ion localization studies by X-ray microanalysis. *J. Electron Microsc. Tech.* (in press).

Hajibagheri, M.A., Flowers, T.J., Collins, J.C. & Yeo, A.R. (1988). A comparison of X-ray microanalysis, compartmental analysis and longitudinal ion profiles to estimate cytoplasmic ion concentrations in two maize varieties. *J. Exp. Bot.*, **39**, 279–90.

Hajibagheri, M.A., Gilmour, D.J., Collins, J.C. & Flowers, T.J. (1986). X-ray microanalysis and ultrastructural studies of cell compartments of *Dunaliella parva*. *J. Exp. Bot.*, **37**, 1725–32.

Hajibagheri, M.A., Harvey, D.M.R. & Flowers, T.J. (1987). Quantitative ion distribution within maize root cells in salt-sensitive and salt-tolerant varieties. *New Phytol.*, **105**, 367–79.

Harvey, D.M.R., Flowers, T.J. & Kent, B. (1984). Improvement in quantitation of biological X-ray microanalysis. *J. Microsc.*, **107**, 1890–8.

Harvey, D.M.R., Hall, J.L. & Flowers, T.J. (1976). The use of freeze-substitution in the preparation of plant tissues for ion localization studies. *J. Microsc.*, **107**, 189–98.

Harvey, D.M.R., Hall, J.L., Flowers, T.J. & Kent, B. (1981). Quantitative ion localization within *Suaeda maritima* leaf mesophyll cell. *Planta*, **151**, 555–60.

Harvey, D.M.R., Stelzer, R., Brandtner, R. & Kramer, D. (1985). Effects of salinity on ultrastructure and ion distributions in roots of *plantago coronopus*. *Physiol. Plant.*, **66**, 328–38.

Katz, A. & Avron, M. (1985). Determination of intracellular osmotic volume and sodium concentration in *Dunaliella*. *Plant Physiol.*, **78**, 817–20.

Pallaghy, C.K. (1973). Electron probe microanalysis of potassium and chloride in freeze substituted leaf sections of *Zea mays*. *Aust. J. Biol. Sci.*, **26**, 1015–34.

Pick, U., Karni, L. & Avron, M. (1986). Determination of ion content and ion fluxes in the halotolerant alga *Dunaliella salina*. *Plant Physiol.*, **81**, 92–6.

Spurr, A.R. (1969). A low viscosity epoxy resin embedding medium for electron microscopy. *J. Ultrastructure Res.*, **26**, 31–43.

Stelzer, R. (1981). Ion localization in the leaves of *Puccinellia peisonis*. *Z. Pflanzenphysiol.*, **103**, 27–36.

Stelzer, R., Kuo, J. & Koyro, H.-W. (1988). Substitution of Na⁺ by K⁺ in tissues and root vacuoles of barley (*Hordeum vulgare* L. cv. Aramir). *J. Plant Physiol.*, **132**, 671–7.

Stelzer, R. & Läuchli, A. (1978). Salt and flooding tolerance of *Puccinellia peisonis*. III. Distribution and localization of ions in the plant. *Z. Pflanzenphysiol.*, **88**, 437–48.

Storey, R., Pitman, M.G. & Stelzer, R. (1983). X-ray microanalysis of cells and cell components of *Atriplex spongiosa*. II. Roots. *J. Exp. Bot.*, **34**, 1196–206.

Van Steveninck, R.F.M., Van Steveninck, M.E. & Läuchli, A. (1982). Profiles of chloride content of vacuoles in lupin root cells as shown by electron probe X-ray microanalysis. *Z. Pflanzenphysiol.*, **108**, 215–22.

Wyn Jones, R.G., Brady, C.J. & Speirs, J. (1979). Ionic and osomotic relations in plant cells. In *Recent advances in the biochemistry of cereals*, ed. D.L. Laidman & R.G. Wyn Jones, pp. 63–103. New York: Academic Press.

Yeo, A.R. & Flowers, T.J. (1980). Salt tolerance in the halophyte *Suaeda maritima* L. Dum: evaluation of the effect of salinity upon growth. *J. Exp. Bot.*, **31**, 1171–83.

Yeo, A.R., Läuchli, A. & Kramer, D. (1977). Ion measurements by X-ray microanalysis in unfixed, frozen, hydrated plant cells of species differing in salt tolerance. *Planta*, **134**, 35–8.

Zmiri, A. & Ginzburg, B.Z. (1983). Extracellular space and cellular sodium content in pellets of *Dunaliella parva* (Dead Sea, 75). *Plant Science Lett.*, **30**, 211–18.

14 Electron probe X-ray microanalysis of diffusible ions in cells and tissues from invertebrate animals

Brij L. Gupta

A Introduction

Invertebrates constitute 90% or more of all the living species in the Animal Kingdom. The class Insecta alone has more than one million recorded species. Invertebrate species can be found flourishing in virtually every ecological environment from the deepest of oceans in the vicinity of hydrothermal vents to the highest of mountains. This process of speciation during the course of evolution has not only generated a bewildering range of morphological form, size and shape but has also led to an exploitation of every conceivable physico-chemical principle in the physiological adaptation. The celebrated Danish zoophysiologist August Krogh once said that there is always some organism which has maximally expressed a particular biological phenomenon and therefore would be ideally suited for investigating that principle. Similarly, while promoting the use of insects in the study of general physiology, Wigglesworth (1948) has said: 'Insects ... are so varied in form, so rich in species, and adapted to such diverse conditions of life that they afford unrivalled opportunities for physiological study'. Furthermore, numerous invertebrate species can often be obtained in large number and maintained in the laboratory at a relatively low cost.

A perusal of any modern text-book of cell and developmental biology, comparative physiology etc. would show that over the past century most of the fundamental biological concepts were founded by investigating material from some invertebrate species. The diffusible ions Na, K, Cl, Mg, Ca, Zn and also H, OH are now implicated in virtually all the cell functions. A great deal of pioneering electron probe X-ray microanalysis (EPXMA) work in order to study ions was also done on invertebrate cells and tissues (Le Furgey, 1988; Gupta, 1991), especially in monitoring the heavy metal pollution in the environment, and in this field invertebrates continue to provide the most useful material (Morgan, 1984; Morgan & Winters, 1987). In most biological EPXMA laboratories these days the invertebrate species have not been exploited to any great extent in the study of cell biology and general physiology. In this chapter I have tried to summarise all the available information on the quantitative EPXMA of the diffusible ions in cells and tissues of invertebrates and project the data in a relevant biological context. It is clear that the contribution of invertebrate material in the studies conducted so far is by no means trivial, and the scope for future studies is considerable.

B   Measurements of ionic concentration: some technical considerations

In order to gain a full insight into the role of ions in cell and tissue functions, one needs to have the following quantitative information.

(a) The total concentration of each ion in mmol/kg wet mass inside the cell and in different cell-types and in tissues in a given physiological state.

(b) The distribution of ions within the cell in the ground cytoplasm (cytosol) and in organelles.

(c) The ionic composition of the microenvironment around the cells and in extracellular matrices such as glycocalyx, basement membrane, cell-membrane invaginations, mucus layer etc.

(d) The fractions of water and dry mass in each of the corresponding components or compartments.

(e) The corresponding free concentration (activity) of each ion in mmol/l water.

(f) How these quantities change with the change in the physiological activity, during cell-cycle, during cell-differentiation and in cell-transformation.

EPXMA is the only established method at present that can provide information for (a)–(d) simultaneously for all the elements with the atomic number $Z = 11$ or higher, and possibly down to $Z = 5$, in a relatively non-destructive fashion *in situ* within a microscopically identifiable image. EPXMA measures the total concentration of an element whether structural, 'bound' or ionised but it cannot measure protons, hydroxyls and other organic electrolytes. Activities of ions can be obtained by electrophysiological methods but without any reliable information on the subcellular localisation *in situ* or by using ion-sensitive dyes such as Fura-2 for Ca^{2+} in a low resolution light microscopical image. Both the latter methods are used on living cells and therefore afford continuous measurements of dynamic events. Other conventional methods used in physiology for measuring the concentrations and distribution of ions have not proved reliable for multi-cellular systems. Measurements of ionic composition in components and organelles biochemically isolated from cells have proved unreliable and totally misleading irrespective of the methods used, even when the isolated structures such as nuclei, secretion granules, nematocysts etc. may *appear* to perform normally *in vitro*. The ionic composition of such isolated systems has been often found to differ drastically from the composition when measured in the 'native' state *in vivo*.

EPXMA measures elements as they are in the specimen under the electron beam. The X-ray data thus obtained are critically dependent on how the specimens are handled during the entire procedure from their native state *in vivo* to the moment of analysis. The specimen may also change under the electron beam itself. It is also now generally agreed that for reliable EPXMA of diffusible elements, the fresh specimens must be *directly* cryofixed as fast as possible. Such specimens may then be analysed: (*a*) in bulk-frozen, hydrated form, (*b*) after cryosectioning, (*c*) after freeze-drying, embedding and sectioning, or (*d*) after freeze-substitution, embedding and sectioning. The complete methodology for EPXMA of biological specimens including the relative merits of different approaches has been described and reviewed in detail by the author (Gupta, Hall & Moreton, 1977; Gupta & Hall, 1982; Hall & Gupta, 1983; Gupta, 1991) and by many others (see this volume). Warley & Gupta (1991) have provided step-by-step practical protocols for the analysis of cryosections including an example of a complex, secretory, invertebrate tissue – the salivary glands of cockroach. Zierold (1991) has reviewed the methods for cryopreserving ultra-fast, millisecond events such as the discharge of trichocysts in *Paramecium* (Schmitz & Zierold, 1989) or of nematocysts in Cnidaria (Zierold, Gerke & Schmitz, 1989; Tardent *et al.*, 1990). In all the past experience with methodology and analysis, invertebrate cells and tissues have proved much more favourable from every point of view (except, of course, for specific bio-medical and pathological questions). The conversion of X-ray data into mmol/kg wet mass, mmol/kg dry mass, mmol/l of the specimen area and mmol/l water has been discussed by Hall & Gupta (1982, 1983, 1986).

If EPXMA is performed on frozen–hydrated specimens and the data quantified by Hall's continuum method, one obtains elemental concentrations directly in mmol/kg wet mass for every area analysed. In the usual practice, the specimen (cryosection) is then dried within the analytical microscope and reanalysed in a frozen–dried state for obtaining values in mmol/kg dry mass in the corresponding fields. By comparing the two concentration values for every element one can obtain a mean local dry mass and hence water fraction for every analysed component (see Gupta & Hall, 1979; Warley & Gupta, 1991). The water fraction can then be used for converting the values from mmol/kg wet mass into mmol/l water. The last value can then be compared with a similar value obtained by other methods for measuring free ionic concentrations. A close agreement between the two values would show that all the measured elements are fully dissolved in all the measured water, as is commonly assumed in the conventional physiological wisdom. Significant

Table 14.1. *Concentration values of elements in the cytoplasm and in the secretion granules of the zymogen cells in cockroach salivary glands measured by EPXMA of 1 μm thick frozen–hydrated cryosections*

Specimen	Concentration values						
Frozen-hydrated	mmol/kg wet mass						
	Na	K	Cl	Mg	Ca	P	S
cytoplasm	9	126	41	8	2.3	120	41
secretion granules	16	76	26	4	17	23	111
Frozen-dried	mmol/kg dry mass						
cytoplasm	43	484	212	25	12	647	174
secretion granules	33	181	61	13	35	43	241
Calculated values[a]	mmol/l water						
cytoplasm	12	164	63	10	3	157	53
secretion granules	30	141	48	6	31	43	207

Dry mass as a fraction of wet mass: mean (SD)[b]
cytoplasm = 0.23 (0.030)
sec. gran. = 0.46 (0.070)

Values are means of 25 different measurements. Data from Gupta & Hall (unpublished) and Warley & Gupta (1991).
[a] The convertion from mmol/kg wet mass is based on the assumption that all the measured elements are fully dissolved in all the measured water fraction in each compartment. This is certainly not true for P and S, which are essentially components of macromolecules. It is also not likely to be true for other elements (see text).
[b] Mean dry mass fraction was estimated by comparing the mmol/kg wet mass and mmol/kg dry mass values for every measured element in each component (see Warley & Gupta, 1991).

differences between the two values can provide information on the binding of ions to polyelectrolyte macromolecules affecting the local activity coefficients of ions. Such interactions form the basis for ionic and osmotic gradients. Such gradients, both within the cells and around the cells in tissues, have now been observed in virtually all quantitative EPXMA studies, both in invertebrates and in vertebrates (see Gupta, 1991).

In most laboratories EPXMA is carried out in frozen–dried cryosections, and the data are generally obtained in mmol/kg dry mass by the Hall method. Such data are of limited value in biology except for comparing the elemental ratios. The use of these mmol/kg dry mass values either for comparing different subcellular components or for comparing cells in different physiological states can be very misleading, as has been recently discussed by Bostrom *et al.* (1991). A simple example from an invertebrate specimen is provided in Table 14.1. It can be seen that the ratio of the concentration of every element in the secretion granules and in the adjacent cytoplasm (paired measurements) is significantly different in each set of data because the secretion granules have almost twice the dry mass of that in the cytoplasm. If one were comparing the relative concentration of Ca, an element sequestered in all secretion granules, the values of secretion granule/cytoplasm would be 7.39 in frozen–hydrated specimen, only 2.92 in frozen–dried specimen, and 10.33 in mmol/l local water. The same holds for other ions which are also important in the osmotic packaging of secretory products in granules and during the exocytotic discharge. (All secretion granules seem to have electrogenic H^+ pumps in their membranes (see below) which provide the electro-motive force for the regulation of other ions.) There is therefore considerable scope for making calculations based on various assumptions (see, for example, Walz & Somlyo, 1984; Baumann *et al.*, 1991). As a rule, in such a situation the measured concentration values in mmol/kg dry mass are exaggerated in the component with low dry mass and suppressed in the component with a high dry mass. Unless it can be firmly established that the local dry mass fractions are the same, data in mmol/kg dry mass alone can not be used for measuring ionic gradients.

The mmol/kg dry mass values obtained by EPXMA of frozen–dried specimens only can also be converted into mmol/kg wet mass or into mmol/l water if a reliable estimate of the local dry mass per unit of the analysed area can be obtained either by using a mass standard in the data-conversion programme or by using a 'peripheral' standard *in situ*, or by some other independent method. Such conversions have recently been applied to the analysis of ∼ 0.2 μm thick frozen–dried cryosections, for example by Zierold & Wessing (1990). In order to facilitate comparisons, I have converted all the EPXMA values published in mmol/kg dry mass only into mmol/kg wet mass by using a nominal dry mass fraction for the component, available from some other source. The sources are indicated in the tables. As is noted below, such conversions have exposed many anomalies that seem to exist in some published data but are not so apparent in the values in mmol/kg dry mass.

C   Biological information

1   Intracellular ions

(i) CYTOPLASM

Before the application of quantitative EPXMA in biology, there was very little reliable information on the intracellular concentrations of Na, K and Cl, which are the major ions of interest in many aspects of cell and tissue physiology (especially in transepithelial transport). Since the first measurements by Gupta & Hall in 1974 in the Malpighian tubules of *Calliphora* (see Gupta, 1991), EPXMA has been used to address many different biological questions in a variety of invertebrates. The information for the cytoplasmic concentrations of Na, K and Cl has been collated in Table 14.2. Many of the data in this table are incidental and do not command detailed comments, because the studies were not aimed at a detailed investigation of intracellular ions under defined physiological conditions. Only Gupta, Hall & co-workers on transporting epithelia in insects, Saubermann & co-workers on the central nervous ganglia of leech, Walz & Somlyo (1984) on the leech photoreceptors and Coles & Rick (1985) on the honey-bee photoreceptors address their studies to this specific question. Some generalities emerge from Table 14.2. The data support the general notion that cells have evolved to maintain within them a high concentration of K and a low concentration of Na and Cl. Gupta *et al.* (1976) first found that in the Malpighian tubules of *Rhodnius prolixus* Na, K and Cl were not uniformly distributed inside the cells but had distinct basal–apical gradients. They also found that nuclei have, contrary to the belief then prevalent, an ionic composition similar to that in the perinuclear cytoplasm (see table in Gupta & Hall, 1982). These observations have now been confirmed in a variety of other tissues both from invertebrates and vertebrates (see Gupta, 1991, for references).

In fresh-water protozoa, the EPXMA results are in line with the previous estimates (mmol/l cell) of Na = 1–20, K = 25–30, Cl = 1–10, intracellular osmolality = about 100 mOsmol (Prusch, 1977). The measurements in Cnidaria are confined to nematocytes and may not apply to other cells in these animals. The work by Zierold *et al.* (1989) and Tardent *et al.* (1990) was aimed at a general survey of the ionic contents of nematocysts (see Table 14.6). In general, the concentration values in their data seem to be much lower than the comparable values from Gupta & co-workers. The results from Zierold's laboratory in the past may have suffered from inadequate corrections for the contributions from extraneous sources to the continuum X-ray quanta (Steinbrecht & Zierold, 1989).

Wherever the estimates have been made by EPXMA, the dry mass fractions of cells (cytoplasm) tend to be about 20–25 % of the wet mass (data not shown) in all the invertebrates and the vertebrates under base-line physiological conditions. Only Saubermann & Scheid (1985) have obtained a dry mass fraction of 45 % for the neuronal cytoplasm of leech, although Walz & Somlyo (1984) found only 25 % dry mass in the photoreceptor cells. Dry mass in the range of 35–45 % is generally found only in highly condensed components such as secretion granules, nematocyst-capsules, sperm-heads etc. Possible inadequacies in Saubermann's method of analysing 'frozen–hydrated' sections have been discussed elsewhere (Gupta & Hall, 1981; Gupta, 1991). Nevertheless Saubermann's work seems to be the only extensive EPXMA study on the ionic regulation in a central nervous ganglion. Walz & Somlyo (1984) and Coles & Rick (1985) have converted their EPXMA values from mmol/kg dry mass into mmol/l water using 75 % as the water fraction and found that resulting values are in good agreement

Table 14.2. *Mean cytoplasmic concentrations of Na, K and Cl of invertebrate cells measured with EPXMA of cryoprepared samples under baseline conditions: mmol/kg wet mass*

Specimen	Na	K	Cl	Reference
PROTOZOA				
Amoeba proteus	nd	62	42	Zierold & Schafer (1988)[a]
Paramecium caudatum				
main cytoplasm	nd	25	6	Schmitz & Zierold
sub-pellicle cytoplasm	11	16	14	(1989)[a]
CNIDARIA:HYDROZOA – Nematocytes				
Hydra attenuata	nd	25	8	Zierold *et al.* (1989)[a]
Aurilia aurita	45	39	114	Tardent *et al.* (1990)[a]
CNIDARIA:SCYPHOZOA – Nematocytes				
Podocryne carnea				
medusa	nd	78	45	Tardent *et al.* (1990)[a]
polyp	11	37	48	Tardent *et al.* (1990)[a]
CNIDARIA:ANTHOZOA – Nematocytes				
Anthopleura elegantissima				
acrorhagi	52	138	152	Lubbock, Gupta & Hall (1981)[b]
acrorhagi	nd	37	70	Tardent *et al.* (1990)[a]
tentacles	9	55	54	Tardent *et al.* (1990)[a]
Actinia equina				
acrorhagi	nd	47	57	Tardent *et al.* (1990)[a]
tentacle	nd	47	51	Tardent *et al.* (1990)[a]
Anemonia viridis				
acrorhagi	11	57	54	Tardent *et al.* (1990)[a]
Calliactis parasitica				
tentacle	7	89	59	Tardent *et al.* (1990)[a]
acontia	32	37	74	Tardent *et al.* (1990)[a]
Rhodactis rhodostoma				
acrorhagi:nematocyte	43	108	106	Lubbock *et al.* (1981)[b]
acrorhagi:columnar cell	32	98	101	Lubbock *et al.* (1981)[b]
ANNELIDA				
Hirudo medicinalis				
photoreceptor cell	33	98	54	Walz & Somlyo (1984)[c]
Macrobdella decora: ventral cord ganglia				
neuronal cell body	56	148	31	Saubermann & Schied (1985)[d]
glial cells[e]	133	29	95	Saubermann & Schied (1985)[d]
neuronal cell body	25	190	50	Saubermann & Stockton (1988)[d]
glial cells	111	51	115	Saubermann & Stockton (1988)[d]

Table 14.2. *Cont.*

Specimen	Na	K	Cl	Reference
INSECTA				
Malpighian tubules				
Calliphora erythrocephala	26	143	—	Gupta & Hall (1974)[f, b]
Rhodnius prolixus : in vivo	12	124	45	Gupta *et al.* (1976)
cricket : *in vivo*	—	101	41	Marshall (1980)[g]
Drosophila hydei : larvae				
posterior tubule	nd	154	19	Zierold & Wessing (1990)
anterior tubule	nd	103	14	Zierold & Wessing (1990)
after pupation	27	55	31	Zierold & Wessing (1990)
Salivary glands				
C. erythrocephala	15	125	23	Gupta *et al.* (1978)[b]
Periplaneta americana				
fluid secreting cells	11	123	23	Gupta & Hall (1983)[b]
Zymogen cells	11	124	39	Gupta & Hall (unpublished)[b]
Intestine				
C. erythrocephala				
rectal papillae	23	85	28	Gupta *et al.* (1980)[b]
Schistocerca gregaria				
midgut caeca	11	133	25	Dow, Gupta & Hall (1981)[b]
Menduca sexta : larval midgut				
in vitro : goblet cells	nd	130	21	Dow *et al.* (1984)[b]
in vitro : columnar cells	nd	129	12	Dow *et al.* (1984)[b]
in vivo : goblet cells	5	91	16	Dow *et al.* (1984)[b]
in vivo : columnar cells	5	127	19	Dow *et al.* (1984)[b]
Neuronal axon				
P. americana : in vivo	—	95	14	Forrest & Marshall (1980)[g]
P. americana : in vitro	—	99	42	Forrest & Marshall (1980)[g]
Photoreceptor (PR)				
Gryllus domesticus				
PR cells	63	187	—	Burovina *et al.* (1978)[h]
pigment cells	60	185	—	Burovina *et al.* (1978)[h]
cone cells	22	175	—	Burovina *et al.* (1978)[h]
Apis mellifera : drone				
PR cytoplasm	12	139	38	Coles & Rick (1985)[i]
PR subrhabdomeric	18	130	43	Coles & Rick (1985)[i]
outer pigment cell	55	113	62	Coles & Rick (1985)[i]
PR cytoplasm	9	164	17	Baumann *et al.* (1991)[c, j]
Olfactory sensilla				
auxl cell cytoplasm	nd	67	4	Steinbrecht & Zierold (1989)[a]
recpt cell cytoplasm	nd	70	5	Steinbrecht & Zierold (1989)[a]

Table 14.2. *Cont.*

Specimen	Na	K	Cl	Reference
Male germ-cells				
Nephrostoma ferruginea				
follicle wall cells	< 10	122	15	Gupta, Forer & Hall (unpublished)[b]
spermatocyte prophase	6	104	25	Gupta, Forer & Hall (unpublished)[b]
meta- and anaphase	8	104	43	Gupta, Forer & Hall (unpublished)[b]
prophase:aster	14	126	31	Forer, Gupta & Hall (1980)[k]
pro-metaphase	12	101	24	Forer, Gupta & Hall (1980)[k]
ECHINODERMATA				
Strogylocentrutus pupuratus				
sperm-head	175	221	167	Cantino, Schachmann & Johnson (1983)[l]

[a] Zierold and his co-workers perform their EPXMA on frozen–dried cryosections in a high resolution STEM. The original data given in mmol/kg dry mass is converted into mmol/kg wet mass by using a nominal value of 0.2 for the dry mass fraction of the cytoplasmic wet mass.

[b] EPXMA carried out directly on about 1 μm thick frozen-hydrated sections in an SEM microanalyser.

[c] Somlyo and his co-workers perform their EPXMA on frozen–dried ultrathin cryosections in a TEM/STEM and obtain concentration values in mmol/kg dry mass. A dry mass fraction of 0.25 suggested by the authors has been used to convert the published values into mmol/kg wet mass.

[d] EPXMA data obtained by an indirect analysis of 0.5 μm thick frozen–hydrated sections in an SEM.

[e] A mean of values from three different types of glia given by the authors.

[f] Quoted from Gupta (1976).

[g] Data based on direct EPXMA of frozen–hydrated bulk samples in an SEM.

[h] Data were obtained from EPXMA of sections from cryofixed, frozen–dried material embedded in epoxy resin.

[i] EPXMA data were obtained from 1 μm thick frozen–dried sections in an SEM: mmol/kg wet mass values were calculated by using characteristic X-ray quanta only (see text).

[j] Published values in mmol/kg dry mass are converted into mmol/kg wet mass by using 0.2 as the dry mass fraction.

[k] EPXMA data were obtained from frozen–dried cryosections only. Light microscopy was used for estimating the % dry mass and this value was used for converting the data from mmol/kg dry mass into mmol/kg wet mass.

[l] EPXMA data were obtained from thin frozen–dried cryosections in an TEM/STEM. The published values in mmol/kg dry mass have been converted into mmol/kg wet mass by using a dry mass fraction of 0.45 as estimated for the sperm-heads by Steinbach & Dunham (1961).

nd = not detected.

Fig. 14.1.   For legend see facing page.

with the previous estimates from other physiological methods. In this conversion the authors have assumed that all the measured elements in the cells are in free solution. Using a similar conversion, Saubermann & co-workers found a great excess of K in the neuronal cytoplasm and have suggested that both the glia and neurons have a large store of excess K which is used for maintaining normal levels of intracellular concentrations under adverse ionic stresses.

(ii) CELL DIVISION

Invertebrate animals, especially insects, have been the favourite material for the study of cytogenetics, mitosis and meiosis over the past century. Very little EPXMA work has been done in this field on the cryoprepared samples. Forer *et al.* (1980) studied the spermatocytes from the cranefly testes in order to examine the role of Ca^{2+} in meiotic spindle but the data were very limited. We repeated this work in 1982 and some of the results (unpublished) are summarised in Fig. 14.1 and Tables 14.3 and 14.4. Data in Table 14.3 are from the direct analysis of frozen–hydrated sections (Fig. 14.1a,b). The ionic concentrations for the follicle cells and primary spermatocytes are in line with the other cells and tissues from invertebrates (Table 14.2). These data also illustrate the validity of analysing 1 µm thick frozen–hydrated sections. The values for the spermatocyte-cytoplasm obtained by the continuum method are compared with the 'characteristic-count-only' method (see Hall & Gupta, 1982; 1986). Excellent agreement shows that the sections must have remained fully hydrated during the analysis. However, poor image details in frozen–hydrated sections precludes analysis of structures in dividing cells. These were analysed in frozen–dried sections.

Table 14.4 summarises the elemental composition of the major components in cells from pro-metaphase (Fig. 14.1g) to late anaphase (Fig. 14.1h). The most notable features are as follows.

(*a*) The meiotic chromosomes have the highest K and P, in a ratio of 1:1. This correlation has previously been noted in the mitotic chromosomes from plants (Cameron, Hunter & Smith, 1984) and in condensed chromatin (Zglinicki, Ziervogel & Bimmler, 1989).

Fig. 14.1.   Dark-field/bright-field STEM images of 1 µm thick cryosections from cranefly testes. (a) One whole cryosection of a testis in frozen–hydrated state ($\times 300$); (b) a primary spermatocyte in a frozen–hydrated cryosection as in (a) ($\times 4500$); (c), two cryosections from a 'ribbon' in a frozen–dried state ($\times 110$); (d, f) primary spermatocytes in early prophase I (d, $\times 2000$; f, $\times 3300$); (e) a follicular epithelial cell ($\times 7000$); (g) a spermatocyte in pro-metaphase from a frozen–dried cryosection; dark rod-like structures surrounding the spindle are mitochondria ($\times 4500$); (h) as in (g), but in late anaphase ($\times 4500$). Labelling: cy = cytoplasm; ch = chromosome; n = nucleus; sp = spindle-pole; marked rectangular field in (h) encloses a spindle-fibre. (Gupta, Forer & Hall, unpublished).

Table 14.3. *Mean concentrations of main elements measured by EPXMA in 1 μm thick frozen–hydrated cryosections of the cranefly testis*

Component	n	% dry mass	Na[a]	K	Mg	Ca[a]	Cl	P	S
Dextran-Ringer[b]									
known	—	19.5	112	4	2.4	2.4	113	8	0
EPXMA	47	18.4 (1)	102 (7)	9 (4)	nd	2 (0.6)	112 (2)	8 (0.5)	nd
Testicular 'fluid'[c]	10	9 (0.4)	91 (13)	22 (7)	16 (4)	12 (6)	101 (11)	12 (1)	16 (1)
Follicular cells	6	17 (2)	<10	122 (7)	9 (7)	2.2 (1)	15 (3)	177 (18)	45 (5)
Primary spermatocyte[a]									
nucleus	7	17 (2)	nd	100 (8)	8 (3)	1.6 (4)	22 (5)	73 (9)	25 (5)
cytoplasm[e]	7	21 (2)	6 (11)	104 (9)	10 (6)	2.4 (1.4)	25 (5)	95 (10)	28 (4)
cytoplasm[f]	7	—	8	111	9	4	25	94	28

Values are given as mmol/kg wet mass (SE).
Data from Gupta, Forer & Hall (unpublished).
[a] Measured with X-ray wavelength dispersive spectrometers (WDS).
[b] Used as a peripheral standard cut with each specimen.
[c] Measured fields must have included cytoplasmic blebs etc., not visible in the images.
[d] Cells were either in interphase or early prophase I. Nuclei and cytoplasm were measured in each cell.
[e] Concentration values obtained by continuum method used in LINK QUANTEM/FLS.
[f] Concentration values obtained by characteristic counts only ('CA' method) but using the known mmol/kg wet mass values for Dextran-Ringer.

Table 14.4. *Mean concentrations of main elements measured by EPXMA in*
1 μm thick frozen–dried cryosections of the cranefly testis

Components	N	% dry mass	Na[a]	K	Mg[b]	Ca[a]	Cl	P	S
Spermatocytes in meta- and anaphase									
Cytoplasm[c]	14	20	8	104	6.4	1.6	43	81	16
Mitochondria[c]	14	21	10	106	6.9	0.6	34	93	15
Chromosomes	29	23	9	166	7.6	1.8	49	171	27
Spindle fibres	47	20	10	156	10.0	2.0	45	118	22
Interzones[d]	4	20	26[e]	119	5.0	5.4	27[e]	76	23
Spindle pole	29	20	16[e]	107	6.7	4.0	33[e]	78	18

Original values in mmol/kg dry mass have been converted into mmol/kg wet mass
using the dry mass fractions given here. SE values have been omitted. Various
components are labelled in Fig. 14.1. Data from Gupta, Forer & Hall
(unpublished).
[a] Measured with X-ray wavelength dispersive (WDS) spectrometers.
[b] EDS data had very high standard deviations.
[c] Cytoplasm and filamentous mitochondria were measured in pair usually at the
extreme periphery of the meiotic spindle (see Fig. 14.1).
[d] This field represents the zone between the spindle fibres and is likely to contain
elements of smooth ER.
[e] The original data in mmol/kg dry wt had very high standard deviations. Reason
is not clear.

(b) 'Spindle-fibres' (about 1 μm thick) have high K and Mg but much less P.
(c) 'Spindle-pole' and the 'interzone' between the spindle-fibres have the
highest levels of Ca. This is likely to be in the membrane system which
remains associated with the spindle (Hyams & Brinkley, 1989) and is said
to be involved in the regulation of free cytosolic Ca^{2+} as discussed
previously by Forer *et al.* (1980).

(iii) SPERM-NUCLEI

Cantino *et al.* (1983) used EPXMA both on frozen–dried spermatozoa and in
frozen–dried cryosections of sperm from the sea-urchin *Strongylocentrutus
pupurata*. In cryosections the published concentrations from the sperm-nuclei
converted into mmol/kg wet mass (dry mass 45 %) are: Na = 175, K = 220,
Cl = 167, Mg = 36, Ca = 3, P = 720, S = 5. The values for Na and K can be
compared with Na = 13, K = 242 in *Arbacia punctata* measured by conventional methods (Steinbach & Dunham, 1961). Very high Na and Cl in
Cantino's results are from sea-water due to incomplete resolution of sperm-
heads even in thin sections. K:P ratio in this maximally packed chromatin is
only 0.3. Presumably, highly basic protamines replace K.

(iv) INTRACELLULAR Ca STORES

In the second-messenger hypothesis for signal-response coupling, Ca^{2+} ions have a key role. The cytosolic levels of Ca^{2+} are maintained in nanomolar range. Therefore virtually all the Ca measured by EPXMA in cells is sequestered mostly in some membrane-bound organelle such as mitochondria, endoplasmic-reticulum (ER), nuclear envelope etc. A major unresolved question is the identity of the Ca store which is mobilised by the universal second-messenger IP3 first discovered in the salivary glands of *Calliphora* in 1980 by Berridge (see Berridge & Irvine, 1989). EPXMA is uniquely suited for this purpose.

It was first noted in *Calliphora* salivary glands (Gupta *et al.*, 1978; Gupta & Hall, 1978) that the highest concentration of Ca (5–10 mmol/kg wet mass) is found not in mitochondria but in and around the nuclear membrane (ER). Current work with other methods for monitoring cytosolic Ca^{2+} in other cells confirms this location (see Gupta, 1991). Little further work has been done on this question in any other invertebrate tissue except in photoreceptors. Baumann *et al.* (1991) exploited for EPXMA the large ER-sacs associated with the photoreceptors in the honey-bee *Apis mellifera*. They noted that in the control samples ER had (mmol/kg dry mass) Na = 106, K = 1335, Cl = 131, Mg = 66, Ca = 48, P = 688, S = 1225. Assuming a 10% dry mass, these values convert into mmol/kg wet mass: Na = 11, K = 135, Cl = 13, Mg = 6.6, Ca = 4.8, P = 69, S = 122. After photo-stimulation the corresponding values were (mmol/kg wet mass): Na: 18, K = 112, Cl = 14, Mg = 8.3, Ca = 2.2, P = 61, S = 105. These authors propose that on stimulation 50% of Ca in ER is released into cytosol and is replaced with an equal amount of Mg for maintaining electroneutrality. It should be noted that the converted elemental concentrations in control ER here are very similar to the interzone in Table 14.3 which is also presumed to be mainly ER. A rise in intracellular Na and a drop in K on cell-stimulation is consistent with the data on other cell-types (see table 2 in Gupta, 1984).

In numerous invertebrates Ca, Zn and other elements are also sequestered in concretion bodies, vacuoles, granules etc. The role of these stores in the regulation of intracellular ions is unclear (see Gupta, 1991; Morgan & Morgan, 1989; Zierold & Wessing, 1990, addendum).

2   Nematocysts and trichocysts

Nematocysts in Cnidaria are structurally the most complex secretory products of a single cell. They come in a bewildering variety and any one organism usually has several types, each deployed for a different purpose. These 'poison-darts' are fired at different speeds in response to specific stimuli. During the past 200 years numerous hypotheses have been proposed for the mechanism of their assembly and exocytotic discharge. Currently the most

Table 14.5. *Mean concentrations of elements (mmol/kg wet mass) and % dry mass in* **R.** *rhodostoma nematocysts during formation, after maturation and in fully discharged states*

Component	Stage	Na	K	Cl	Ca	Mg	Cu[a]	Zn[a]	P	S	% dry mass
Cytoplasm[b]	early nematoblast	43	108	106	9	32	nd	nd	38	135	18
	mature nematocyte	16	100	99	9	32	nd	nd	30	121	16
	after stimulation	117	116	167	11	33	nd	nd	46	98	19
Capsule fluid	early nematoblast	29	137	90	17	nd	nd	nd	17	156	18
	mature, undischarged	32	11	17	606	33	nd	nd	nd	54	32
	discharging	194	7	105	189	18	nd	< 20	nd	18	22
	fully discharged	350	9	401	35	45	nd	nd	nd	37	11
Capsule wall	mature, undischarged[c]	93	62	89	21	28	nd	nd	5	181	52
	discharging nematocyst	188	59	125	18	31	nd	nd	6	136	42
	fully discharged	140	61	181	15	29	nd	< 10	5	162	50
Filament	tube in nematoblast	39	105	147	8	7	nd	nd	49	118	19
	mature uneverted[c]	106	22	15	42	21	~ 30	~ 80	3	142	42
	fully everted	201	18	130	92	17	~ 30	~ 10	2	151	31
Vacuole[a]	discharging nematocyst	234	11	389	531	24	nd	< 20	2	18	19
21/25 sea water + dextran		328	7	382	7	38	nil	nil	0	20	19

SE values given in the original data (Lubbock *et al.*, 1981; Gupta & Hall, 1984) have been omitted here.

[a] These values are approximate: exact quantification was not done.

[b] Cytoplasm contained numerous dense granules loaded with copper and zinc. These granules were excluded from the measured fields.

[c] Data are based on measurements which did not show high Ca given by Lubbock *et al.* (1981): the latter is considered as due to capsule fluid trapped in pockets (inner folds).

[d] Vacuole refers to the space between the nematocyst capsule and the nematocyte membrane.

Table 14.6. *Mean concentrations of major cationic elements in the capsular fluid (matrix) of Cnidarian nematocysts determined by EPXMA of cryosections: conc. = mmol/kg wet mass*

Species	Element	Conc.	Reference
Anthozoa			
Anthopleura elegantissima			
acrorhagi	Ca	542	Lubbock *et al.* (1981)
acrorhagi	Ca	352	Tardent *et al.* (1990)[a]
tentacle	Mg	537	Tardent *et al.* (1990)[b]
Actinia equina			
acrorhagi	Ca	380	Tardent *et al.* (1990)
tentacle	Mg	590	Tardent *et al.* (1990)[b]
Anemonia viridis			
acrorhagi	Mg	992	Tardent *et al.* (1990)
tentacle	Mg	483	Tardent *et al.* (1990)[b]
Calliactis parasitica			
acontia	K	734	Tardent *et al.* (1990)
tentacle	K	735	Tardent *et al.* (1990)
Rhodactis rhodostoma			
acrorhagi	Ca	606	Lubbock *et al.* (1981)
Hydrozoa			
Hydra attenuata			
undischarged	K	881	Zierold *et al.* (1989)[a]
discharged	K	385	Zierold *et al.* (1989)[c]
Podocryne carnea			
medusa	K	505	Tardent *et al.* (1990)
Scyphozoa			
Aurelia aurita			
polyp	K	653	Tardent *et al.* (1990)

[a] These workers analysed frozen–dried cryosections only and the original data are given in mmol/kg dry mass. In this table the original values have been converted into mmol/kg wet mass by using a nominal dry mass fraction of 0.4 for the capsular fluid as measured by Tardent *et al.* (1990).
[b] Very high concentrations of sulphur were also present.
[c] A dry mass fraction of 0.15 for the residual mass in the discharged capsule (Lubbock *et al.*, 1991) was used for this conversion.

popular hypothesis suggests that the contents of the capsule in undischarged nematocysts have a very high osmotic potential which is used for discharge but the ionic basis for such a mechanism has remained controversial (see Gupta & Hall, 1984; Weber, 1989; Tardent *et al.*, 1990). Measurements of the elemental compositions in different components of the nematocysts *in situ* in a large number of species have illustrated the potential of EPXMA.

Lubbock *et al.* (1981) and Gupta & Hall (1984) used EPXMA on frozen–hydrated cryosections of acrorhagal tissue in *Anthopleura elegan-*

Table 14.7. *Average concentrations of main elements in the basement membranes* (= *basal laminae*) *of some tubular epithelia from insects measured by EPXMA in 1 μm thick frozen–hydrated sections*

Element	basal cytoplasm	basement membrane	serosal fluid[a]
		mmol/kg wet mass	
Na	10–20	80–120	~ 120
K	100–140	20–120	5–20
Cl	20–40	~ 100	140
Ca	~ 1.0	up to 10	2.0
Mg	~ 5.0	up to 10	2.0
P	~ 200	~ 50	< 10
S	~ 50	~ 50	~ 0
% dry mass	20–30 %	up to 20 %	< 2 %

[a] Serosal fluid would be haemolymph *in vivo* but insect-Ringer in preparations *in vitro*.
Data from Gupta *et al.* (1976, 1977, 1978); Dow *et al.* (1981); Gupta (1989); and Gupta *et al.* (unpublished).

tissima and *Rhodactis rhodostoma*. Their complete data are summarised in Table 14.5. Their main conclusions are as follows.

(a) During the formation within a secretory vacuole (ER-Golgi product) the capsule acquires some highly polyanionic polymer (dry mass = 35–45 %) which sequesters some 500 mmol/kg wet mass of Ca. The complex and highly folded 'filament' (= tubule) has little Ca but has ~ 30 mmol of Cu and ~ 80 mmol of Zn.

(b) Upon exocytotic discharge the capsular contents lose dry mass (residual mass 10–15 %) and Ca but acquire Na and Cl presumably from sea water. During eversion the filament loses Zn but not Cu.

These authors propose that the release of Ca is somehow concerned with the osmotic phenomenon. Zn stabilises the folded structure of the filament, perhaps by 'zipping up' the barbs. Its replacement (either by Na^+ or H^+) releases the stored mechanical energy and causes the filament to unravel.

In an extensive survey, Tardent *et al.* (1990) confirmed the presence of high Ca in the capsule-fluid of the acrorhagial nematocysts in *A. elegantissima* but found Mg instead in the nematocysts from the oral tentacles. In other species either Ca or Mg or K were found but *never* Na. Inordinately high concentrations of K were also found in the nematocysts of a fresh-water *Hydra* which discharge at an ultrahigh speed of a millisecond (Zierold *et al.*, 1989). By comparing the concentrations of the capsule-fluid in mmol/kg dry mass they did not find any change in K between the resting state and after stimulated discharge (also see Tardent *et al.*, 1990). However, it is clear from

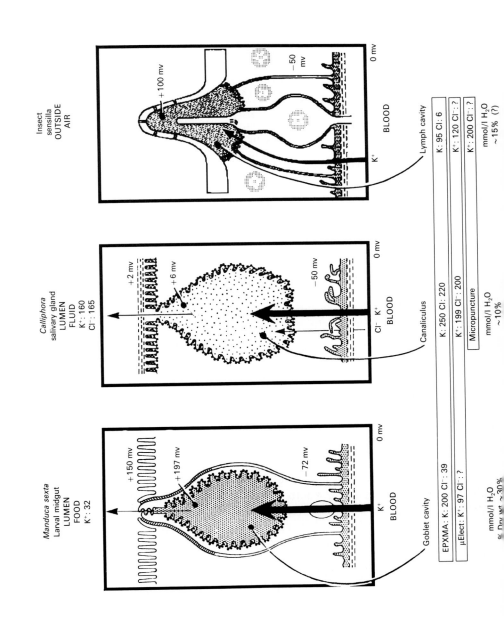

Manduca sexta
Larval midgut
LUMEN
FOOD
K⁺: 32

+150 mv
+197 mv
−72 mv
0 mv
K⁺
BLOOD
Goblet cavity

EPXMA: K: 200 Cl: 39
µElect: K⁺: 97 Cl⁻: ?
mmol/l H₂O
% Dry wt ~30%

Calliphora
salivary gland
LUMEN
FLUID
K⁺: 160
Cl⁻: 165

+2 mv
+6 mv
−50 mv
0 mv
Cl⁻   K⁺
BLOOD
Canaliculus

K: 250 Cl: 220
K⁺: 199 Cl⁻: 200
Micropuncture
mmol/l H₂O
~10%

Insect
sensilla
OUTSIDE
AIR

+100 mv
−50 mv
0 mv
K⁺
BLOOD
Lymph cavity

K: 95 Cl: 6
K⁺: 120 Cl⁻: ?
K⁺: 200 Cl⁻: ?
mmol/l H₂O
~15% (?)

Fig. 14.2. A schematic representation of EPXMA data compared with information obtained using other techniques, as reported in the literature. In the goblet cells of *Manduca sexta* (Dow *et al.*, 1984) and in insect-sensilla (Steinbrecht & Zierold, 1989), as well as other similarly constructed sensory organs, an apical electrogenic pump secretes K^+ into an electrically insulated cavity. This contains a high concentration of polyanionic glycoproteins acting as 'fixed' negative charges for maintaining electroneutrality. There is little short-circuiting of the generated transmembrane potential. The efficiency of the matrix in reducing the activity of K is seen in the difference between the EPXMA values and those from micro-electrodes (μElect.). As noted in the text, the values in Steinbrecht & Zierold's EPXMA data are probably 50% of the real values because of technical problems.

In a very similar geometrical arrangement in the salivary glands of *Calliphora* (Gupta *et al.*, 1978) the apical canaliculi have much less matrix and additionally a regulated Cl^- uniport in the apical membrane. The transmembrane potential generated by the apical pump is almost completely short-circuited but the cavity now forms the site of solute–solvent coupling for secreting an isotonic KCl solution at a high rate. (Further discussions in Dow *et al.*, 1984; Gupta *et al.*, 1985; Gupta, 1989.)

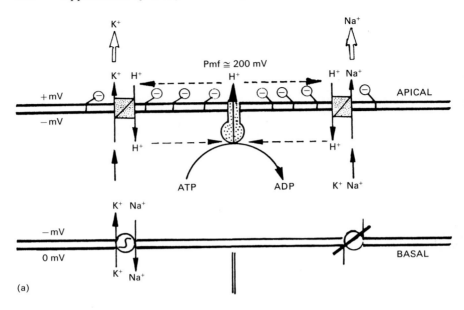

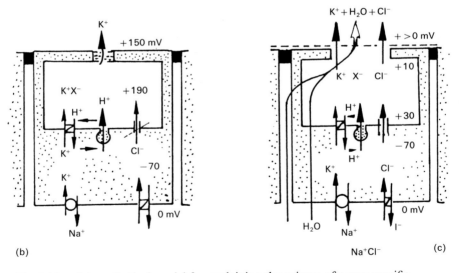

Fig. 14.3. A hypothetical model for explaining the enigma of a non-specific, apically located 'cation'-pump which dominates many insect-epithelia and seems to pump either K^+ or Na^+ out of cell into the lumen *in vivo* and *in vitro* depending on the major cation present in the cell (see Gupta *et al.*, 1976, 1978; Gupta & Hall, 1979). It now seems that the electrogenic pump might be a vacuolar-type $H–^+$-ATPase (Wieczorek *et al.*, 1989). Protons are essentially trapped by the fixed negative charges

the published images that the discharged capsules have lost most of the mass, as was measured by Lubbock *et al.* (1981). Allowing for this loss of mass, the values converted into mmol/kg wet mass (Table 14.6) show that the discharged capsules have lost some 500 mmol of K. Since no other ion was detected, it may have been replaced by H^+ for electro-neutrality!

Trichocysts are analogous structures found in *Paramecium* and in many other protozoa. In a technically elegant study, Schmitz & Zierold (1989) surprisingly found some 177 mmol/kg dry mass of Na ($= 80$ mmol/kg wet mass assuming 45% dry mass) in the main body of the trichocyst. No Na was found in the trichocyst-ghosts (stimulated discharge). These authors discuss EPXMA findings for evaluating previously proposed mechanisms for trichocyst discharge.

3   Pericellular compartments

In the study of animal epithelia transporting ions and water, a major achievement of EPXMA has been to measure the ionic composition of the fluid interspaces involved in solute–solvent coupling (Gupta, 1984) and additionally of the acidic glycoprotein/glycosaminoglycan structures such as basement membranes, mucus layers and matrices in the invaginated cavities etc. The role of these matrices has until recently been ignored in transepithelial physiology. It has now emerged that these matrices preferentially sequester ions such as K^+, Ca^{2+} and H^+, exclude anions such as Cl^-, and therefore maintain a critical microenvironment around single cells and around cells in tissues by acting as Donnan-systems (Gupta, 1989, 1991). A similar Donnan-function for the acidic extracellular matrices has been proposed by Walz & Somlyo (1984) in leech photoreceptors and by Godde & Krefting (1989) in the labellar taste-hairs of *Protophormia terraenovae*.

(i) BASEMENT MEMBRANES (BM)

Gupta *et al.* (1976, 1977, 1978) noted that in the tubular epithelia of insects which freely bathe in the haemolymph, the relatively tough and complex BM (up to 1 μm), have an ionic composition distinct from the other adjacent

---

in the glycocalyx in the microvilli of the brush-borders (e.g. Malpighian tubules) and recycled via H^+/K^+ and/or H^+/Na^+ antiports. In epithelia without an apical, invaginated cavity the net transport of a isotonic salt-solution will result if a parallel Cl^--uniport is present. In principle the same pump could appear as an active transport of Cl^-. (a) The general model for the insect-type 'cation'-pump; (b) epithelial cells with an electrically insulated cavity as in the goblet cells and in insect-sensilla in Fig. 14.2; (c) postulated ion-pump and ion-channel arrangements for epithelial cells secreting salt solutions as in the salivary glands of *Calliphora* but have apical, invaginated cavities (canaliculi) as shown in Fig. 14.2.

compartments. The collective EPXMA data from BM in insect tissues are shown in Table 14.7. As a rule there seems to be 4–6 times more K and Ca but less Na and Cl in BM than in the bathing medium. The exact composition of BM can differ from site to site along the same epithelium even in the same preparation. Fessler & Fessler (1989) have recently reviewed the extracellular matrices (essentially BM) in *Drosophila* and have discussed their key role in cell–cell interactions during differentiation and morphogenesis. They report a new acidic glycoprotein called 'glutactin' in *Drosophila* BM. Of the residues in glutactins, 44 % are glutamic acid, glutamate and 4-*O*-sulphated tyrosines. Glutactins preferentially bind Ca^{2+} in the presence of excess Mg^{2+}, thus providing a biochemical basis for EPXMA observations. Laminin, the major glycoprotein of BM in all animals, self-aggregates in the presence of Ca^{2+}, also suggesting Ca^{2+}-binding function. Gupta (1989) has suggested that these pericellular acidic glycosaminoglycans/glycoproteins have a key role in ionic homeostasis, especially in conserving and recycling K^+ leaking out of the cells. In amoeba, Hedril (1971) reported much earlier that the concentration of monovalent cations in glycocalyx is at least 2.1 times more than in the external medium. Prusch (1980) proposed that during the induction of endocytosis in amoeba, Ca^{2+} bound to the glycocalyx is displaced and enters the cell. Even in onion-roots, Cameron *et al.* (1984) found that the lateral cell-wall has nearly twice the amount of K and 10 to 20-fold higher Ca than in cells.

Gupta (1989) has also discussed the role of acidic polysaccharide matrices in large apical invaginations ('cavities', 'canaliculi', 'vacuoles') in several insect epithelia. The principle is illustrated in Fig. 14.2 and has emerged from EPXMA measurements of the ionic composition in these apical invaginations. In all the examples, the net ion-transport is dominated by an electrogenic, apically located cation-pump which normally pumps K^+ out of the cell into a highly acidic polymeric matrix in the cavity. Then, depending on the concentration of the polymer in the cavity and on the antiport, symport and uniport type ionic channel-proteins in the apical membranes, the system can sustain either a net transepithelial potential of > 100 mV (lumen + ve) or secrete at a high rate an isotonic salt solution (usually KCl). Further discussion can be found in Dow *et al.* (1984), Gupta *et al.* (1985) and Gupta (1989).

The general structure of these cation-pumps was discovered by Gupta & Berridge (1966) in the rectal papillae of *Calliphora* in the form of ~ 10 nm stalked particles very similar to the 'lollipops' on the inner membrane of mitochondria. These authors suggested that these stalked particles might be proton-pumps. It has now emerged that similar particles on the cytoplasmic face of the apical cell-membrane lining the 'goblet' cavities in the larval midgut of *Manduca sexta* (Fig. 14.2) are indeed V-type H^+-ATPases (Wieczorek *et al.*, 1989). A unifying model based on H^+-ATPases for the transepithelial transport of ions or fluid in insects is presented in Fig. 14.3. V-type H^+-ATPases have long been recognised as the major pumps in bacteria, fungi and plants. It now seems that they might also be the dominant electrogenic ion-pumps in animals (including vertebrates (Forgac, 1989)).

References

Baumann, O., Walz, B., Somlyo, A.V. & Somlyo, A.V. (1991). Electron microanalysis of calcium release and magnesium uptake by endoplasmic reticulum in bee photoreceptors. *Proceedings of the National Academy of Sciences U.S.A.*, **88**, 741–4.

Berridge, M.J. & Irvine, R.F. (1989). Inositol phosphates and cell signalling. *Nature*, **341**, 197–205.

Bostrom, T. E., Field, M. J., Gyory, A. Z., Dyne, M. & Cockayne, D.J.H. (1991). Electron probe X-ray microanalysis of intracellular elemental concentrations in cryosections in the presence of changes in cell volume. *Journal of Microscopy*, **162**, 319–33.

Burovina, I.V., Gribakin, F.G., Petrosyan, A.M., Pivovarova, N.B. & Pogorelova, A.G. (1978). Ultrastructural localization of potassium and calcium in an insect omatidium as demonstrated by X-ray microanalysis. *Journal of Comparative Physiology*, A**127**, 245–53.

Cameron, I.L., Hunter, K.E. & Smith, N.K.R. (1984). The subcellular concentration of ions and elements in thin cryosections of onion root meristam cells. *Journal of Cell Biology*, **72**, 295–306.

Cantino, M.E., Schackmann, R.W. & Johnson, D.E. (1983). Changes in subcellular elemental distributions accompanying the acrosome reaction in sea urchin sperm. *Journal of Experimental Zoology*, **226**, 255–68.

Coles, A.J. & Rick, R. (1985). An electron microprobe analysis of photoreceptors and outer pigment cells in the retina of the honeybee drone. *Journal of Comparative Physiology*, A**156**, 213–22.

Dow, J.A.T., Gupta, B.L. & Hall, T.A. (1981). Microprobe measurements of Na, K, Cl, P, S, Ca, Mg and water in frozen–hydrated sections of anterior caeca of the locust *Schistocerca gragaria*. *Journal of Insect Physiology*, **27**, 629–39.

Dow, J.A.T., Gupta, B.L., Hall, T. & Harvey, W.R. (1984). X-ray microanalysis of elements in frozen–hydrated sections of an electrogenic K^+ transport system: the posterior midgut of tobacco hornworm (*Menduca sexta*) *in vivo* and *in vitro*. *Journal of Membrane Biology*, **77**, 223–41.

Fessler, J.H. & Fessler, L.I. (1989). *Drosophila* extracellular matrix. *Annual Reviews of Cell Biology*, **5**, 309–39.

Forer, A., Gupta, B.L. & Hall, T.A. (1980). Electron probe X-ray microanalysis of calcium and other elements in meiotic spindles in frozen sections of spermatocytes from cranefly testes. *Experimental Cell Research*, **126**, 217–26.

Forgac, M. (1989). Structure and functions of the vacuolar class of ATP-driven proton pumps. *Physiological Reviews*, **69**, 765–96.

Forrest, O.G. & Marshall, A.T. (1980). Electron-probe X-ray microanalysis of fully frozen–hydrated bulk-specimens of axons from connective. *Micron*, **11**, 399–400.

Godde, J. & Krefting, E.-R. (1989). Ions in the receptor lymph of the labellar taste hairs of the fly *Protophormia terraenovae*. *Journal of Insect Physiology*, **35**, 107–18.

Gupta, B.L. (1976). Water movement in cells and tissues. In *Perspectives in experimental biology*, Vol. 1, ed. P. Spencer-Davies, pp. 25–42. Oxford: Pergamon Press.

(1984). Models of salt and water flow across epithelia: an evaluation by the electron probe X-ray microanalysis. In *Osmoregulation in esturine and marine animals*, ed. A. Pequeux, R. Gilles & L. Bolis, pp. 191–211. Berlin: Springer-Verlag.

(1989). The relationship of mucoid substances and ions and water transport with new data on intestinal goblet cells and a model for gastric secretion. *Symposia of the Society for Experimental Biology*, **43**, 81–110.

(1991). Ted Hall and the science of biological microprobe X-ray analysis: a historical perspective of methodology and biological dividends. *Scanning Microscopy*, **5**, 379–426.

Gupta, B.L. & Berridge, M.J. (1966). A coat of repeating subunits on the cytoplasmic surface of the plasma membrane in the rectal papillae of blowfly, *Calliphora erythrocephala* (MEIG) studied *in situ* by electron microscopy. *The Journal of Cell Biology*, **29**, 376–82.

Gupta, B.L., Berridge, M.J., Hall, T.A. & Moreton, R.B. (1978). Electron microprobe and ion-selective microelectrode studies of fluid secretion in salivary glands of *Calliphora. Journal of Experimental Biology*, **72**, 261–84.

Gupta, B.L., Dow, J.A.T., Hall, T.A. & Harvey, W.R. (1985). Electron probe X-ray microanalysis of the effect of *Bacillus thurengiensis Var kurstaki* crystal protein insecticide on the ions in an electrogenic K^+-transporting epithelium of the larval midgut in the lepidopteran, *Meduca sexta, in vitro. Journal of Cell Science*, **74**, 137–52.

Gupta, B.L. & Hall, T.A. (1978). Electron probe microanalysis of calcium. *Annals of New York Academy of Sciences*, **307**, 28–51.

(1979). Quantitative electron probe X-ray microanalysis of electrolyte elements within epithelial tissue compartments. *Federation Proceedings, Federation of American Societies for Experimental Biology*, **38**, 144–53.

(1981). The X-ray microanalysis of frozen–hydrated sections in scanning electron microscopy: an evaluation. *Tissue & Cell*, **13**, 623–43.

(1982). Electron probe X-ray microanalysis. In *Techniques in cellular physiology*, Vol. 128, ed. P.F. Baker, pp. 1–52. County Clare: Elsevier.

(1983). Ionic distribution in dopamine stimulated NaCl-secreting cockroach salivary gland. *American Journal of Physiology*, **244**, R176–86.

(1984). Role of high concentration of Ca, Cu, and Zn in the maturation and discharge in situ of sea anemone nematocysts as shown by X-ray microanalysis of cryosections. In *Toxins, drugs and pollutants in marine animals*, ed. L. Bolis, J. Zadunaisky & R. Giles, pp. 77–95. Berlin: Springer-Verlag.

Gupta, B.L., Hall, T.A., Maddrell, S.H.P. & Moreton, R.B. (1976). Distribution of ions in fluid transporting epithelium examined by electron probe X-ray microanalysis. *Nature*, **264**, 284–7.

Gupta, B.L., Hall, T.A. & Moreton, R.B. (1977). Electron probe X-ray microanalysis. In *Transport of ions and water in animals*, ed. B.L. Gupta, R.B. Moreton, J.L. Oschman & B.J. Wall, pp. 83–143. London: Academic Press.

Gupta, B.L., Wall, B.J., Oschman, J.L. & Hall, T.A. (1980). Direct Microprobe evidence of local concentration gradients and recycling of electrolytes during fluid absorption in the rectal papillae of *Calliphora. Journal of Experimental Biology*, **88**, 21–47.

Hall, T.A. & Gupta, B.L. (1982). Quantitation for the X-ray microanalysis of cryosections. *Journal of Microscopy*, **126**, 333–45.

Hall, T.A. & Gupta, B.L. (1983). The localization and assay of chemical elements by microprobe methods. *Quarterly Reviews of Biophysics*, **16**, 279–339.

(1986). EDS application in biology. In *Principles of analytical electron microscopy*, ed. D.C. Joy, A.D. Romig & J.I. Goldstein, pp. 219–48. New York: Plenum Publishing Corporation.

Hedril, K.B. (1971). Ion exchange properties of glycocalyx of amoeba *Chaos chaos* and its relation to pinocytosis. *Comptes rendus des travaux du Laboratoire Carlsberg*, **38**, 187–1211.

Hyams, J.S. & Brinkley, B.R. (1989). *Mitosis: molecules and mechanisms*. New York, London: Academic Press.

Le Furgey, A. (1988). Frontiers in electronprobe microanalysis: application to cell physiology. *Ultramicroscopy*, **24**, 185–220.

Lubbock, R., Gupta, B.L. & Hall, T.A. (1981). Novel role of calcium in exocytosis: mechanism of nematocyst discharge as shown by X-ray microanalysis. *Proceedings of National Academy of Sciences U.S.A.*, **78**, 3624–8.

Marshall, A.T. (1980). Quantitative X-ray microanalysis of frozen–hydrated bulk-biological specimens. *Scanning Electron Microscopy*, **1980/II**, 335–48.

Morgan, A.J. (1984). The localization of heavy metals in tissues of terrestrial invertebrates by electron microprobe X-ray analysis. *Scanning Electron Microscopy*, **1984/IV**, 1847–65.

Morgan, J.E. & Morgan, A.J. (1989). Zinc sequestration by earthworm (Annelida: Oligochaeta) chloragocytes. An *in vivo* investigation using fully quantitative electron probe X-ray micro-analysis. *Histochemistry*, **90**, 405–11.

Morgan, A.J. & Winters, C. (1987). The contribution of electron probe X-ray microanalysis (EPXMA) to pollution studies. *Scanning Microscopy*, **1**, 133–57.

Prusch, R.D. (1977). Protozoan osmotic and ionic regulation. In *Transport of ions and water in animals*, ed. B.L. Gupta, R.B. Moreton, J.L. Oschman & B.J. Wall, pp. 363–77. London: Academic Press.

Prusch, R.D. (1980). Endocytic sucrose uptake in *Amoeba proteus* induced with calcium ionophore A 23187. *Science*, **209**, 691–2.

Saubermann, A.J. & Scheid, V.L. (1985). Elemental composition and water contents of neuron and glial cells in the central nervous system of the North American medicinal leech (*Macrobdella decora*). *Journal of Neurochemistry*, **44**, 825–34.

Saubermann, A.J. & Stockton, J.D. (1988). Effects of increased extracellular K on the elemental composition and water content of neuron and glial cells in Leech CNS. *Journal of Neurochemistry*, **51**, 1797–807.

Schmitz, M. & Zierold, K. (1989). X-ray microanalysis of ion changes during fast processes in cells as exemplified by trichocyst exocytosis of *Paramecium caudatum*. In *Electron microscopy of subcellular dynamics*, ed. H. Plattner, pp. 325–39. Boca Raton, Fl: CRC Press.

Steinbach, H.B. & Dunham, P.B. (1961). Ionic gradients in some invertebrate spermatozoa. *The Biological Bulletin of the Marine Biological Laboratory, Woodshole*, **120**, 411–19.

Steinbrecht, R.A. & Zierold, K. (1989). Electron probe X-ray microanalysis in silk-moth antenna – problem of quantification in ultrathin sections. In *Electron probe microanalysis: application in biology and medicine*, ed. K. Zierold & H. Hagler, pp. 87–98. Berlin: Springer-Verlag.

Tardent, P., Zierold, K., Klug, M. & Weber, J. (1990). X-ray microanalysis of elements present in the matrix of cnidarian nematocysts. *Tissue & Cell*, **22**, 629–43.

Walz, B. & Somlyo, A. P. (1984). Quantitative electronprobe microanalysis of leech photoreceptors. *Journal of Comparative Physiology*, A**154**, 81–7.

Warley, A. & Gupta, B.L. (1991). Quantitative biological X-ray microanalysis. In *Electron microscopy of tissue, cells and organelles: a practical approach*, ed. J.R. Harris, pp. 243–81. Oxford: I.R.L. Oxford University Press.

Weber, J. (1989). Nematocysts (stinging capsules of *Cnidaria*) as Donnan potential dominated osmotic systems. *European Journal of Biochemistry*, **184**, 465–76.

Wigglesworth, V.B. (1948). The insect as a medium for the study of physiology. *Proceedings of the Royal Society London*, B**135**, 430–46.

Wieczorek, H., Weerth, S., Schindleck, M. & Klein, U. (1989). A vacuolar-type proton pump in a vesicle fraction enriched with potassium transporting plasma membranes from tobacco hornworm midgut. *Journal of Biological Chemistry*, **264**, 11 143–8.

Zglinicki, T., Ziervogel, H. & Bimmler, M. (1989). Binding of ions to nuclear chromatin. *Scanning Microscopy*, **3**, 1231–9.

Zierold, K. (1991). Cryofixation methods for ion localization in cells by electron probe microanalysis: a review. *Journal of Microscopy*, **161**, 357–66.

Zierold, K., Gerke, I. & Schmitz, M. (1989). X-ray microanalysis of fast exocytotic process. In *Electron probe microanalysis: application in biology and medicine*, ed. K. Zierold & H. Hagler, pp. 281–92. Berlin: Springer-Verlag.

Zierold, K. & Schafer, D. (1988). Preparation of cultured and isolated cells for X-ray microanalysis. *Scanning Microscopy*, **2**, 1175–790.

Zierold, K. & Wessing, A. (1990). Mass dense vacuoles in *Drosophila* Malpighian rubules contain zinc, not sodium. A reinvestigation by X-ray microanalysis of cryosections. *European Journal of Cell Biology*, **53**, 222–6.

Addendum

Recently Wessing, Zierold & Hevert (1992) have shown by EPXMA of frozen–dried sections that in *Drosophila* Malpighian tubules two types of luminal concretions are found which differ in their distribution along the length of the tubules and in different morphogenetic stages. Type-I have a matrix of proteoglycans and accumulate large quantities of Ca and Mg. Type-II concretions are deficient in matrix and accumulate K. By experimentally changing the elemental composition of diet, Wessing & Zierold (1992) have found that each type of concretion shows high selectivity in the ions it sequesters. These results on concretions have interesting parallels with the ion binding and sequestering selectivity found in extracellular matrices and in nematocysts discussed in the main text.

Wessing, A. & Zierold, K. (1992). Metal-salt feeding causes alterations in concretions in *Drosophila* larval Malpighian tubules as revealed by X-ray microanalysis. *Journal of Insect Physiology*, **38**, 623–32.

Wessing, A. & Zierold, K. & Hevert, F. (1992). Two types of concretions in *Drosophila* Malpighian tubules as revealed by X-ray microanalysis: a study on urine formation. *Journal of Insect Physiology*, **38**, 543–54.

15 X-ray microanalysis in pollution studies

J.A. Nott

A   Introduction

X-ray microanalysis (XRMA) is a young science, but in the past 20 years it has produced major advances in the material and biological sciences. This is particularly the case for the study of the uptake of pollutant heavy metals by animals and plants.

Before the arrival of XRMA it had been established that marine animals were capable of taking-up metals in large amounts and concentrating them in particular tissues. These effects were investigated by atomic absorption spectrophotometry and to a limited extent by histochemistry. However, structural resolution was limited and most specimen preparation procedures either removed the metals from the tissues or produced some degree of translocation. These restrictions did not assist investigation into the cytology of metal metabolism. This situation changed dramatically with the arrival of XRMA which coupled the structural resolution of the electron microscope with the element identification of the energy dispersive X-ray detector. Early applications to marine organisms were described by Nott & Parkes (1975) and Walker *et al.* (1975a,b).

The metals that are found in tissues can be grouped biologically into those essential for metabolism and those which are non-essential. Most metals are toxic when they occur in excess amounts. Chemically the metals can be grouped into the hard acid type (for example Na, Mg, K and Ca) which bind electrostatically with a preference for $O > N > S$ donating ligands, and the soft acid type (for example Cu, Cd, Hg and Ag) which bind covalently with a preference for $S > N > O$. Other metals including Mn, Fe, Co, Ni and Zn show intermediate properties. In terms of generalised biochemistry the ligands for the hard acid metals are carbonate and phosphate anions which occur in mineralised intracellular granules. The carbonate granules bind only magnesium and calcium but the phosphate variety bind, also, a wide range of intermediate metals. The ligands for the soft acid metals are proteins which can have a high sulphur content; these occur in the cytosol and in residual lysosomes. Intracellular accumulations of complexed metals are readily detectable by XRMA.

Table 15.1. *References to XRMA work on metals in marine crustaceans* (*R = review*)

Reference(s)	Metal(s)	Animal	Tissue(s)
Al-Mohanna & Nott, 1986a,b, 1987, 1989	Cu, Zn, Au, Th, Fe	Shrimp	Gut
Chassard-Bouchaud, 1985	Cu, Zn, Mn, Fe, Fr, Cd, Ag etc.	Shrimp	Gut
Hopkin & Nott, 1979, 1980	Pb, Ca, P, Au, Th, S	Crab	Gut
Icely & Nott, 1980, 1985 (R)	Au, Fe, Th	Amphipod	Gut
Koulish, 1976	Zn, P	Barnacle	Gut, body wall
Nott & Mavin, 1986	Ca, Mg, P, S	Shrimp	Gut, blood
Nott *et al.*, 1985	Pb (aryl sulphatase)	Copepod	Gut
Rainbow, 1987 (R)	Cu, Zn, Mn, Fe, etc.	Barnacle	Gut parenchyma
Thomas & Ritz, 1986	Zn	Barnacle	Gut parenchyma
Walker, 1977	Cu, Zn	Barnacle	Gut parenchyma
Walker *et al.*, 1975a,b	Zn, Mn, Fe, Z	Barnacle	Gut parenchyma

Tables 15.1–15.5 are reproduced, with permission, from *Scanning Microscopy*, **5**(1) (1991). Tables 15.1–15.3 are sources of data in Figs 15.1–15.3 respectively.

B   Methodology

The developments of XRMA techniques over 20 years have produced major advances in instrumentation and in specimen preparation.

Initially the solid state detectors were cumbersome and could only be attached to the large specimen chamber of the scanning electron microscope (SEM). Later, with more refined engineering, they were attached to transmission electron microscopes (TEMs). Now the TEM and SEM are designed to accommodate X-ray detectors and the control systems of the microscopes and analysers can be fully interactive. Also the image and data processing systems are highly sophisticated.

The development of scanning transmission systems (STEM) for the TEM increased the effectiveness of XRMA by decreasing the spot size of the incident beam and by providing images from biological sections in excess of 1 μm thickness. This is of particular benefit for the examination of intracellular minerals which tend to fall out of ultrathin sections.

Specimen preparation for pollution studies has been transformed by the development of rapid and productive cryosystems. These have replaced the original chemical techniques of fixation and dehydration which, in the case of cadmium, could remove the bulk of the metal from the tissue (George, Pirie & Coombs, 1976). The modified fixatives which were designed to precipitate the intracellular metals as insoluble pyroantimonate or sulphide were

Table 15.2. *References to XRMA work on metals in marine bivalves* (*R = review*)

Reference(s)	Metal(s)	Animal(s)	Tissue(s)
Ballan-Dufrancais et al., 1982, 1985	Metals	Scallop	Gut
Carmichael & Fowler, 1981	Cd	Scallop	Kidney
Carmichael et al., 1979	Ca, Mg, P, Zn, Mn	Scallop	Kidney
Chassard-Bouchaud, 1983	U, P	Mussel	Blood, kidney, gut
Chassard-Bouchaud & Hallegot, 1984	La, P	Mussel	Gut macrophages, gill
Chassard-Bouchaud et al., 1985, 1986, 1989	Ag, Pb, Cr, P, S	Mussel, oyster	Gut, gill, blood, muscle, kidney, byssus
Coombs & George, 1978 (R)	Fe, Pb, Cu, Zn, P, S	Mussel, oyster	Gill, amoebocytes, kidney
Doyle et al., 1978	Ca, Mg, P, Mn, Fe, Cu, Zn	Clams	Kidney
Fowler et al., 1975	Fe, Hg	Clam	Mantle
George & Pirie, 1979, 1980	Mg, Ca, P, Zn, Fe, Cd	Mussel	Kidney, blood
George et al., 1976, 1977, 1978, 1980	Fe, Zn, Mn, Ca, P, Cu, S	Mussel, scallop, oyster	Gill, blood, kidney, gut, amoebocytes
Marshall & Talbot, 1979	Cd, Pb	Mussel	Gill
Martoja & Martin, 1985	Cd, Zn, Cu	Oyster	Amoebocytes
Martoja et al., 1985	Ag	Scallop	Gut
Mauri & Orlando, 1982	Mn	Wedge shell	Kidney
Pirie et al., 1984	Cu, Zn	Oyster	Blood
Schulz-Baldes, 1977	Pb, P, S	Mussel	Blood, kidney
Thomson et al., 1985	Cu, Zn, Fe	Oyster	Blood, kidney, mantle, gill

non-specific and introduced additional peaks for antimony and sulphur, respectively.

For cryopreparation, fresh tissue is quenched rapidly in liquid ethane or propane which is cooled by liquid nitrogen (Ryan & Purse, 1984; Ryan et al., 1987, 1990) and analysed in the hydrated or dehydrated state. Hydrated specimens present the ultimate preparation but requirements for cryoultramicrotomy and EM cold stages tend to reduce the volume of output of X-ray data. Instead, for routine pollution studies the quenched tissue blocks are dehydrated either by freeze-substitution for observations of fine structure or

Table 15.3. *References to XRMA work on metals in marine gastropods* ($R =$ review)

Reference(s)	Metal(s)	Animal(s)	Tissue(s)
Bouquegneau & Martoja, 1982	Cu, Cd, Zn	Gastropods	Gut
Bouquegneau et al., 1984 (R)	Metals	Molluscs	Gut
Greaves et al., 1984 (R)	Mn	Snail	Gut
Martoja et al., 1980	Cu	Winkle	Kidney, gut
Martoja et al., 1985	Cu, Ag	Whelk, winkle	Gut
Mason & Nott, 1980	Zn, Mn, Fe	Winkle	Gut, kidney
Mason & Nott, 1981 (R)	Metals	Gastropods	Blood, gut, mantle, kidney, shell
Mason & Simkiss, 1982	Mn	Snail	Gut
Mason et al., 1984	Mn, Fe, Zn, Cu	Winkle	Kidney, gut, gill
Nott & Langston, 1989	Cd, P, S	Winkle	Gut
Nott & Nicolaidou, 1989a,b, 1990	Metals, Zn, Mn	Gastropods, whelks	Gut, faeces
Simkiss, 1979, 1984 (R)	Metals/ granules, Mn	Molluscs/ invertebrates	Gut
Simkiss & Mason, 1983, 1984 (R)	Metals, Cu, Fe, Mn, Zn	Molluscs, winkle	Blood, gut, kidney, gill
Taylor et al., 1988	Mn	Snail	Gut

by freeze-drying for XRMA (Nott & Nicolaidou, 1989b). Thus, specimens dried by both methods are subsequently embedded in resin and sectioned for observation and analysis by TEM and STEM. Some intracellular mineralised concretions can be prepared by squashing fresh tissue on a graphite stub, air drying and then analysing by XRMA in the SEM. These specimens do not have to be coated, especially when examined with a backscattered electron detector (Nott & Nicolaidou, 1990).

Resin embedded specimens have the limitation that some elements account for less than *c.* 0.1 % of the mass and are, therefore, below the level of detectability by XRMA. The mass fraction of inorganic cell constituents can be increased by removing the organic material from sections by microincineration. The resulting inorganic ash produces a recognisable image of the cells in the form of a spodograph (Mason & Nott, 1980; Al-Mohanna & Nott, 1986a). This technique has made it possible to detect zinc which is dispersed in the cytosol of the nephrocytes from the winkle (Mason & Nott, 1980). This zinc is not detectable in normal resin embedded sections but would probably be detected in frozen–dried cryosections.

In marine organisms pollutant metals are usually accumulated within intracellular compartments. Quantitatively, these can be analysed directly

Table 15.4. *Review papers on the cytology of metals which cite X-ray microanalytical work*

Review	Metal(s)	Animal(s)
Brown, 1982	Metal containing granules	Invertebrates
Bryan, 1984	Metal pollution	Marine organisms
Fowler, 1987	Intra-cellular metals	Aquatic organisms
Fowler *et al.*, 1981	Intra-cellular metals	Estuarine organisms
George, 1980	Metal metabolism	Marine bivalves
George, 1982	Metal detoxification	Aquatic animals
George & Viarengo, 1985	Metal detoxification	Mussels
Martoja *et al.*, 1975	Ecological metals	Various organisms
Rainbow, 1988	Metal accumulations	Decapods
Ray & McLeese, 1987	Cadmium	Marine organisms
Simkiss, 1976	Biomineralisation	Invertebrates
Simkiss *et al.*, 1982	Metal detoxification	Molluscs
Taylor & Simkiss, 1984	Biomineralisation	Invertebrates
Taylor *et al.*, 1986	Intracellular granules	Invertebrates
Viarengo, 1989	Metal metabolism	Marine invertebrates

Other reviews appear in Tables 15.1–15.3 marked (R).

(George, Pirie & Coombs, 1980) or indirectly by analysis of tissues which have been solubilised, spiked with cobalt, and sprayed onto EM grids as microdroplets (Nott & Mavin, 1986). Also, XRMA has been productive in the area of semi-quantitative analysis. The ratios of elements can be calculated with accuracy from the peak integrals when the relative collection efficiencies of the detector for different atomic numbers is established. The angle of tilt for the TEM specimen and the general area of grid analysed must be monitored because both parameters affect the collection efficiency for low energy X-rays. Semi-quantitative analyses can determine the ratios of various pollutant metals to phosphorus and sulphur which are the two intracellular XRMA markers for phosphate and sulphur ligands (see Tables 15.1–15.4 for references).

C   Metals in marine organisms

In marine organisms, pollutant metals are accumulated in particular tissues and XRMA has shown that they are restricted to particular cells. Within the cells, the metals can be dispersed in the cytosol or, more generally, they are compartmentalised in vesicles of the lysosomal system or in mineralised concretions. The contents of these compartments are not homogeneous; the mineralised concretions have concentric layers and the residual lysosomes

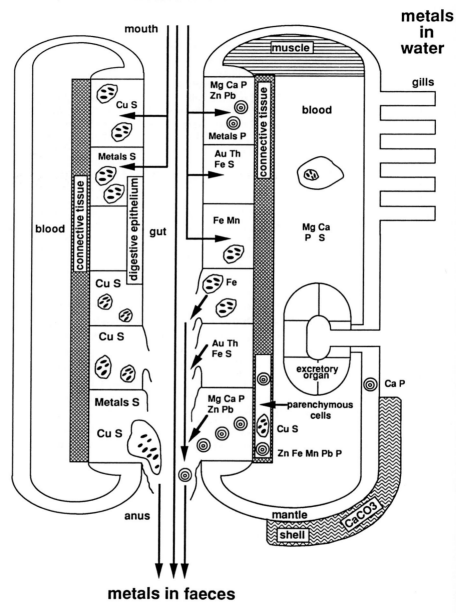

Fig. 15.1.   For legend see page 265.

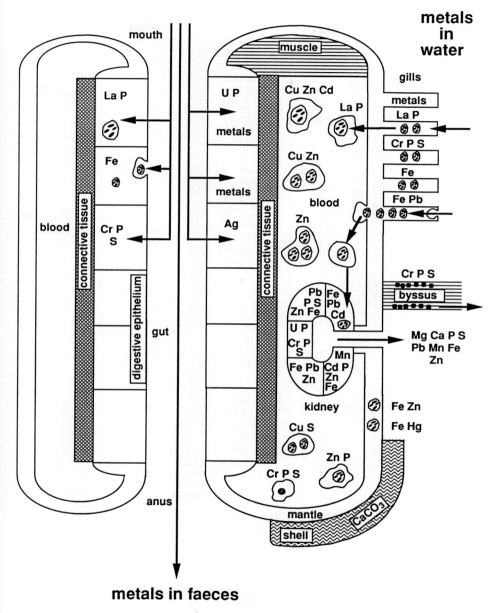

Fig. 15.2. For legend see page 265.

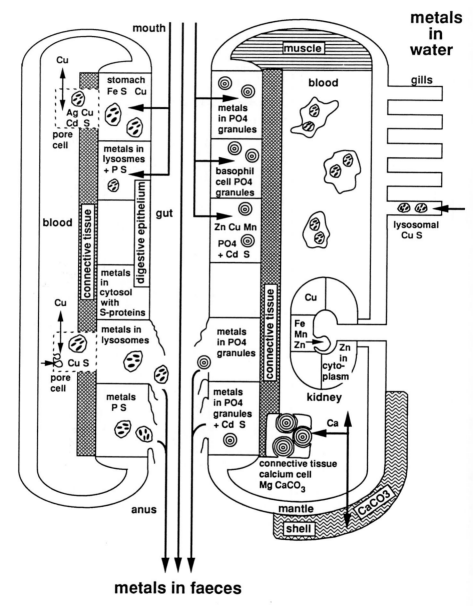

Fig. 15.3. For legend see facing page.

Table 15.5. *Additional papers which utilise XRMA for studying metals in marine organisms*

Reference	Metal(s)	Animal(s)/tissue(s)
Bell *et al.*, 1982	Vanadium & sulphur	Ascidian blood cells
Bone *et al.*, 1987	Ionic analysis	Chaetognath body fluids
Gibbs & Bryan, 1980	Cu	Polychaete jaws
Gibbs & Bryan, 1984	$CaPO_4$	Polychaete skeleton
Gibbs *et al.*, 1981	Cu	Polychaete
Gooday & Nott, 1982	$BaSO_4$, Sr (Intracellular)	Deep-sea protozoans
Gupta & Hall, 1984	Ca, Cu, Zn	Sea anemone
Nott & Parkes, 1975	Ca secretion	Polychaete
Pirie & Bell, 1984	Vanadium and sulphur	Ascidian blood
Pirie *et al.*, 1985	Cu, Zn	Polychaete
Rowley, 1982	Vanadium	Ascidian blood cells
Southward, 1982	Zn, Cu, Ca, P	Pogonophora
Pedersen & Roomans, 1983[a]	Br, I	Brown seaweed

[a] For seaweed references.

show heterogeneous density patterns. The metals in the cytosol are bound to high sulphur metalloproteins. These are probably degraded in the lysosomes wherein the metals remain associated with sulphur (see Table 15.4 for references). The various combinations of intracellular binding and compartmentalisation serve as detoxification mechanisms which protect the biochemical processes of normal metabolism from dislocation by reactive pollutant metals.

Figs 15.1.–15.3. Schematic diagrams of marine organisms to show accumulations of metals in cells and the blood as detected by XRMA. Sources of the data in Figs 15.1–15.3 can be found by making cross references to the metals and tissues listed in Tables 15.1–15.3 respectively. Reproduced, with permission, from *Scanning Microscopy*, **5**(1) (1991).

Fig. 15.1. Marine crustaceans. Sites of metal accumulation occur mainly in the epithelial and sub-epithelial cells of the gut. The gills, blood and excretory organs are active in the uptake, transport and excretion of metals (for example, Bryan, Hummerstone & Ward, 1986) but no accumulation in these tissues have been detected by XRMA.

Fig. 15.2. Marine bivalves. Accumulations of metals take place at sites of uptake in the gut and gills. However, in these animals the kidney is the dominant tissue for accumulation and excretion of metals.

Fig. 15.3. Marine gastropods. Sites of metal accumulation occur in the gills and kidney, although most recordings have been made on intracellular inclusions in the gut diverticulum or hepatopancreas where metals are taken up and excreted.

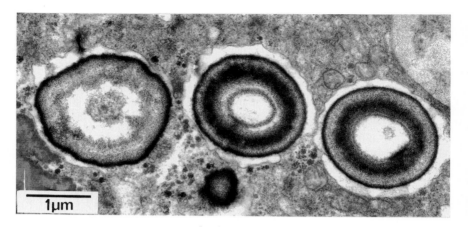

Fig. 15.4.   Cerithium vulgatum: section of digestive gland showing mineralised
phosphate granules. Reproduced, with permission, from Nott &
Nicolaidou, 1989b.

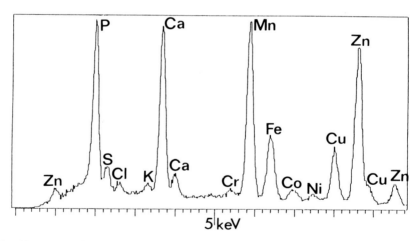

Fig. 15.5.   Cerithium vulgatum: X-ray analysis of phosphate granules containing
heavy metals. (Compare with Figs 15.9 and 15.10 which show
phosphate granules without heavy metals.) Reproduced, with
permission, from Nott & Nicolaidou, 1989b.

XRMA investigations in marine biology have concentrated on the bivalve
molluscs, particularly mussels and oysters, the gastropod molluscs, including
winkles and whelks, and crustaceans, namely crabs, shrimps and lobsters.
This work is summarised in Figs 15.1–15.3 which show sites of uptake,
translocation, storage and excretion of metals in schematic diagrams of the

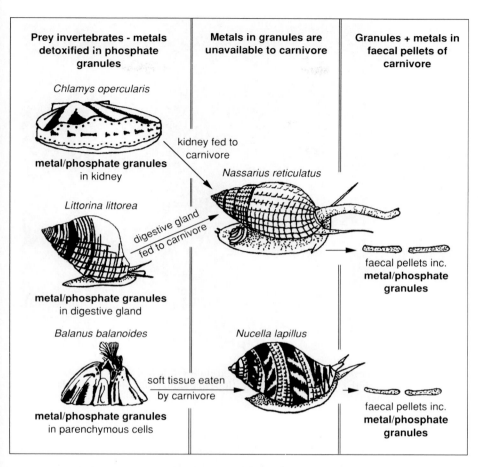

Fig. 15.6. Predator/prey food chain. Metals are incorporated in intracellular
          phosphate granules in the prey (left column). The soft tissues containing
          the granules are ingested by the carnivore (centre column). The tissues
          are digested by the carnivore and the residues including granules are
          egested in the faecal pellets (right column). Reproduced, with
          permission, from Nott & Nicolaidou, 1990.

whole animals. The source references are given in Tables 15.1–15.3. XRMA
work on other marine organisms is listed in Table 15.5.

The biology of metals requires the study of a dynamic phenomenon
because the metal content of an organism is a variable net balance of
continual uptake and loss. In marine animals, metals are taken up from the
water, food and sediments via the gut, gills and other body surfaces. The
uptake of iron into pinocytic vesicles in the gills and gut of mussels has been
demonstrated by XRMA (George *et al.*, 1976). At sites of uptake the metals

Fig. 15.7.   *Nassarius reticulatus*: masses of granules in faecal pellet squashed on graphite stub. Reproduced, with permission, from Nott & Nicolaidou, 1990.

are bound to ligands and may be transported via the blood. Both copper and zinc have been identified by XRMA in the amoebocytic blood cells of oysters (see the entries for blood in Table 15.2 for references). The metals are accumulated in the kidney of bivalves and in the gut of gastropods and crustaceans where individual metals can exceed 1 % of the dry mass of the tissue. They are excreted via the gut, kidney and other tissues often by the complete disintegration of the storage cell (Nott & Nicolaidou, 1989b).

D   Marine pollution effects

A nickel smelting plant on the shore of the Aegean Sea contaminates the local environment with heavy metals and it has been established by atomic absorption spectrophotometry that these are accumulated by marine organisms. The animals include three species of gastropod mollusc (snail) which take up different metals in widely different amounts (Nott & Nicolaidou, 1989b). The tower shell *Cerithium vulgatum* feeds on sediments and takes up

Fig. 15.8. *Nassarius reticulatus*: detail of granules in Fig. 15.7. The smaller dense granules are the phosphate type. Reproduced, with permission, from Nott & Nicolaidou, 1990.

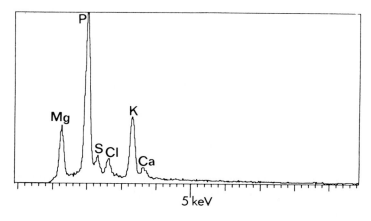

Fig. 15.9. *Littorina littorea* (winkle): analysis of phosphate granule in digestive gland (compare with Fig. 15.10). Figs 15.9–15.14 are reproduced, with permission, from *Scanning Microscopy*, **5**(1) (1991).

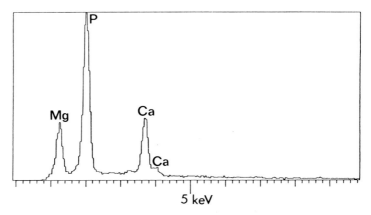

Fig. 15.10. Nassarius reticulatus (carnivorous whelk): analysis of phosphate granule in faecal pellet: granule is derived from winkle and has passed through gut of the whelk (compare with Fig. 15.9).

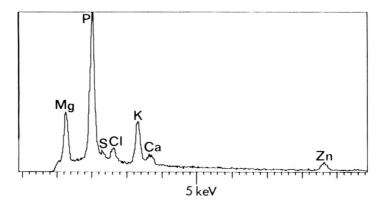

Fig. 15.11. Littorina littorea: analysis of phosphate granule containing zinc (compare with Fig. 15.12).

the highest levels of metals; the top shell, *Monodonta articulata* grazes on algae and takes up metals to a lesser extent; the carnivorous whelk, *Murex trunculus*, which includes *Cerithium vulgatum* in its diet, takes up the least amounts. The metals are accumulated in the digestive gland where they are shown by XRMA to be incorporated in the form of phosphate compounds in concentrically structured, mineralised granules (Figs 15.4 and 15.5).

It appears from this work that the detoxification system that operates in *Cerithium* continues to protect the carnivore, in that *Murex* does not extract the metals from the tissues of *Cerithium*. This concept of detoxification transfer along a food chain was tested by an XRMA investigation on animals

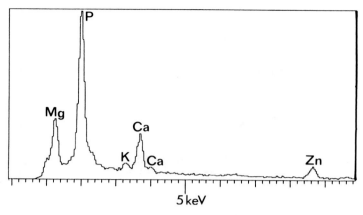

Fig. 15.12. Nassarius reticulatus: analysis of phosphate granule in faceal pellet: the zinc incorporated in the pellet by the winkle has not been removed by the digestive system of the whelk (compare with Fig. 15.11).

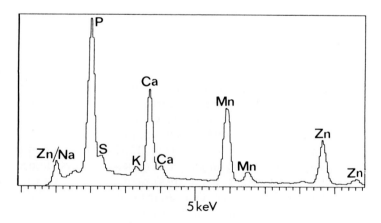

Fig. 15.13. Chlamys opercularis (scallop): analysis of phosphate granule in kidney (compare with Fig. 15.14).

taken from the Plymouth area (Nott & Nicolaidou, 1990) (Fig. 15.6). The grazing periwinkle *Littorina littorea* produces masses of intracellular magnesium/calcium phosphate granules in the digestive gland. This tissue was fed to the carnivorous whelk *Nassarius reticulatus* and subsequently the faecal pellets were collected. Samples of digestive gland from the periwinkle and faecal pellets from the whelk were squashed thinly on SEM graphite specimen stubs and air-dried. Granules were particularly numerous in the faecal pellets where they formed the bulk of the sample (Figs 15.7 and 15.8). The granules in the two samples produced XRMA spectra (Figs. 15.9 and 15.10) typical of

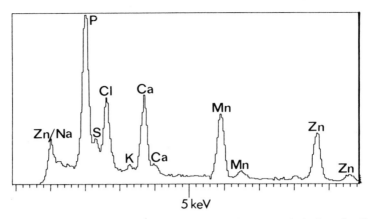

Fig. 15.14. Nassarius reticulatus: analysis of phosphate granule in faceal pellet: granule is derived from scallop kidney and has passed through the gut of the whelk (compare with Fig. 15.13).

magnesium/calcium phosphate except that the digestive system of the carnivore had removed the more soluble S, Cl and K (Fig. 15.10). This was taken as strong circumstantial evidence that granules synthesised in the winkle had passed through the gut of the carnivore. To test the transfer of detoxified metals between animals some winkles were dosed with 1 p.p.m. Zn in seawater for 16 days and the metal appeared in the phosphate granules in the digestive gland (Fig. 15.11). This tissue was fed to the carnivore and the faecal pellets were examined by XRMA. The granules in the pellets produced a similar spectrum with a peak for zinc (Fig. 15.12). This experiment was repeated with manganese and again the metal appeared to be retained in the granules from the faecal pellets. From these qualitative analyses it can be proposed that the metals have been rendered insoluble and unavailable in the winkle and that they have remained in this form during passage through the gut of the carnivore. The system of detoxification has protected two stages of the food chain.

The queen scallop *Chlamys opercularis* is typical for a bivalve mollusc in that it accumulates heavy metals in phosphate granules in the kidney. The granules were examined by XRMA and the spectra showed that they were contaminated with manganese and zinc (Fig. 15.13). The kidney tissue was fed to the carnivore, *Nassarius reticulatus* and, again, the granules passed through the gut and into the faecal pellets (Fig. 15.6). Qualitative XRMA analyses showed that manganese and zinc peaks were retained in the granules (Fig. 15.14) which suggested that the metals were transferred in detoxified form.

Barnacles produce granules in parenchymous tissue which surrounds the gut (Walker *et al.*, 1975a,b) and they are the natural prey of the dogwhelk,

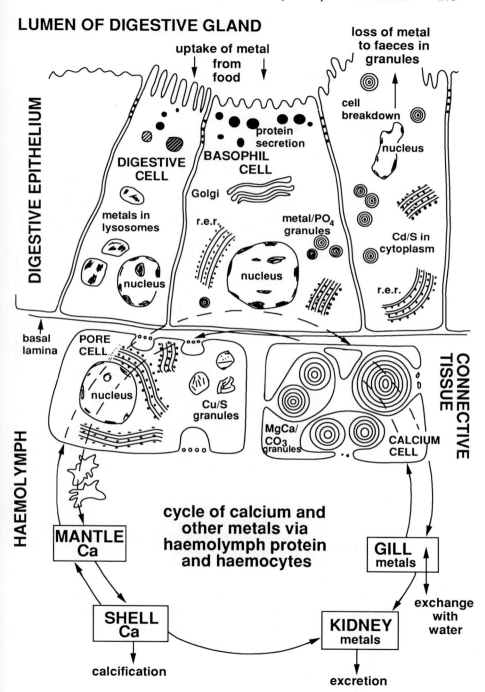

Fig. 15.15. Sites of uptake and excretion of magnesium, calcium and other metals in *Littorina littorea* with routes of translocation between tissues. Reproduced, with permission, from McGraw-Hill Yearbook 1991.

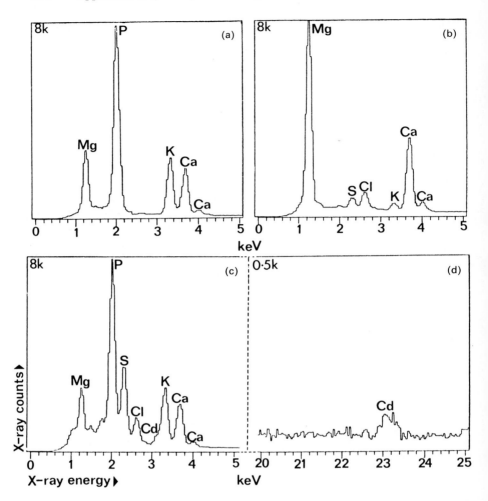

Fig. 15.16. Littorina littorea. (a) Typical X-ray microanalytical spectrum from basophil cell phosphate granule in control animal. Section similar to Fig. 15.4. (b) Typical spectrum from connective tissue calcium cell carbonate granule in same section as (a). (c, d) X-ray microanalysis of margin of phosphate granule from animal treated with cadmium. Note peaks for cadmium at 3.13 keV (Lα) and 23.2 keV (Kα) and associated peaks for sulphur and chlorine. Analysis of the centre of the granule is the same as (a). (c) and (d) are different regions from the same spectrum; in (d) the vertical scale has been expanded to 0.5 K. Reproduced, with permission, from Nott & Langston, 1989.

Nucella lapillus. Granules from the barnacle and from the whelk faecal pellets both produce spectra with a peak for zinc. This suggests that the metal is not available to the whelk.

These results are qualitative because the XRMA spectra from the solid spherical granules could not be processed quantitatively. The surface geometry of the spheres produced significant variations in peak height, particularly at the low energy end of the spectrum. To quantify the results, fresh samples have been quench frozen, freeze-dried, embedded and sectioned for analysis by STEM/XRMA at 200 kV. The sections have flat surfaces and the granules give reproducible spectra. Integrals are recorded for the elemental peaks (P) and selected regions of background (b) and the P–b values are calculated by direct subtraction.

XRMA can give some cytological basis to the synergistic and antagonistic reactions between different metals in the tissues of the periwinkle. Large inputs of heavy metals into the phosphate granules of the basophil cells (Figs 15.5 and 15.14) substitute for magnesium which disappears from the spectrum (Nott & Nicolaidou, 1989a). However, the tissue levels of magnesium increase (Nott & Nicolaidou, 1989b), which suggests that it is shunted to the $Mg/CaCO_3$ granules which occur in the connective tissue (Fig. 15.15). Unlike the phosphate granules, the carbonate granules do not contain a measurable peak for phosphorus and they never contain any heavy metals (Fig. 15.16). They appear to form an integral part of a buffering system for the body fluids (Mason & Nott, 1981). Deposits of copper and sulphur in the pore cells are concerned with the metabolism of haemocyanin and they do not contain other metals (Mason *et al.*, 1984). Pollutant metals are accumulated together with phosphorus and sulphur in the residual lysosomes of the digestive cells (Nott & Nicolaidou, 1989b).

XRMA combined with the cryopreparation of tissue and the resolution of STEM has located cadmium and sulphur in association with the membrane which encloses each phosphate granule (Fig. 15.16) (Nott & Langston, 1989). Cadmium does not occur in the mineralised granule itself.

Thus, XRMA has contributed substantially to work on the uptake, transport, accumulation and excretion of heavy metals in marine organisms. The combined observations on fine structure and microanalysis can resolve the various intracellular sites of inorganic chemical activity. Indeed, XRMA is customised for work on the metabolism of pollutant heavy metals.

Acknowledgements

The author acknowledges contributions from many investigators whose names appear in the reference list. Recently, there has been collaborations with Artemis Nicolaidou, Linda Mavin and Keith Ryan. David Nicholson provided assistance with photographic work.

References

Al-Mohanna, S.Y. & Nott, J.A. (1986a). The accumulation of metals in the hepatopancreas of the shrimp *Penaeus semisulcatus* De Haan (Crustacea: Decapoda) during the moult cycle. In *Proc. First Arabian Gulf Conf. on Environment and Pollution*, ed. D. Clayton & M. Behbehani, pp. 195–209. Kuwait University, Faculty of Science, P.O. Box 5969, Kuwait.

(1986b). B-cells and digestion in the hepatopancreas of *Penaeus semisulcatus* (Crustacea: Decapoda). *Journal of the Marine Biological Association of the United Kingdom*, **66**, 403–14.

(1987). R-cells and the digestive cycle in *Penaeus semisulcatus* (Crustacea: Decapoda). *Marine Biology*, **95**, 129–37.

(1989). Functional cytology of the hepatopancreas of *Penaeus semisulcatus* (Crustacea: Decapoda) during the moult cycle. *Marine Biology*, **101**, 535–44.

Ballan-Dufrancais, C., Jeantet, A.-Y., Feghali, C. & Halpern, S. (1985). Physiological features of heavy metal storage in bivalve digestive cells and amoebocytes: EPMA and factor analysis of correspondences. *Biology of the Cell*, **53**, 283–92.

Ballan-Dufrancais, C., Jeantet, A.-Y. & Halpern, S. (1982). Localisation intracellulaire par microanalyse X de métaux et de métalloïdes dans la glande digestive d'un Mollusque Bivalve *Pecten maximus*. Implications des processus de digestion. *Comptes Rendus Hebdomadaires des Séances de l'Académie des Sciences, Paris*, **294**, 673–8.

Bell, M.V., Pirie, B.J.S., McPhail, D.B., Goodman, B.A., Falk-Petersen, I.-B. & Sargent, J.R. (1982). Contents of vanadium and sulphur in the blood cells of *Ascidia mentula* and *Ascidiella aspersa*. *Journal of the Marine Biological Association of the United Kingdom*, **62**, 709–16.

Bone, Q., Brownlee, C., Bryan, G.W., Burt, G.R., Dando, P.R., Liddicoat, M.I., Pulsford, A.L. & Ryan, K.P. (1987). On the differences between the two 'indicator' species of chaetognath *Sagitta setosa* and *S. elegans*. *Journal of the Marine Biological Association of the United Kingdom*, **67**, 545–60.

Bouquegneau, J.M. & Martoja, M. (1982). La teneur en cuivre et son degré de complexation chex quatre Gastéropodes marines. Données sur le cadmium et le zinc. *Oceanologica Acta*, **5**, 219–28.

Bouquegneau, J.M., Martoja, M. & Truchet, M. (1984). Heavy metal storage in marine animals under various environmental conditions. In *Toxins, drugs and pollutants in marine animals*, ed. L. Bolis, J. Zadunaisky & R. Gilles, pp. 147–60. Berlin, Heidelberg: Springer-Verlag.

Brown, B.E. (1982). The form and function of metal-containing 'granules' in invertebrate tissues. *Biological Reviews*, **57**, 621–67.

Bryan, G.W. (1984). Pollution due to heavy metals. *Marine Ecology*, **5**, 1289–431.

Bryan, G.W., Hummerstone, L.G. & Ward, E. (1986). Zinc regulation in the lobster *Homarus gammarus*: importance of different pathways of absorption and excretion. *Journal of the Marine Biological Association of the United Kingdom*, **66**, 175–99.

Carmichael, N.G. & Fowler, B.A. (1981). Cadmium accumulation and toxicity in the kidney of the Bay Scallop *Argopecten irradians*. *Marine Biology*, **65**, 35–43.

Carmichael, N.G., Squibb, K.S. & Fowler, B.A. (1979). Metals in the molluscan kidney: a comparison of two closely related bivalve species (Argopecten) using X-ray microanalysis and atomic absorption spectroscopy. *Journal of the Fisheries Research Board of Canada*, **36**, 1149–155.

Chassard-Bouchaud, C. (1983). Rôle des lysosomes et des sphérocristaux dans le phénomène de concentration de l'uranium chez la Moule *Mytilus edulis* (L.)

Microanalyse par spectrographie des rayons X. *Comptes Rendus Hebdomadaires des Séances de l'Academie des Sciences, Paris,* **296**, 581–6.

(1985). Bioaccumulation de métaux stables et radioactifs par les organismes benthiques de la baie de Seine: aspects structuraux, ultrastructuraux et microanalytiques. *Cahiers de Biologie Marine,* **26**, 63–85.

Chassard-Bouchaud, C., Boutin, J.F., Hallegot, P. & Galle, P. (1989). Chromium uptake, distribution and loss in the mussel *Mytilus edulis*: a structural, ultrastructural and microanalytical study. *Diseases of Aquatic Organisms,* **7**, 117–36.

Chassard-Bouchaud, C., Fiala-Medioni, A. & Galle, P. (1986). Etude microanalytique de *Bathymodiolus* sp. (Mollusque Lamellibranche Mytilidae) provenant des sources hydrothermales de la Ride du Pacifique oriental. Donées préliminaires. *Comptes Rendus Hebdomadaires des Séances de l'Academie des Sciences, Paris,* **302**, 117–24.

Chassard-Bouchaud, C., Galle, P. & Escaig, F. (1985). Mise en évidence d'une contamination par l'argent et le plomb de l'Huitre *Crassostrea gigas* et de la Moule *Mytilus edulis* dans les eaux côtiéres françaises. Etude microanalytique par émission ionique secondaire. *Comptes Rendus Hebdomadaires des Séances de l'Academie des Sciences, Paris,* **300**, 3–8.

Chassard-Bouchaud, C. & Hallegot, P. (1984). Bioaccumulation de lanthane par des Moules *Mytilus edulis* (L.) récoltées sur les côtes francaises. Microanalyse par spectrographie des rayons X et par émission ionique secondaire. *Comptes Rendus Hebdomadaires des Séances de l'Academie des Sciences, Paris,* **298**, 567–72.

Coombs, T.L. & George, S.G. (1978). Mechanisms of immobilization and detoxication of metals in marine organisms. In *Physiology and behaviour of marine organisms,* ed. D.S. McLusky & A.J. Berry, pp. 179–87. Oxford: Pergamon Press.

Doyle, L.J., Blake, N.J., Wood, C.C. & Yevich, P. (1978). Recent biogenic phosphorite: concretions in mollusc kidneys. *Science,* **199**, 1431–3.

Fowler, B.A. (1987). Intracellular compartmentation of metals in aquatic organisms: roles in mechanisms of cell injury. *Environmental Health Perspectives,* **71**, 121–8.

Fowler, B.A., Carmichael, N.G., Squibb, K.S. & Engel, D.W. (1981). Factors affecting trace metal uptake and toxicity to estuarine organisms. II. Cellular mechanisms. In *Biological monitoring of marine pollutants,* ed. F.J. Vernberg, A. Calabrese, F.P. Thurberg & W.B. Vernberg, pp. 145–63. New York: Academic Press.

Fowler, B.A., Wolfe, D.A. & Hettler, W.F. (1975). Mercury and iron uptake by cytosomes in mantle epithelial cells of quahog claims (*Mercenaria mercenaria*) exposed to mercury. *Journal of the Fisheries Research Board of Canada,* **32**, 1767–75.

George, S.G. (1980). Correlation of metal accumulation in mussels with the mechanisms of uptake, metabolism and detoxification: a review. *Thalassia Jugoslavica,* **16**, 347–65.

(1982). Subcellular accumulation and detoxication of metals in aquatic animals. In *Physiological mechanisms of marine pollutant toxicity,* ed. W.B. Vernberg, A. Calabrese, F.P. Thurberg & F.J. Vernberg, pp. 3–52. New York: Academic Press.

George, S.G. & Pirie, B.J.S. (1979). The occurrence of cadmium in sub-cellular particles in the kidney of the marine mussel, *Mytilus edulis,* exposed to cadmium. The use of electron microprobe analysis. *Biochemistry and Biophysica Acta,* **580**, 234–44.

(1980). Metabolism of zinc in the mussel *Mytilus edulis* (L.): a combined ultrastructural and biochemical study. *Journal of the Marine Biological Association of the United Kingdom,* **60**, 575–90.

George, S.G., Pirie, B.J.S., Cheyne, A.R., Coombs, T.L. & Grant, P.T. (1978). Detoxication of metals by marine bivalves: an ultrastructural study of the compartmentation of copper and zinc in the oyster *Ostrea edulis*. *Marine Biology*, **45**, 147–56.

George, S.G., Pirie, B.J.S. & Coombs, T.L. (1976). The kinetics of accumulation and excretion of ferric hydroxide in *Mytilus edulis* (L.) and its distribution in the tissues. *Journal of Experimental Marine Biology and Ecology*, **23**, 71–84.

(1977). Absorption, accumulation and excretion of iron-protein complexes by *Mytilus edulis* (L.). In *International Conference on Heavy Metals in the Environment*, vol. 2, ed. T.C. Hutchinson, pp. 887–900. University of Toronto.

(1980). Isolation and elemental analysis of metal-rich granules from the kidney of the scallop *Pecten maximus* (L.). *Journal of Experimental Marine Biology and Ecology*, **42**, 143–56.

George, S.G. & Viarengo, A. (1985). A model for heavy metal homeostasis and detoxication in mussels. In *Marine pollution and physiology: recent advances*, ed. F.J. Vernberg, F.P. Thurberg, A. Calabrese & W.B. Vernberg, pp. 125–39. The Belle W. Baruch Library in Marine Science, No. 13. University of Southern California Press.

Gibbs, P.E. & Bryan, G.W. (1980). Copper – the major component of glycerid polychaete jaws. *Journal of the Marine Biological Association of the United Kingdom*, **60**, 205–14.

(1984). Calcium phosphate granules in muscle cells of *Nephthys* (Annelida, Polychaeta) – a novel skeleton? *Nature*, **310**(5977), 494–5.

Gibbs, P.E., Bryan, G.W. & Ryan, K.P. (1981). Copper accumulation by the polychaete *Melinna palmata*: an antipredation mechanism? *Journal of the Marine Biological Association of the United Kingdom*, **61**(3), 707–22.

Gooday, A.J. & Nott, J.A. (1982). Intracellular barite crystals in two Xenophyophores, *Aschemonella ramuliformis* and *Galatheammina* sp. (Protozoa: Rhizopoda) with comments on the taxonomy of *A. ramuliformis*. *Journal of the Marine Biological Association of the United Kingdom*, **62**, 595–605.

Greaves, G.N., Simkiss, K., Taylor, M. & Binsted, N. (1984). The local environment of metal sites in intracellular granules investigated by using X-ray absorption spectroscopy. *Biochemical Journal*, **221**, 855–68.

Gupta, B.L. & Hall, T.A. (1984). Role of high concentrations of Ca, Cu and Zn in the maturation and discharge *in situ* of sea anemone nematocysts as shown in X-ray microanalysis of cryosections. In *Toxins, drugs and pollutants in marine animals*, ed. L. Bolis, J. Zadunaisky & R. Gilles, pp. 77–95. Berlin, Heidelberg: Springer-Verlag.

Hopkin, S.P. & Nott, J.A. (1979). Some observations on concentrically structured, intracellular granules in the hepatopancreas of the shore crab *Carcinus maenas* (L.). *Journal of the Marine Biological Association of the United Kingdom*, **59**, 867–77.

(1980). Studies on the digestive cycle of the shore crab *Carcinus maenas* (L.) with special reference to the B-cells in the hepatopancreas. *Journal of the Marine Biological Association of the United Kingdom*, **60**, 891–907.

Icely, J.D. & Nott, J.A. (1980). Accumulation of copper within the 'hepatopancreatic' caeca of *Corophium volutator* (Crustacea: Amphipoda). *Marine Biology*, **57**, 193–9.

(1985). Feeding and digestion in *Corophium volutator* (Crustacea: Amphipoda). *Marine Biology*, **89**, 183–95.

Koulish, S. (1976). Organization of 'special' parenchymal cells underlying the midgut in some barnacles. *Journal of Experimental Marine Biology and Ecology*, **23**, 155–70.

Marshall, A.T. & Talbot, V. (1979). Accumulation of cadmium and lead in the gills of *Mytilus edulis*: X-ray microanalysis and chemical analysis. *Chemical and Biological Interactions*, **27**, 111–23.

Martoja, M., Bouquegneau, J.M., Truchet, M. & Martoja, R. (1985). Recherche de l'argent chez quelques mollusques marins, dulcicoles et terrestres. Formes chemiques et localisation histologique. *Vie Milieu*, **35**(1), 1–13.

Martoja, M., Tue, V.T. & Elkaim, B. (1980). Bioaccumulation du cuivre chez *Littorina littorea* (L.) (Gastéropode Prosobranche): signification physiologique et écologique. *Journal of Experimental Marine Biology and Ecology*, **43**, 251–70.

Martoja, R., Alibert, J., Ballan-Dufrancais, C., Jeantet, A.Y., Lhonore, D. & Truchet, M. (1975). Microanalyse et ecologie. *Journal de Microscopie et de Biologie Cellulaire*, **22**, 441–8.

Martoja, R. & Martin, J.-L. (1985). Recherche des mécanismes de détoxication du cadmium par l'Huitre *Crassostrea gigas* (Mollusque, Bivalve). I. Mise en evidence d'une protéine sulfhydrilée de complexation du métal dans les amoebocytes à zinc et cuivre. *Comptes Rendus Hebdomadaires des Séances de l'Academie des Sciences, Paris*, **300**(15), 549–54.

Mason, A.Z. & Nott, J.A. (1980). A rapid, routine technique for the X-ray microanalysis of microincinerated cryosections: an SEM study of inorganic deposits in tissues of the marine gastropod *Littorina littorea* (L.). *Journal of Histochemistry and Cytochemistry*, **28**, 1301–11.

(1981). The role of intracellular biomineralized granules in the regulation and detoxification of metals in gastropods with special reference to the marine prosobranch *Littorina littorea*. *Aquatic Toxicology*, **1**, 239–56.

Mason, A.Z. & Simkiss, K. (1982). Sites of mineral deposition in metal-accumulating cells. *Experimental Cell Research*, **139**, 383–91.

Mason, A.Z., Simkiss, K. & Ryan, K.P. (1984). The ultrastructural localization of metals in specimens of *Littorina littorea* collected from clean and polluted sites. (Menai Bridge [clean] and Restronguet Creek on Fal Estuary [polluted]). *Journal of the Marine Biological Association of the United Kingdom*, **64**, 699–720.

Mauri, M. & Orlando, E. (1982). Experimental study on renal concretions in the wedge shell *Donax trunculus* L. *Journal of Experimental Marine Biology and Ecology*, **63**, 47–57.

Nott, J.A., Corner, E.D.S., Mavin, L.J. & O'Hara, S.C.M. (1985). Cyclical contributions of the digestive epithelium to faecal pellet formation by the copepod *Calanus helgolandicus*. *Marine Biology*, **89**, 271–9.

Nott, J.A. & Langston, W.J. (1989). Cadmium and the phosphate granules in *Littorina littorea*. *Journal of the Marine Biological Association of the United Kingdom*, **69**, 219–27.

Nott, J.A. & Mavin, L.J. (1986). Adaptation of a quantitative programme for the X-ray analysis of solubilized tissue as microdroplets in the transmission electron microscope: application to the moult cycle of the shrimp *Crangon crangon* (L.). *Histochemical Journal*, **18**, 507–18.

Nott, J.A. & Nicolaidou, A. (1989a). Metals in gastropods – metabolism and bioreduction. *Marine Environmental Research*, **28**, 201–5.

(1989b). The cytology of heavy metal accumulations in the digestive glands of three marine gastropods. *Proceedings of the Royal Society of London*, **B237**, 347–62.

(1990). Transfer of metal detoxification along marine food chains. *Journal of the Marine Biological Association of the United Kingdom*, **70**, 905–12.

Nott, J.A. & Parkes, K.R. (1975). Calcium accumulation and secretion in the serpulid polychaete *Spirorbis spirorbis* L. at settlement. *Journal of the Marine Biological Association of the United Kingdom*, **55**, 911–23.

Pedersen, M. & Roomans, G.M. (1983). Ultrastructural localization of bromine and iodine in the stipes of *Laminaria digitata* (Huds.) Lamour, *Laminaria saccharina* (L.) and *Laminaria hyperborea* (Gunn.) Fosl. *Botanica Marina*, **26**, 113–18.

Pirie, B.J.S. & Bell, M.V. (1984). The localization of inorganic elements, particularly vanadium and sulphur, in haemolymph from the ascidians *Ascidia mentula* (Muller) and *Ascidiella aspersa* (Muller). *Journal of Experimental Marine Biology and Ecology*, **74**, 187–94.

Pirie, B.J.S., Fayi, L. & George, S. (1985). Ultrastructural localisation of copper and zinc in the polychaete *Nereis diversicolor*, from a highly contaminated estuary. *Marine Environmental Research*, **17**, 197–8.

Pirie, B.J.S., George, S.G., Lytton, D.G. & Thomson, J.D. (1984). Metal-containing blood cells of oysters: ultrastructure, histochemistry and X-ray microanalysis. *Journal of the Marine Biological Association of the United Kingdom*, **64**, 115–23.

Rainbow, P.S. (1987). Heavy metals in barnacles. In *Barnacle biology* (Crustacean issues 5), ed. A.J. Southward, pp. 405–17. Rotterdam: A.A. Balkema.

(1988). The significance of trace metal concentrations in decapods. *Symposium of the Zoological Society of London*, **59**, 291–313.

Ray, S. & McLeese, D.W. (1987). Biological cycling of cadmium in marine environment. In *Cadmium in the aquatic environment*, ed. J.O. Nriagu & J.B. Sprague, pp. 199–220. John Wiley & Sons.

Rowley, A.F. (1982). The blood cells of *Ciona intestinalis*: an electron probe X-ray microanalytical study. *Journal of the Marine Biological Association of the United Kingdom*, **62**(3), 607–20.

Ryan, K.P., Bald, W.B., Newmann, K., Simonsberger, P., Purse, D.H. & Nicholson, D.N. (1990). Cooling rate and ice-crystal measurement in biological specimens plunged into liquid ethane, propane and Freon 22. *Journal of Microscopy*, **158**, 365–78.

Ryan, K.P. & Purse, D.H. (1984). Rapid freezing: specimen supports and cold gas layers. *Journal of Microscopy*, **136**, RP5–6.

Ryan, K.P., Purse, D.H., Robinson, S.G. & Wood, J.W. (1987). The relative efficiency of cryogens used for plunge-cooling biological specimens. *Journal of Microscopy*, **145**, 89–96.

Schulz-Baldes, M. (1977). Lead transport in the common mussel *Mytilus edulis*. In *Proc. 12th European Marine Biology Symposium*. ed. D.S. McLusky & A.J. Berry, pp. 211–18. Pergamon Press.

Simkiss, K. (1976). Intracellular routes in biomineralization. *Symposia of the Society of Experimental Biology*, **30**, 423–44.

(1979). Metal ions in cells. *Endeavour* (New Series), **3**, 2–6.

(1984). The karyotic mineralization window (KMW). *American Zoologist*, **24**, 847–56.

Simkiss, K. & Mason, A.Z. (1983). Metal ions: metabolic and toxic effects. In *The Mollusca*, vol. 2, ed. P.W. Hochachka, pp. 101–64. Academic Press.

(1984). Cellular responses of molluscan tissues to environmental metals. *Marine Environmental Research*, **14**, 103–18.

Simkiss, K., Taylor, M. & Mason, A.Z. (1982). Metal detoxication and bioaccumulation in molluscs. *Marine Biology Letters*, **3**, 187–201.

Southward, E.C. (1982). Bacterial symbionts in Pogonophora. *Journal of the Marine Biological Association of the United Kingdom*, **62**, 889–906.

Taylor, M. & Simkiss, K. (1984). Inorganic deposits in invertebrate tissues. *Environmental Chemistry*, **3**, 102–38.

Taylor, M., Simkiss, K. & Greaves, G.N. (1986). Amorphous structure of intracellular mineral granules. *Biochemical Society Transactions*, **14**, 549–52.

Taylor, M., Simkiss, K., Greaves, G.N. & Harries, J. (1988). Corrosion of intracellular granules and cell death. *Proceedings of the Royal Society of London*, **B234**, 463–76.

Thomas, P.G. & Ritz, D.A. (1986). Growth of zinc granules in the barnacle *Eliminius modestus*. *Marine Biology*, **90**, 255–60.

Thomson, J.D., Pirie, B.J.S. & George, S.G. (1985). Cellular metal distribution in the Pacific oyster *Crassostrea gigas* (Thun.) determined by quantitative X-ray microprobe analysis. *Journal of Experimental Marine Biology and Ecology*, **85**, 37–45.

Viarengo, A. (1989). Heavy metals in marine invertebrates: mechanisms of regulation and toxicity at the cellular level. *Reviews in Aquatic Sciences*, **1**(2), 295–317.

Walker, G. (1977). 'Copper' granules in the barnacle *Balanus balanoides*. *Marine Biology*, **39**, 343–9.

Walker, G., Rainbow, P.S., Foster, P. & Crisp, D.J. (1975a). Barnacles: possible indicators of zinc pollution? *Marine Biology*, **30**, 57–65.

Walker, G., Rainbow, P.S., Foster, P. & Holland, D.L. (1975b). Zinc phosphate granules in tissue surrounding the midgut of the barnacle *Balanus balanoides*. *Marine Biology*, **33**, 161–6.

16 X-ray microanalysis in biomaterials research

H.K. Koerten, R.A. Dalmeyer, I. van den Brink,
M. Okumura and C.A. van Blitterswijk

A Introduction

Biomaterials include all materials that are used as surgical implants, dental materials and prosthetic implants. The applicability of a material as a biomaterial is determined by its chemical and mechanical properties, by its behaviour in contact with body fluids, cells and tissues, and by the eventual release of toxic components during biodegradation.

Biomaterials science is, therefore, concerned with tissue reactions that occur when artificial, manufactured devices are brought into contact with the living body.

Various microscopical techniques, including light microscopy, transmission electron microscopy and scanning electron microscopy are applied to study these tissue reactions. X-ray microanalysis is of special value in studies on phenomena occurring during biodegradation, i.e. the release and accumulation of chemical trace elements, and also to study the process of calcium phosphate deposition in bone bonding implants. In this chapter, the role of X-ray microanalysis in biomaterials research is discussed, specifically in reference to the biological evaluation of materials in a tissue culture system and post-implantation in experimental animals.

B The total alloplastic middle ear

Implants routinely consist of various types of biomaterial, as shown, for example, by the artificial Total Alloplastic Middle Ear implant (TAM) (Bakker, 1988).

The TAM was developed to reconstruct in one operation the canal wall, the tympanic membrane and the ossicle chain in patients that suffer from chronic otitis media. It consists of a hydroxyapatite canal wall as suspension system to which a hydroxyapatite ossicular chain is connected. As an intermediate between these two components an alloplastic tympanic membrane, made of an elastomeric polymer, is applied.

The canal wall is made of macroporous hydroxyapatite. Tissue ingrowth

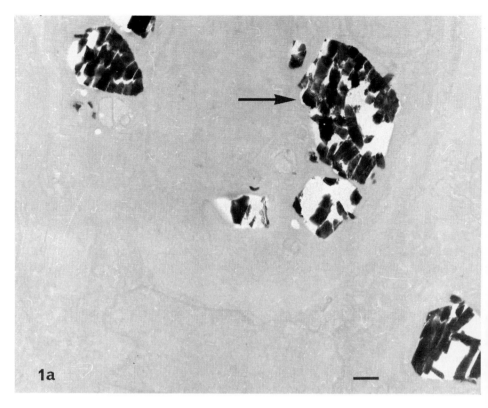

1a

Fig. 16.1a.   For legend see facing page.

and the formation of new bone, especially in the pores, is one of the properties of this biomaterial. Biodegradation of macroporous hydroxyapatite is, to a certain extent, also considered to be possible. The ossicles are made of dense hydroxyapatite. The size of the pores in this material is such that tissue ingrowth in the pores is highly improbable. Furthermore, implants made of dense hydroxyapatite are assumed to be very poorly degradable in a biological system. The tympanic membrane consists of *POLYACTIVE* (HC Implants BV, Leiden, The Netherlands), a biodegradable polymer with bone bonding capacities. *POLYACTIVE* can be produced both as a dense and as a macroporous biomaterial.

Before an implant like the TAM can be applied clinically, the composing materials have to be tested for their chemical and physical properties, such as elasticity and mechanical strength. Materials that pass this test are subsequently tested biologically, first *in vitro* in a tissue culture system to test the reactions of a specific cell or type of tissue to the material under study. Then

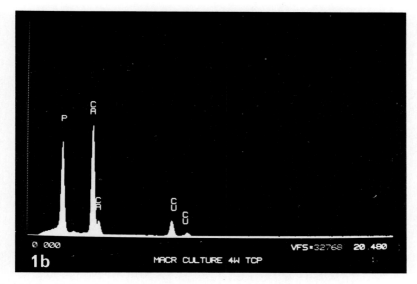

Fig. 16.1.   (a) Transmission electron micrograph of electron dense inclusions in the cytoplasm of macrophages cultured in the presence of calcium phosphate ceramic granules at the interval of one week of culture. Fixation: glutaraldehyde only. Unstained Epon section. Bar: 1 μm. (b) Spectrum of a X-ray microanalytical spot analysis of an inclusion in (a) (arrow). The peaks for calcium and phosphorus confirm the calcium phosphate ceramic nature of this inclusion. The Cu peak is an extraneous peak from the grid.

the materials have to be investigated *in vivo* to establish the reactions of a complete functioning organism to their presence. Properties like biocompatibility, biodegradation, and bone bonding capacity are thus evaluated.

C   *In vitro* biodegradation tests

Since macrophages are assumed to be the cells that are predominantly involved in degradation of biomaterials, tests to verify the rate of biodegradation are performed by adding suspensions of particulates of the materials to macrophages in culture.

In one of our recent studies (Koerten *et al.*, 1991), macrophages were cultured in the presence of five different types of calcium phosphate, i.e. hydroxyapatite, tri-calcium phosphate, tetra-calcium phosphate, fluoroapatite, and magnesium Whitlockite. After a culture period of one week the macrophages were processed for electron microscopical evaluation.

Electron dense inclusions were observed in the cytoplasm of the

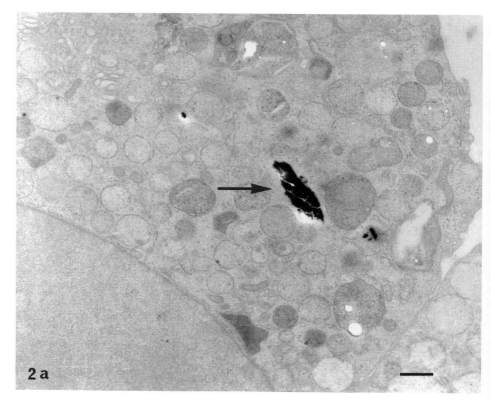

2 a

Fig. 16.2a. For legend see facing page.

macrophages (Fig. 16.1a) and X-ray microanalysis showed that these inclusions were composed of the characteristic elements of the biomaterials under study (Fig. 16.1b). Thus it was demonstrated that endocytosis of the biomaterial occurs *in vitro*. At the longer intervals 2 to 12 weeks, electron dense inclusions, rich in other elements than Ca and P, i.e. aluminium, silicon, chromium, nickel, iron and titanium, were detected. Generally, a number of these elements were simultaneously detected in one inclusion (Fig. 16.2a,b). Since these elements could not be detected by X-ray microanalysis in the raw materials before injection, it has to be concluded that they were released from the materials by biodegradation and accumulated in the lysosomes of the macrophages.

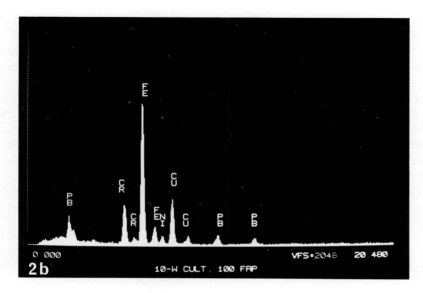

Fig. 16.2. (a) Transmission electron micrograph of an electron dense inclusion (arrow) detected in the cytoplasm of a macrophage at four weeks of culture in the presence of Mg-Whitlockite. Fixation: glutaraldehyde, OsO_4. Epon section stained with lead hydroxide. Bar: 0.5 μm.
(b) Spectrum of a X-ray microanalytical spot analysis of the indicated inclusion in (a). The presence of Cr, Fe and Ni is demonstrated. The Cu and Pb peaks are extraneous peaks derived from the grid and from the staining procedure.

D *In vivo* biocompatibility tests

Implantation of biomaterials within an experimental animal provides a general test for *in vivo* compatibility. A common method is the implantation of plugs or small sheets of the material by surgical procedures at the site where the implant will be applied. Although this method indeed gives direct information on the tissue reaction to the presence of the material at the implantation site, this method is disadvantageous in experiments where the inflammatory reaction to the implanted material has to be studied. In these studies, the wound reaction that is caused by the surgical implantation will obscure the plain tissue reaction to the implanted material. To avoid a wound reaction, injection of a suspension of granulates into the peritoneal cavity of experimental animals can be used as an alternative (Koerten *et al.*, 1988). Intraperitoneal injection of the same calcium phosphates as used in our *in vitro* studies was applied to study biodegradation of these materials *in vivo*. For this purpose, suspensions of the calcium phosphates were injected and at various intervals, ranging from one week to six months, the peritoneal cavities of the mice were evaluated.

The results of these peritoneal experiments showed that the particulates from the suspensions aggregated within 24 hours and at the longer intervals, i.e. from one week and longer, gave rise to the formation of foreign body granulomas (Fig. 16.3). In these granulomas fibroblasts, collagen and capillaries, as well as macrophages, giant cells, lymphocytes, plasma cells and mast cells were seen. Such granulomas were found after the intraperitoneal administration of all of the five calcium phosphates. Vacuoles containing electron dense inclusions were frequently seen in the macrophages and in the multinucleated giant cells. X-ray microanalysis showed that the majority of these inclusions consisted of calcium and phosphorus and therefore the conclusion is justified that they were derived from the injected calcium phosphates.

In agreement with our *in vitro* observations, electron dense inclusions with a variable chemical composition, including calcium, phosphorus, iron, silicon, aluminium, titanium, chromium and magnesium were detected. Also here combinations of the above mentioned elements per inclusion were observed (Fig. 16.4a,b).

As already concluded from our *in vitro* experiments, we assumed that these elements were derived from the injected biomaterials and accumulated by the macrophages during the process of biodegradation.

A number of the retrieved elements, however, are also present in biological organisms as essential (or possibly essential) trace elements and therefore it remained uncertain to what extent the trace elements found *in vivo* were derived from the injected materials or were of a systemic origin. To establish whether or not macrophages are triggered to accumulate systemic trace elements in reaction to the presence of poorly degradable materials, a suspension of crocidolite asbestos fibres was intraperitoneally injected. Crocidolite asbestos was selected because it is generally known that this type of asbestos is a very inert mineral (Zussman, 1979) and therefore cannot be the source of chemical elements in the matrix of macrophage lysosomes. The injected asbestos fibres, like the biomaterials, gave rise to the formation of foreign body granulomas. Within 24 hours of crocidolite injection, asbestos fibres were retrieved in vacuoles of the macrophages in the granulomas. At longer intervals, i.e. two weeks and longer, iron was demonstrated at a relatively high concentration in the lysosomes (Fig. 16.5). Since asbestos is very inert, it has to be concluded that this iron is of a systemic origin.

Trace elements similar to those observed in the present study were also observed in other studies on biodegradation by our group (Blitterswijk *et al.*, 1986; Bakker *et al.*, 1990) and we conclude therefore that the release and accumulation of trace elements from an implant is a general phenomenon.

Enhanced concentrations of certain metallic ions are known to give toxic reactions (Christoffersen & Christoffersen, 1985) and inhibit the function of

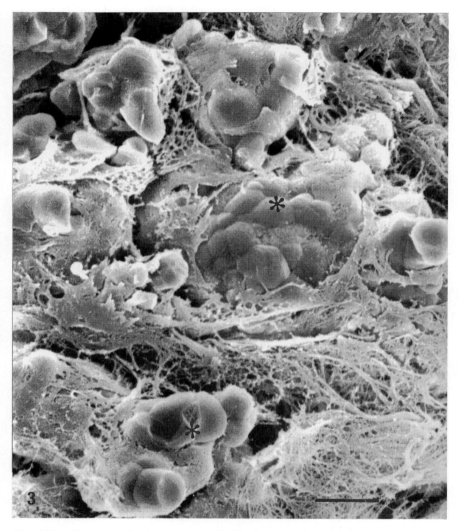

Fig. 16.3. Scanning electron micrograph of a fracture through a granuloma that was formed after the intraperitoneal injection of hydroxyapatite. The HAP particles (asterisks) are surrounded by fibrous tissue and cells. Fixation: glutaraldehyde only. The tissue was dehydrated in alcohol and critical-point dried in CO_2. Bar: 5 μm.

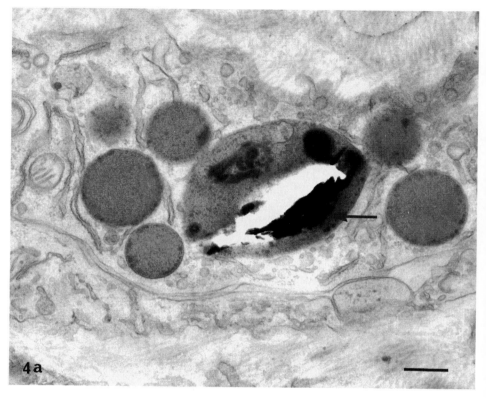

Fig. 16.4a.   For legend see facing page.

receptors at the surface of lymphocytes (Carvalho & de Sousa, 1988). Therefore, it is our opinion that the release and accumulation of trace elements by the surrounding macrophages can contribute to the failure of an implant and thus studies to monitor and control trace element release from implants are important.

E   X-ray microanalysis and bone formation

The proper functioning of an orthopaedic implant depends on its firm attachment to the surrounding bone. Therefore studies to enhance the stability of an implant are of great importance. A better stability can be obtained by the ingrowth of new bone into the pores of the implant and by bonding of the implant material to bone.

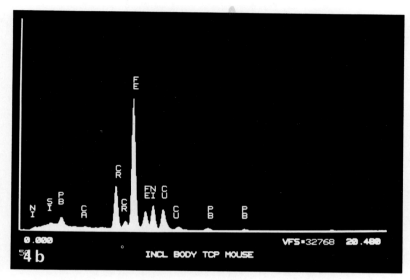

Fig. 16.4.　(a) Transmission electron micrograph of an electron dense inclusion found in a tri-calcium phosphate granuloma. Fixation: glutaraldehyde, OsO_4. Epon sections were stained with lead hydroxide. Bar: 0.5 μm. (b) Spectrum of an X-ray microanalytical spot analysis of the dark inclusion in (a) (arrow). The presence of Ca, Cr, Fe, Ni and Si is demonstrated. Peaks indicating Cu and Pb are extraneous peaks derived from the grid and from the staining procedure.

To study the attachment to bone and new bone formation, macroporous biomaterials were combined with marrow cells and implanted subcutaneously (Okumura *et al.*, 1991). Four weeks after implantation, formation of new bone was observed in the pores of the biomaterials. Implantation of porous *POLYACTIVE* according to this model showed the formation of new bone which was in direct contact with the implantation material. Transmission electron microscopy of the interface between the implant and bone showed that small crystals were present both in the newly formed bone and also in the matrix of the implanted *POLYACTIVE* (Fig. 16.6). X-ray microanalysis demonstrated that these crystals consisted of calcium and phosphorus and thus it was concluded that they were apatite crystals. The exact role of these crystals in the process of implant attachment to bone is still uncertain and the subject of further investigations. Similar observations were made in a study on the implantation of *POLYACTIVE* plugs into the tibiae of rats. At the same period after implantation, scanning electron microscopy showed a direct contact between bone and *POLYACTIVE* without the presence of connective tissue at the interface (Fig. 16.7). X-ray mapping of these implants

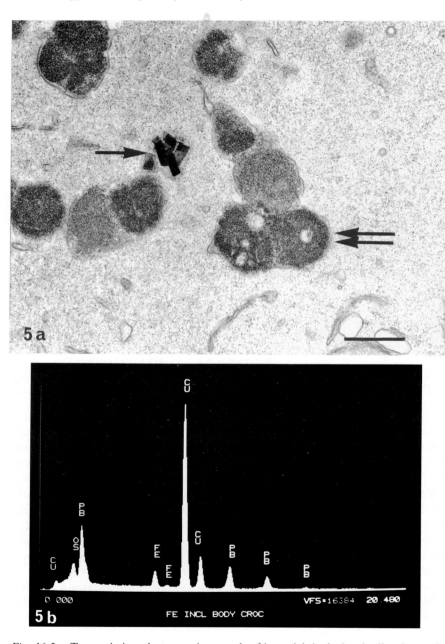

Fig. 16.5. Transmission electron micrograph of iron-rich inclusion bodies detected in a macrophage and an asbestos induced granuloma. Fixation: glutaraldehyde, OsO_4. Epon sections were stained with lead hydroxide. Bar: 0.5 μm. (b) X-ray microanalytical spot analysis of an inclusion body that contained no asbestos fibres in the plane of the section

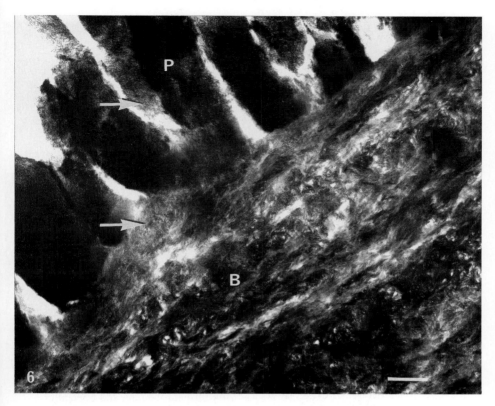

Fig. 16.6. Transmission electron micrograph of the interface between newly formed bone (B) and *POLYACTIVE* (P). Numerous needle-like crystals (arrows) can be seen, both in the implant and in the matrix of the bone. Bar: 0.25 μm.

confirmed the presence of calcium and phosphorus and it was shown by the use of this technique that bone is indeed lining the implant interface.

F   Summary

Experiments on the biocompatibility of calcium phosphates have shown that these substances are subject to biodegradation. It has also been shown that, as a result of the digestive activity of macrophages functioning at the material

---

(double arrow). The spectrum shows that this type of inclusion body is only rich in iron. Peaks indicating Cu, Os and Pb are derived from the grid and from the preparative procedure. The electron dense inclusions (arrow) invariably gave spectra typical of crocidolite asbestos.

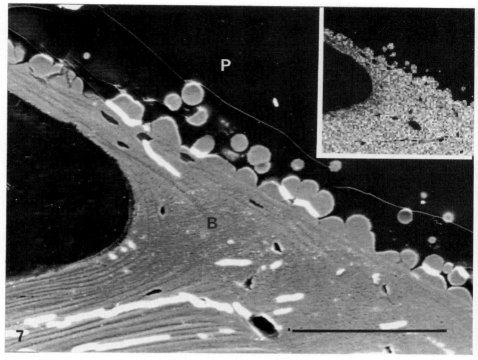

Fig. 16.7.   *POLYACTIVE* plug (P), implanted in the tibia of a rat. The scanning
electron micrograph shows a section through the plug and bone after
embedding in Spurr Ultra Low Viscosity embedding medium. After
sectioning the surface has been polished with aluminium oxide powder.
This micrograph shows that the bone (B) has grown along the
periphery and into the pores of the implant. Inset: X-ray map of the
same area showing the presence of calcium, indicative of bone. The
map confirms the close contact between implant and bone and it shows
that there is no connective tissue at the interface between implant and
bone. Bar: 1000 μm.

interface, trace elements are released from these materials and accumulate in
residual bodies. Toxic reactions resulting from the presence of these trace
elements were not observed in this study. However, since the extent to which
the released elements are transported to other organs where their presence
might cause severe damage (Parkinson, Ward & Kerr, 1981) is not known, we
are of the opinion that biomaterials intended for clinical application should
be extremely pure. Trace element accumulation studies to assess this purity
should therefore be part of the evaluation of newly developed biomaterials.
    Finally, X-ray microanalysis has shown to be valuable in studies on the
formation of new bone. It has been demonstrated that a verdict about the
interfacial deposition of calcium and phosphorus at the sub-microscopical

level is only possible when electron microscopy is used in combination with X-ray microanalysis.

References

Bakker, D. (1988). Alloplastic tympanic membrane. Thesis, Leiden.

Bakker, D., van Blitterswijk, C.A., Hesseling, S.C., Koerten, H.K., Kuijpers, W. & Grote, J.J. (1990). Biocompatibility of a polyether urethane, polypropylene oxide, and a polyether polyester copolymer. A qualitative and quantitative study of three alloplastic tympanic membrane materials in the rat middle ear. *J. Biomedical Materials Research*, **24**, 489.

Blitterswijk, C.A. van, Grote, J.J., Koerten, H.K. & Kuijpers, W. (1986). The biological performance of calcium phosphate ceramics in an infected implantation site. III. Biological performance of β-Whitlockite in the non-infected and infected rat middle ear. *J. Biomedical Materials Research*, **20**, 1197.

Carvalho, C.S. & de Sousa, M. (1988). Iron exerts a specific inhibitory effect on CD2 expression of human PBL. *Immunology Letters*, **19**, 163.

Christoffersen, M.R. & Christoffersen, J. (1985). The effect of aluminium on the rate of dissolution of calcium hydroxyapatite – a contribution to the understanding of aluminium induced bone disease. *Calcified Tissue International*, **37**, 673.

Koerten, H.K., Vogelaar, J., Leenders, H. & van Blitterswijk, C.A. (1992). Inflammatory response to ceramics tested in an *in vivo* model. *Cells and Materials* (submitted).

Koerten, H.K., van Blitterswijk, C.A., Grote, J.J. & Daems, W.Th. (1988). Accumulation of trace elements by macrophages during degradation of biomaterials. In *Implant materials in biofunction*, ed. C. de Putter, G.L. Lange, K. de Groot & A.J.C. Lee, *Advances in Biomaterials*, **8**, 43, Amsterdam: Elseviers Science Publishers.

Okumura, M., van Blitterswijk, C.A., Koerten, H.K. & Ogushi, H. (1991). Osteogenic response of rat bone marrow cells in porous alumina hydroxyapatite and kiel bone. In *Bioceramics. Proceedings of the 4th International Symposium on Ceramics and Medicine,* ed. W. Bonfield, G.W. Hastings & K.E. Tanner, Vol. 4, pp. 3–8. London: Butterworth-Heinemann Ltd.

Parkinson, I.S., Ward, M.K. & Kerr, D.N.S. (1981). Dialysis encephalopathy, bone disease and anaemia: the aluminium intoxication syndrome during regular haemodialysis. *J. Clinical Pathology*, **34**, 1285.

Zussman, J. (1979). The mineralogy of asbestos. In *Asbestos, vol 1, properties, applications, and hazards*, ed. L. Michaels & S.S. Chissick, p. 45. Belfast: The University Press.

17 Applications of X-ray microanalysis in biomedicine: an overview

Godfried M. Roomans

A Summary

X-ray microanalysis has developed into a reliable technique for the determination of the intracellular distribution of elements, due to the use of cryopreparation techniques that preserve the *in vivo* localisation of diffusible elements, and the introduction of quantitative techniques. X-ray microanalysis is now used for investigating a great number of biological problems. The major fields of application are environmental biology/toxicology and physiology/pathology. In human and animal physiology (and pathology), studies have concentrated on the role of calcium in cellular processes; but ion transport in epithelia, under both normal and pathological conditions, is also an important field of study. Furthermore, the possible relation of Na^+ ions to mitogenesis and oncogenesis is a problem where X-ray microanalysis has made interesting contributions.

B Introduction

Biologists have been using electron probe X-ray microanalysis for the past 30 years to complement the morphological information provided by the electron beam with the chemical information provided by the process of X-ray generation in the specimen. The first applications were carried out using a crystal spectrometer, but when energy-dispersive (semiconductor-based) spectrometers became available, these soon became the dominating type of instrument for biological X-ray microanalysis. Energy-dispersive spectrometers were easier to handle, allowed the analysis of all elements simultaneously, and had a better efficiency, which allowed lower beam currents. Only in the analysis of microdroplets (Roinel, 1988) is the wavelength-dispersive spectrometer the instrument of choice. While the cradle of the technique of X-ray microanalysis as such stood in France, where Castaing developed the first spectrometer in the late 1940s, it can be said that biological X-ray microanalysis grew up in Cambridge, round the group of Hall and Gupta. The early history of biological microanalysis has been reviewed by Hall (1986, 1989b) and more recently by Gupta (1991a,b).

These past 30 years have seen a remarkable development of preparative

297

techniques, where conventional preparation techniques for electron microscopy have virtually disappeared from the toolbox of the biological X-ray microanalyst. The need to obtain results that could be communicated to physiologists prompted the development of quantitative techniques. X-ray microanalysis is used in very diverse fields and a survey of all applications is beyond the scope of a single chapter. Prominent fields of application are environmental pollution, cell physiology, muscle and epithelial physiology, and pathology. As examples of areas with potential significance for human pathology, cancer research and epithelial (patho)physiology will be considered in somewhat more detail.

C   Specimen preparation

With regard to preparation of cells and tissues for X-ray microanalysis, one can distinguish four groups of specimens.

1. Specimens needing little or no preparation, other than a conductive (carbon) coating, e.g. hair, nails, gall stones and kidney stones.
2. Specimens that are digested prior to analysis since only the particulate content, not the tissue itself, is important. The reason for considering this category in a review on biological microanalysis is that analysis of particulates in, for example, the lung is quantitatively one of the most important applications of X-ray microanalysis in pathology.
3. Specimens prepared by conventional aldehyde fixation, possibly also osmium fixation, dehydration, embedding and wet sectioning. These specimens are used for analysis of elements that are very tightly bound to tissue structures and where only qualitative analysis is of importance. It is well-known that aldehyde fixation results in loss and redistribution of diffusible ions (Roomans, 1990b).
4. Specimens prepared by cryotechniques: rapid freezing followed by sectioning, or possibly by freeze-substitution or embedding of freeze-dried tissue. Bulk specimens can be analysed in the frozen–hydrated state or freeze-dried.

With regard to the freezing step, the experience of the past 20 years with freezing of animal material has resulted in consensus on a number of issues.

1. Freezing should be carried out as fast as possible to avoid the formation of ice crystals that may displace ions and destroy tissue structure (see contribution by Zierold elsewhere in this volume). Since biological tissue is a poor conductor of heat, the use of small samples is a prerequisite and even then only a small, outer layer of the sample can be frozen without visible ice crystal 'damage' (reviewed by Robards & Sleytr, 1985). Propane and ethane are more suitable liquid coolants than the freons, and much better than liquid nitrogen, in which a thermally insulating layer

surrounding the specimen develops. Even better than liquid coolants may be the use of instruments in which the tissue is frozen against liquid nitrogen- or liquid helium-cooled metal. In some cases use of high-pressure freezing may give superior results (Zierold, Tobler & Müller, 1991).

2. Even in work with experimental animals that can be done under well-controlled conditions, damage to the tissue prior to freezing should be avoided. Dissecting the sample may introduce ion shifts because of anoxia (von Zglinicki, Bimmler & Purz, 1986) even within time periods of a minute. Also, with small samples, loss of water by evaporation may occur. Therefore, methods for *in situ* freezing, such as cooled clamps and cryobioptic needles (von Zglinicki *et al.*, 1986) have been developed.

Optimal freezing conditions can, in practice, only be obtained with a limited number of specimen types, where the investigator has complete control over all relevant experimental conditions. This may apply to work with cell cultures, but limitations are encountered in work with experimental animals, in particular mammals, and even more restrictions apply to analysis of human tissue.

The specific problems associated with obtaining human tissue for micro-analytical studies can be divided into problems occurring before freezing the tissue, and problems occurring during the freezing.

If tissue is to be obtained after the death of the patient, it may not be possible to control the time between death and sampling because of legal or practical reasons. Delays of several hours, up to a day, may easily occur. In such cases, post-mortem changes in the ion distribution have to be taken into account (Kuypers & Roomans, 1980; Roomans & Wroblewski, 1985): the cellular concentrations of Na, Cl and Ca will increase, those of K and Mg will decrease. No systematic studies on human tissue have as yet been carried out to investigate whether such material is still useful for X-ray microanalysis. In a pilot study of animal material, it could be shown that changes in cellular ion content induced before death (mimicking a pathological process) could be distinguished even after a post-mortem period of 24 h (Roomans & Wroblewski, 1985).

If tissue is to be obtained by biopsy, this may have to be performed under local or general anaesthesia out of consideration for the patient; effects of anaesthetics on ion distribution have not been investigated by X-ray microanalysis, but from what is known about the mechanism of action of several anaesthetics it is evident that they may indeed affect ion and water distribution. This does not necessarily preclude the use of biopsies obtained during anaesthesia: in a study on hamster tracheal epithelium we could show that a physiological process such as chloride efflux induced by β-adrenergic stimulation persisted during anaesthesia (Spencer & Roomans, 1989). On the other hand, when tissue is taken without anaesthesia the excitement or tension (adrenalin) might be a complicating factor.

Patients from whom the biopsy is taken may be undergoing drug treatment for the disease investigated or for unrelated diseases or conditions (e.g. contraceptives). Such treatment has to be recorded in the experimental protocol and the possible implications on ion and water distribution considered.

While it may be possible to motivate a patient to undergo a biopsy for diagnostic or research purposes, it may be difficult or unethical to obtain control material when the biopsy procedure is particularly painful or carries some clinical risk. The problem is exacerbated if special requirements such as age- or sex-matching are important.

Also, when freezing human tissue it may not be possible to use *in situ* freezing techniques. This could be because the tissue is inaccessible for an *in situ* method: it would be unethical to use open biopsy procedures, for example, in the case of a liver biopsy. *In situ* freezing techniques also cause damage to the surrounding tissue which may be unacceptable. When freezing is to be carried out in the operating theatre considerations of safety and sterility apply that may make it impossible to use particular freezing methods. The use of inflammable liquids such as propane or ethane may be prohibited, and the use of large equipment may be unpractical. Finally, a well-known problem in ultrastructural pathology is that in many changes in disease processes may be unevenly distributed in the tissue. In morphological studies one therefore always starts with a light microscopical investigation. Similarly, for X-ray microanalysis the use of very small samples may give unrepresentative and erroneous results and the initial investigation of relatively large samples may be necessary.

It may be possible that the use of *in vitro* systems, where the tissue can recover from the trauma of the biopsy, can alleviate some of the problems described above. Use of an *in vitro* system uncouples the biopsy procedure and the freezing procedure both in time and in place and would allow the use of the most sophisticated freezing techniques in the laboratory without any of the restrictions imposed in a clinical environment. However, although such systems may be fully acceptable to the physiologist, the elemental composition of incubated tissue may not be the same as that of tissue *in situ*. However, very little attention has been given to possible improvements of the incubation procedure and much remains to be done. Also the use of cell and organ cultures could be a valuable complement to studies of tissue *in situ* (Wroblewski & Roomans, 1984; Warley, 1987).

In most cases, freezing is followed by cryoultramicrotomy at very low temperature (Roomans, Wei & Sevéus, 1982). There has been discussion about the mechanism by which sections are obtained at temperatures below $-100\ °C$ and Saubermann, Riley & Beeuwkes (1977) have advocated the use of much higher temperatures on the grounds that the process of thin sectioning at low temperature would transfer so much energy to the section that partial melting would be possible. On the other hand, Karp, Silcox &

Somlyo (1982) showed that the increase in temperature during cryosectioning around $-100\ °C$ could not be more than a few degrees, and most groups section below $-100\ °C$. However, Kirk, Knoff & Lee (1991) recently showed that both 'fracturing' and 'cutting' could occur during preparation of thin cryosections, and that the latter process might be associated with superficial melting of the sections. The frozen sections can be transferred to the electron microscope at low temperature, and be freeze-dried in the vacuum of the microscope. According to Hagler & Buja (1986) 'external freeze-drying' in a separate instrument carries the risk of rehydration of the sections and gross redistribution of ions, but von Zglinicki & Uhrik (1988) showed that with careful handling of externally freeze-dried sections artifacts could be avoided. Very thin cryosections cannot be analysed in the frozen–hydrated state (Zierold, 1988); the minimal thickness for analysis of frozen–hydrated sections appears to be 1–2 µm.

The main alternatives to the preparation of thin cryosections are freeze-substitution and sectioning of freeze-dried embedded tissue (Wroblewski, Wroblewski & Roomans, 1988). Comparative studies have given conflicting results. According to Roos & Barnard (1986), freeze-substitution resulted in extensive losses of diffusible elements, but their results may have been due to an unfortunate choice of substitution fluid. More careful studies by Condron & Marshall (1990) and Edelmann (1991) show losses not larger than 30 % for diffusible elements.

Analysis at the cellular level can be carried out on frozen–hydrated bulk tissue (Marshall, 1988), freeze-fractured freeze-dried bulk samples (Zs.-Nagy, 1989) or freeze-dried cryostat sections (Wroblewski *et al.*, 1988; McMillan & Roomans, 1990). Such specimens are relatively easy to prepare and analyse and can also be prepared from tissue that has not been frozen under optimal conditions. This type of specimen appears useful in the field of experimental and clinical pathology.

Fig. 17.1 a–c shows the same tissue (mouse intestinal epithelium) prepared in three different ways: freeze-dried thick cryosections for analysis at the cell level; freeze-dried thin cryosections for analysis at the sub-cellular level; frozen–hydrated bulk samples for analysis of local water content.

D Quantitative analysis

It was realised quite early on, that if X-ray microanalysis were to be applied to problems in physiology and pathology, quantitative techniques would be necessary. This is particularly valid for studies involving endogenous elements, e.g. in studies of the (patho)physiology of ion transport.

The continuum theory of quantitative X-ray microanalysis was developed by Dr T.A. Hall in the late 1960s (Marshall & Hall, 1968; Hall, 1968; reviewed by Hall, 1989a) for thin sections, based on earlier theories on

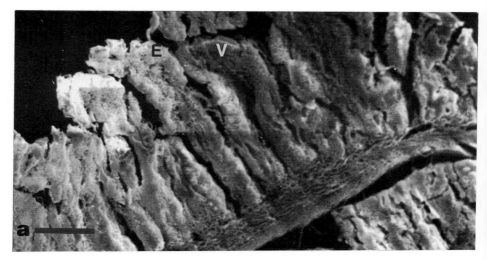

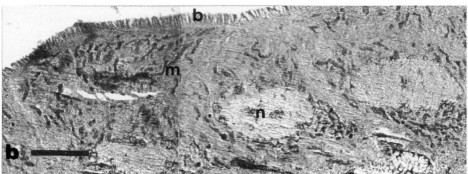

Figure 17.1a and b.   For legend see facing page.

characteristic and continuum X-ray generation, in particular on Kramers's (1923) 'law' on the relation between the continuum intensity and the atomic number of the target. The derivation of the theory was outlined in the now classic papers by Hall (1971) and Hall, Anderson & Appleton (1973). With the advent of microcomputers, a general formalism for quantitative micro-analysis of biological specimens was developed (Roomans & Sevéus, 1976) that can be solved by an iterative procedure

$$C_{sp} = C_{st} \frac{R_{sp}(Z^2/A)_{sp}}{R_{st}(Z^2/A)_{st}}$$

where C is the concentration, R the relative intensity (peak/background),

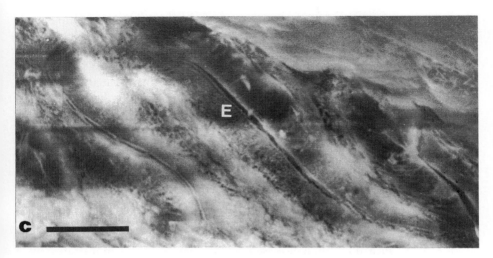

Fig. 17.1. (a) Freeze-dried, thick (16 μm) cryosection of mouse intestinal epithelium; villi (V) and epithelial cells (E) can easily be recognised. Bar = 100 μm, (b) Freeze-dried, thin cryosection of the same tissue, where brush border (b), nucleus (n), and mitochondria can be recognised. Bar = 10 μm. (c) Frozen–hydrated bulk specimen, where the epithelial layer (E) (though not individual cells) can be recognised. Bar = 50 μm.

Z^2/A the weighted mean value of Z(atomic number)^2 divided by A (atomic weight); subscripts sp and st refer to specimen and standard, respectively. C_{sp} and Z^2/A_{sp} are related unknowns (since Z^2/A_{sp} is a function of C_{sp}), and the iterative procedure matches both unknowns. This formalism forms the basis of the commercial programmes that are now routinely used for quantitative biological microanalysis. The formula given above also makes it clear that in quantitative biological microanalysis standards have to be used, in contrast to the case in the materials sciences, where standardless quantitation based upon the determination of elemental ratios is possible. However, biological specimens generally consist of about 95% of elements that cannot be detected by conventional semiconductor detectors (H, C, N and O). Methods for preparation of proper standards have been reviewed by Roomans (1979, 1980) and more recently by Warley (1990).

It has been clear for some time (Hall *et al.*, 1973) that the contribution of extraneous sources to the spectrum is a problem in quantitative analysis. Gupta & Hall (1979) and Roomans & Kuypers (1980) have provided a comprehensive formalism for the correction of extraneous background. In practice, problems may still occur if the extraneous background is large in proportion to the specimen-generated background. The theory used in correcting for extraneous background assumes that all of the extraneous signal is due to electrons hitting the grid bars after a single high-angle scattering event in the section. However, in practice, most of the extraneous background may be generated by multiple scattering events involving specimen holder and pole piece. By keeping geometrical conditions as constant as possible, the errors can be kept systematic (Roomans, 1988), and may cancel out to a large extent if specimen and standard are analysed in the same way. However, the difficulty in adequately correcting for the extraneous background is the major weakness of the continuum method (Steinbrecht & Zierold, 1989). There are alternative methods to determine specimen mass, such as the use of brightfield or darkfield transmission or scanning transmission electron microscopy signals (e.g. Linders *et al.*, 1982; Zierold, 1988) or the zero-loss signals determined by electron energy-loss spectroscopy (Leapman, Fiori & Swyt, 1984) but these methods require separate measurements and/or additional equipment and have not become popular.

The continuum or 'Hall' method is not limited to thin sections. A linear relationship between the peak-to-background ratio and the concentration of an element in a thick biological sample was empirically demonstrated by Cobet & Traub (1971) and Zs.-Nagy *et al.* (1977). Statham & Pawley (1978) and Small *et al.* (1978) introduced the peak-to-background ratio method for inorganic samples with an irregular shape, such as particles. The background was determined in the same energy region as the peak, i.e. the background under the peak was chosen. This was done to provide an 'intrinsic' correction for absorption, which is the most important part of the ZAF-correction, also in biological bulk specimens (Roomans, 1981). The use of the peak-to-background method for biological bulk samples thus became routine (Roomans, 1981; Zs.-Nagy & Casoli, 1990) especially since the conventional ZAF-methods did not work well with biological specimens (Boekestein *et al.*, 1983). The same method can be used with semi-thick specimens (Wroblewski *et al.*, 1983) and with frozen–hydrated specimens (Zs.-Nagy, Lustyik & Bertoni-Freddari, 1982).

Water can be regarded as one of the most important constituents of the cell. Measurement of local water content is necessary to interpret the data obtained by X-ray microanalysis, e.g. in physiological studies of ion transport. Methods for the determination of local water concentration by X-ray microanalysis can be based on measurements on the same sample both in the frozen–hydrated and the frozen–dried state, but also on measurements in the frozen–dried state only. The quantitative X-ray analysis of water in biological specimens has recently been reviewed (Roomans, 1990a).

E   Applications

The number of biological and biomedical problems investigated by X-ray microanalysis is now so large that a single review on this subject cannot possibly cover all fields. X-ray microanalysis is frequently used to localise heavy metals in studies of a toxicological or environmental character. Due to the relatively high local concentration needed to detect a particular element by X-ray microanalysis (around 200–500 p.p.m.), X-ray microanalysis is generally only useful in situations where the metal of interest is accumulated in particular cellular or extracellular structures, a mechanism which is present in a number of invertebrate animals (Nott, 1991; Morgan & Winters, 1991) and is significant in the protection of these animals against high levels of toxic metals in the environment. Applications in human pathology (reviewed by Shelburne *et al.*, 1989) often concern identification of particulate matter in lung (Churg, 1989; Roggli, 1991; Abraham, Burnett & Hunt, 1991) and studies of pathological calcifications. The role of calcium as a regulator of a variety of biological processes, in particular of muscle contraction, is an area of intense research, where X-ray microanalysis is one of the many techniques used for localisation of calcium at the (sub)cellular level. Another important area of research is the localisation of the diffusible ions Na, K, and Cl both in plant cells and in animal cells. A few biological problems where X-ray microanalysis has made or is making important contributions will be considered in somewhat more detail.

1 Cell physiology

X-ray microanalysis has been applied to a large variety of tissue-specific biological problems (for instance, epithelia, smooth muscle or heart muscle), but some lessons have been learned that apply to cells in general. It has been learned that, while mitochondria *can* take up large amounts of calcium, mitochondria *in situ* normally have a low calcium concentration (Gupta, 1991b). Accumulation of calcium by mitochondria is a definite sign of cell damage, caused either by a pathological process or by preparation artifacts (Gupta, 1991b; Krefting *et al.*, 1988; Sevéus, Brdiczka & Barnard, 1978). X-ray microanalysis has also helped to dispel the idea of the cell as a watery bag of ions. Ion gradients and ion binding to cellular structures (von Zglinicki, Ziervogel & Bimmler, 1989) show that a complex relationship exists between the inorganic and the organic constituents of the cell. These complex relationships are not limited to the interior of the cell, but can also be found in extracellular matrix and mucus (Engel & Catchpole, 1989; Gupta, 1989). The X-ray microanalysis data often appear to conflict with microelectrode studies that suggest that all ions in the cell are in free solution (Edelmann, 1990). The interpretation of ion distribution data is still a matter of dispute and theories that totally disagree with the conventional concept of 'pump and channel'-mediated ion transport have been proposed and discussed (e.g. Ling, 1988, 1990).

2   Cell proliferation and cancer

Several lines of evidence indicate that there is a relationship between the intracellular ionic environment and cellular proliferation. Cone (1971) developed a theory proposing that stimulation of quiescent cells to divide was coupled to an influx of Na^+ ions into the cells. The increased cellular Na^+ concentration could be part of the regulatory mechanism of cell proliferation. Specifically, Cone (1971) proposed that cancer cells would have an abnormally high Na^+ concentration, and that this was part of the defective regulation of proliferation in cancer cells. Energy-dispersive X-ray microanalysis has provided an excellent method to test Cone's hypothesis. In one of their initial studies on this subject, Cameron *et al.* (1980) compared a variety of tumour cells with their non-tumour counterparts and found large and significant increases of both Na and Cl concentrations in the tumour cells, whereas the concentrations of other elements measurable by X-ray microanalysis did not differ significantly between the two groups of cells. On the other hand, when a group of rapidly dividing normal (non-tumour) cells was compared with slow-growing counterparts, a smaller difference in Na and Cl was noted but, in addition, the rapidly dividing cells had higher concentrations of Mg, P and K. Elevated Na concentrations were also noted in the S and M phase of HeLa cells (Stephen *et al.*, 1990). Significantly elevated concentrations of Mg, P and K were also found in epidermal cells in affected areas in patients with psoriasis (non-proliferating areas from the same patients were used as controls) (Grundin *et al.*, 1985). It is tempting to speculate that the increased level of P is due to an increased concentration of nucleic acids that are preferably stabilised by K^+ and Mg^{2+} ions. It seems that an increased intracellular Na^+ concentration can be related to mitogenesis, but that the much higher Na^+ concentration in cancer cells suggested an even stronger link between Na and oncogenesis (Cameron & Smith, 1989). In a study of normal, preneoplastic and neoplastic mouse mammary tissue (Smith *et al.*, 1981) it was found that the nuclear concentration of Na and Cl in the neoplastic cells was two- to threefold higher than in normal or preneoplastic cells; in addition, K was significantly increased in the neoplastic cells. In yet another system, colon cells of rats treated with the carcinogen 1,2-dimethylhydrazine (DMH), a significant increase of the intracellular Na concentration was found prior to overt neoplastic transformation of the cells, even though the increase in Na followed the increased proliferative activity of the tissue (Cameron & Smith, 1989). Pieri *et al.* (1983, 1984) found that in Morris hepatoma (as well as in regenerating liver cells, or in liver cells stimulated to proliferate by treatment with tri-iodothyronine) both Na and Cl were significantly higher than in normal liver cells. Interestingly, amiloride, an inhibitor of the Na^+–H^+ antiport mechanism, not only specifically decreased the intracellular Na concentration in a number of tumours *in vivo* (a mammary adenocarcinoma and a hepatoma) but also inhibited the growth of these tumours in a dose-dependent fashion (Sparks *et al.*, 1983). Szolgay-Daniel *et al.* (1991) recently observed that short-term treatment with amiloride decreased the intracellular Na concentration in a glioma cell line and in cultured colon carcinoma cells. Cameron & Hunter (1983) also found that amiloride decreased the intracellular Na concentration in the rapidly proliferating intestinal crypt cells. In recent studies, our group found that chronic amiloride treatment results in a markedly decreased intracellular Na/K ratio in the epithelial cells of the small intestine of mice treated *in utero* with amiloride, while the effect on adult animals is very small (Mörk, von Euler & Roomans, 1991; von Euler, Mörk & Roomans, unpublished results).

Increased intracellular Na/K ratios were established in a number of human tumours *in situ*: invasive urogenital cancers (Zs.-Nagy *et al.*, 1981), thyroid tumours (Zs.-Nagy *et al.*, 1983) and laryngeal tumours (Zs. Nagy *et al.*, 1987), as well as in tumours of the oral mucosa (Wroblewski, Anniko & Wroblewski, 1983). In tumors *in situ* a number of necrotic cells occur; these cells have a high Na and Cl content because their Na^+–K^+-pump is no longer functional. However, necrotic cells can be distinguished on the basis of several criteria from living tumour cells (Zs.-Nagy, 1989). Hence, the increased Na and Cl in tumour cells cannot simply be considered as an artifact due to the presence of necrotic cells; also the fact that similar findings have been made on cultured cells (Szállási *et al.*, 1988) speaks against such a notion. Nevertheless, in studies involving tumours *in situ* attention should always be given to the possibility of changes in intracellular Na/K ratios due to metabolic factors or sampling technique rather than mitogenesis or oncogenesis. As an example, Tvedt *et al.* (1987) showed an increase in Na/K ratio in human prostate neoplasms compared with normal prostate cells. However, this group found that differences in sampling techniques could cause significant differences in intracellular Na and K concentrations (Tvedt *et al.*, 1989) and did not draw any conclusions from their Na and K data. On the other hand, they suggested that an increase in intranuclear calcium concentration with advancing age could be of pathological significance to growth disturbances in the prostate (Tvedt *et al.*, 1987).

Evidently, the exact relationship between intracellular Na concentrations and cell proliferation has not yet been defined. The problem is highly interesting, however, since the hypothesis of Cone (1971) and the accumulated microprobe data confirming the hypothesis would seem to suggest a role for amiloride, or its more potent dimethyl- or diethyl-analogues, in treatment of cancer; according to Zs.-Nagy (1989), clinical tests of amiloride in this context had not yet been carried out. X-ray microanalysis could, in particular, be put to good use in investigations of tumours *in situ* to correlate intracellular ion levels with morphological and chemical characteristics defining the state of the tumour cell.

3   Epithelial ion transport

A recent survey (LeFurgey, Bond & Ingram, 1988) shows that many epithelia have been the subject of physiological studies by X-ray microanalysis. Our own research has centred on transepithelial ion transport, since the congenital hereditary disease cystic fibrosis (CF) is caused by a defective regulation of chloride (and water) secretion by epithelial cells. In patients with this disease, the chloride channel in the apical membrane of the epithelial cells cannot be opened by cAMP. It is thought that this results in the production of viscous mucus blocking the smaller airways, leading to repeated infections, and eventually resulting in serious deterioration of lung function and death. The disease is further complicated by malfunctioning of other epithelial tissues, such as pancreas and intestine, in which water transport is much less than normal. In a study on hamster tracheal epithelium (Spencer & Roomans, 1989) it was shown that β-adrenergic stimulation of the animals by an intraperitoneal injection of isoproterenol resulted in a significant decrease of the cellular chloride content, indicating that the chloride efflux exceeds the chloride intake across the basolateral membrane. This decrease in chloride content after stimulation with cAMP was also observed in normal cultured human respiratory epithelial cells, but not in respiratory epithelial cells from patients with cystic fibrosis, which confirms that chloride efflux in CF cells cannot be activated by cAMP (Sagström *et al.*, 1990b). It has been debated whether the defect in chloride transport in CF epithelial cells can also be found in other cells, such as fibroblasts. Since fibroblasts are much easier to culture than most epithelial cells, they would provide a very useful experimental system. However, our recent data (von Euler & Roomans, 1991) show that cAMP stimulates chloride efflux from CF fibroblasts to about the same extent as from normal fibroblasts.

The diagnostic criterion for CF usually is the elevated concentration of Na and Cl in the sweat. Although this is generally done by chemical analysis, X-ray microanalysis can be used both for direct analysis of sweat droplets (Quinton, 1978) and for indirect analysis of sweat dried onto the nail surface (Roomans *et al.*, 1978). This elevated concentration of Na and Cl is due to a defective reabsorption of these ions in the duct. In sweat glands from Cf patients, the Na and Cl concentrations in duct cells did not increase on stimulation (Wilson *et al.*, 1988), in contrast to what happened in normal glands; this result is consistent with a defective reabsorption.

It would be interesting to have an animal model for cystic fibrosis where the sequence of pathological tissue changes, starting from the defective trans-epithelial chloride transport to the full-blown clinical symptoms, could be studied. Since it is difficult to inhibit the chloride channel itself, we have attempted to reproduce the human disease in experimental animals by inhibiting the transepithelial chloride transport by long-term treatment with diuretics. Treatment with furosemide or amiloride for up to three months was indeed found to cause changes in the elemental content of pancreas and submandibular gland acinar cells (Scarlett *et al.*, 1988; Sagström *et al.*, 1990a, Mörk *et al.*, 1991). In particular, the intracellular chloride concentration decreased; this reduces or abolishes the gradient driving chloride efflux through the apical membrane. Physiological experiments confirmed that fluid transport was partially inhibited. However, the changes in tissue structure were relatively minor, even if treatment was continued for a period of up to 13 months. Treatment *in utero*, however, appeared to cause more marked changes in the structure of the pancreas, as well as a decrease in cellular chloride content (von Euler, Mörk & Roomans, 1992).

Early microprobe studies on the intestinal epithelium were carried out by Gupta, Hall & Naftalin (1978). This tissue is particularly interesting since a disturbance of ion and water transport is of great clinical significance. If cystic fibrosis is due to an inhibition of water transport, secretory diarrhoea presents the opposite problem. The diarrhoea in cholera, for example, is caused by a continued activation of chloride channels, because the breakdown of cAMP is inhibited by the bacterial toxin, and the chloride and water efflux cannot be shut off. Hence we investigated by X-ray microanalysis the activation of chloride efflux by various agents. Analysis of resting cells showed ion gradients (measured on a dry weight basis) between the apical part (brush border) to the basal part (von Zglinicki & Roomans, 1989a). While the high Na and Cl levels in the brush border may be partially due to the inclusion of some extracellular luminal fluid in the measured area, the difference between the terminal web area and the basal cytoplasm is significant. Vaso-intestinal peptide (VIP) is a known stimulator of fluid secretion in the intestine and VIP stimulation significantly decreases the chloride and potassium content of the crypt cells, while Na is not significantly affected (von Zglinicki & Roomans, 1990). The effectiveness of β-adrenergic stimulation of fluid transport in the intestine is controversial. However, stimulation of the animals with isoproterenol caused a decrease of the cellular chloride concentration in crypt cells (von Zglinicki & Roomans, 1989b). Moreover, this decrease could be inhibited by alloxan, which is an inhibitor of the enzyme adenylate cyclase which produces cAMP. This indicates that isoproterenol stimulation of secretion in the intestine could be mediated by cAMP, similar to the case in the respiratory epithelium.

Absorptive diarrhoea, on the other hand, is assumed to be caused by inhibition of water influx in the villus tip cells. However, a recent study by Spencer *et al.* (1990) showed that diarrhoea induced by rotavirus-infection is a very complicated process affecting both villus tip and villus base cells. Apparently, the infection results in the shedding of villus cells, followed by regeneration. Interestingly, there seems to be a sodium gradient rising from the base of the villi to their tips, although the concentration difference found by Spencer *et al.* (1990) is less than that observed by Sjöqvist & Beeuwkes

(1989), who used freeze-dried, paraffin-embedded thick sections, rather than thin cryosections.

References

Abraham, J.L., Burnett, B.R. & Hunt, A. (1991). Development and use of a pneumoconiosis database of human pulmonary inorganic particulate burden in over 400 lungs. *Scanning Microscopy*, **5**, 95–108.
Boekestein, A., Stadhouders, A.M., Stols, A.L.H. & Roomans, G.M. (1983). Quantitative biological X-ray microanalysis of bulk specimens: an analysis of inaccuracies involved in ZAF-correction. *Scanning Electron Microscopy*, **1983/II**, 725–36.
Cameron, I.L. & Hunter, K.E. (1983). Effect of cancer cachexia and amiloride treatment on the intracellular sodium content in tissue cells. *Cancer Research*, **43**, 1074–8.
Cameron, I.L. & Smith, N.K.R. (1989). Applications of electron probe X-ray microanalysis to the study of ionic regulation of growth in normal and cancer cells. In *Microprobe analysis in medicine*, ed. P. Ingram, J.D. Shelburne & V.L. Roggli, pp. 291–302. New York: Hemisphere.
Cameron, I.L., Smith, N.K.R., Pool, T.B. & Sparks, R.L. (1980). Intracellular concentration of sodium and other elements as related to mitogenesis and oncogenesis *in vivo*. *Cancer Research*, **40**, 1493–500.
Churg, A. (1989). Quantitative methods for analysis of disease induced by asbestos and other mineral particles using the transmission electron microscope. In *Microprobe analysis in medicine*, ed. P. Ingram, J.D. Shelburne & V.L. Roggli, pp. 79–95. New York: Hemisphere.
Cobet, U. & Traub, F. (1971). Untersuchungen an speziellen biologischen Geweben mit der Elektronenstrahlmikrosonde. *Experimentelle und Technische Physik*, **19**, 479–80.
Condron, R.J. & Marshall, A.T. (1990). A comparison of three low temperature techniques of specimen preparation for X-ray microanalysis. *Scanning Microscopy*, **4**, 439–47.
Cone, C.D. Jr. (1971). Unified theory on the basic mechanism of normal mitotic control and oncogenesis. *Journal of Theoretical Biology*, **30**, 151–81.
Edelmann, L. (1990). The physical state of potassium in frog skeletal muscle studied by ion-sensitive microelectrode and by electron microscopy: interpretation of seemingly incompatible results. *Scanning Microscopy*, **3**, 1219–30.
(1991). Freeze-substitution and preservation of diffusible ions. *Journal of Microscopy*, **161**, 217–28.
Engel, M.B. & Catchpole, H.R. (1989). Microprobe analysis of element distribution in bovine extracellular matrices and muscle. *Scanning Microscopy*, **3**, 887–94.
Grundin, T.G., Roomans, G.M., Forslind, B., Lindberg, M. & Werner, Y. (1985). X-ray microanalysis of psoriatic skin. *Journal of Investigative Dermatology*, **85**, 378–80.
Gupta, B.L. (1989). 1 µm thick frozen hydrated/dried sections for analysing pericellular environment in transport epithelia. New results from old data. In *Electron probe microanalysis: applications in biology and medicine*, ed. K. Zierold & H.K. Hagler, pp. 199–212. Berlin: Springer.
(1991a). Theodore Alvin Hall: a biographical sketch and personal appreciation. *Scanning Microscopy*, **5**, 369–78.

(1991b). Ted Hall and the science of biological microprobe X-ray analysis. A historical perspective of methodology and biological dividends. *Scanning Microscopy*, **5**, 379–426.

Gupta, B.L. & Hall, T.A. (1979). Quantitative electron probe microanalysis of electrolyte elements within epithelial compartments. *Federation Proceedings*, **38**, 144–53.

Gupta, B.L., Hall, T.A. & Naftalin, R.J. (1978). Microprobe measurement of Na, K and Cl concentration profiles in epithelial cells and intracellular spaces of rabbit ileum. *Nature*, **272**, 70–3.

Hagler, H.K. & Buja, L.M. (1986). Effect of specimen preparation and section transfer techniques on the preservation of ultrastructure, lipids and elements in cryosections. *Journal of Microscopy*, **141**, 311–17.

Hall, T.A. (1968). Some aspects of the microprobe analysis of biological specimens. In *Quantitative electron microprobe analysis*, ed. K.F.J. Heinrich, NBS Special Technical Publication, 298, pp. 269–99.

(1971). The microprobe assay of chemical elements. In *Physical techniques in biochemical research*, Vol. 1A, ed. G. Oster, pp. 157–275. New York: Academic Press.

(1986). The history and current status of biological electron probe X-ray microanalysis. *Micron and Microscopica Acta*, **17**, 91–100.

(1989a). Quantitative electron probe X-ray microanalysis in biology. *Scanning Microscopy*, **3**, 461–6.

(1989b). The history of electron probe microanalysis in biology. In Electron Probe Microanalysis. In *Applications in biology and medicine*, ed. K. Zierold & H.K. Hagler, pp. 1–15. Berlin: Springer.

Hall, T.A., Anderson, H.C. & Appleton, T. (1973). The use of thin specimens for X-ray microanalysis in biology. *Journal of Microscopy*, **99**, 177–82.

Karp, R.D., Silcox, J.C. & Somlyo, A.V. (1982). Cryoultramicrotomy: evidence against melting and the use of a low temperature cement for specimen orientation. *Journal of Microscopy*, **125**, 157–65.

Kirk, R.G., Knoff, L. & Lee, P. (1991). Surfaces of cryosections: is cryosectioning 'cutting' or 'fracturing'. *Journal of Microscopy*, **161**, 445–53.

Kramers, H.A. (1923). On the theory of X-ray absorption and the continuous X-ray spectrum. *Philosophical Magazine*, **46**, 836–71.

Krefting, E.R., Höhling, H.J., Felschmann, M. & Richter, K.D. (1988). Strontium as a tracer to study the transport of calcium in the epiphyseal growth plate (electron probe microanalysis). *Histochemistry*, **88**, 321–6.

Kuypers, G.A.J. & Roomans, G.M. (1980). Post-mortem elemental redistribution in rat studied by X-ray microanalysis and electron microscopy. *Histochemistry*, **69**, 145–56.

Leapman, R.D., Fiori, C.E. & Swyt, C.R. (1984). Mass thickness determination by electron energy loss for quantitative X-ray microanalysis in biology. *Journal of Microscopy*, **33**, 239–53.

LeFurgey, A., Bond, M. & Ingram, P. (1988). Frontiers in electron probe microanalysis: application to cell physiology. *Ultramicroscopy*, **24**, 185–220.

Linders, P.W.J., Stols, A.L.H., van de Vorstenbosch, R.A. & Stadhouders, A.M. (1982). Mass determination of thin biological specimens for use in quantitative electron probe X-ray microanalysis. *Scanning Electron Microscopy*, **1982/IV**, 1603–15.

Ling, G.N. (1988). A physical theory of the living state: application to water and solute distribution. *Scanning Microscopy*, **2**, 899–914.

(1990). The physical state of potassium in the cell. *Scanning Microscopy*, **4**, 737–68.

Marshall, A.J. (1988). Progress in quantitative X-ray microanalysis of frozen–hydrated bulk biological samples. *Journal of Electron Microscopy Technique*, **9**, 57–64.

Marshall, D.J. & Hall, T.A. (1968). Electron-probe X-ray microanalysis of thin films. *British Journal of Applied Physics*, **1**, 1651–6.

McMillan, E.B. & Roomans, G.M. (1990). Techniques for X-ray microanalysis of intestinal epithelium using bulk specimens. *Biomedical Research (India)*, **1**, 1–10.

Mörk, A.C., von Euler, A. & Roomans, G.M. (1991). Effects of chronic amiloride treatment on epithelial cells determined by X-ray microanalysis. *Biomedical Research (India)*, **2**, 182–90.

Morgan, A.J. & Winters, C. (1991). Diapause in the earthworm, *Apporrectodea longa*: morphological and quantitative X-ray microanalysis of cryosectioned chloragogenous tissue. *Scanning Microscopy*, **5**, 219–28.

Nott, J.A. (1991). Cytology of pollutant metals in marine invertebrates: a review of microanalytical applications. *Scanning Microscopy*, **5**, 191–206.

Pieri, C., Giuli, C. & Bertoni-Freddari, C. (1983). X-ray microanalysis of monovalent electrolyte contents of quiescent, proliferating as well as tumor rat hepatocytes. *Carcinogenesis*, **4**, 1577–81.

Pieri, C., Lustyik, G., Giuli, C. & Bertoni-Freddari, C. (1984). Amiloride inhibition of tri-iodothyronine stimulated hepatocyte proliferation *in vivo* and involvement of the intracellular Na^+ content in mitoric regulation: an X-ray microanalytic study. *Cytobios*, **41**, 71–83.

Quinton, P.M. (1978). SEM-EDS X-ray analysis of fluids. *Scanning Electron Microscopy*, **1978/II**, 391–7.

Robards, A.W. & Sleytr, U.B. (1985). *Low temperature methods in biological electron microscopy*. Amsterdam: Elsevier.

Roggli, V.L. (1991). Scanning electron microscopic analysis of mineral fiber content of lung tissue in the evaluation of diffuse pulmonary fibrosis. *Scanning Microscopy*, **5**, 71–84.

Roinel, N. (1988). Quantitative X-ray analysis of biological fluids: the microdroplet technique. *Journal of Electron Microscopy Technique*, **9**, 45–56.

Roomans, G.M. (1979). Standards for X-ray microanalysis of biological specimens. *Scanning Electron Microscopy*, **1979/II**, 649–57.

(1980). Quantitative X-ray microanalysis of thin sections. In *X-ray microanalysis in biology*, ed. M.A. Hayat, pp. 401–53. Baltimore: University Park Press.

(1981). Quantitative electron probe X-ray microanalysis of biological bulk specimens. *Scanning Electron Microscopy*, **1981/II**, 345–56.

(1988). The correction for extraneous background in quantitative X-ray microanalysis of biological thin sections: some practical aspects. *Scanning Microscopy*, **2**, 311–18.

(1990a). The Hall method in the quantitative X-ray microanalysis of biological specimens: a review. *Scanning Microscopy*, **4**, 1055–63.

(1990b), X-ray microanalysis. In *Biophysical electron microscopy*, ed. P.W. Hawkes & U. Valdrè, pp. 347–411. London: Academic Press.

Roomans, G.M., Afzelius, B.A., Forslind, B. & Kollberg, H. (1978). Electrolytes in nails analysed by X-ray microanalysis in electron microscopy. Considerations on a new method for the diagnosis of cystic fibrosis. *Acta Paediatrica Scandinavica*, **67**, 89–94.

Roomans, G.M. & Kuypers, G.A.J. (1980). Background determination in X-ray microanalysis of biological thin sections. *Ultramicroscopy*, **3**, 81–3.

Roomans, G.M. & Sevéus, L.A. (1976). Subcellular localization of diffusible ions in the yeast *Saccharomyces cerevisiae*: quantitative microprobe analysis of thin frozen–dried sections. *Journal of Cell Science*, **21**, 119–27.

Roomans, G.M., Wei, X. & Sevéus, L. (1982). Cryoultramicrotomy as a preparative method for X-ray microanalysis in pathology. *Ultrastructural Pathology*, **3**, 65–84.

Roomans, G.M. & Wroblewski, J. (1985). Post-mortem storage of tissue for X-ray microanalysis in pathology. *Scanning Electron Microscopy*, **1985/II**, 681–6.

Roos, N. & Barnard, T. (1986). Preparation methods for quantitative electron probe X-ray microanalysis of rat exocrine pancreas: a review. *Scanning Electron Microscopy*, **1986/II**, 703–11.

Sagström, S., McMillan, E., Marijianowski, M., Mulders, H. & Roomans, G.M. (1990a). Changes in rat and mouse salivary glands and pancreas after chronic treatment with diuretics: a potential animal model for cystic fibrosis. *Scanning Microscopy*, **4**, 161–70.

Sagström, S., Roomans, G.M., Keulemans, J.L.M. & Bijman, J. (1990b). X-ray microanalysis of cultured respiratory epithelial cells from patients with cystic fibrosis. In *Proceedings of the XIIth International Congress for Electron Microscopy*, Vol. 2, ed. L.D. Peachey & D.B. Williams, pp. 352–3. San Francisco: San Francisco Press.

Saubermann, A.J., Riley, W.D. & Beeuwkes, R. III (1977). Cutting work in a thin section cryomicrotomy. *Journal of Microscopy*, **111**, 39–49.

Scarlett, S.M., Sagström, S., Sagulin, G.B. & Roomans, G.M. (1988). Effects of chronic furosemide treatment on rat exocrine glands. *Experimental and Molecular Pathology*, **48**, 206–15.

Sevéus, L., Brdiczka, D. & Barnard, T. (1978). On the occurrence and composition of dense particles in mitochondria in ultrathin frozen dry sections. *Cell Biology International Reports*, **2**, 155–62.

Shelburne, J.D., Tucker, J.A., Roggli, V.L. & Ingram, P. (1989). Overview of applications in medicine. In *Microprobe analysis in medicine*, ed. P. Ingram, J.D. Shelburne & V.L. Roggli, pp. 55–77. New York: Hemisphere.

Sjöqvist, A. & Beeuwkes, R. III (1989). Villous sodium gradient associated with volume absorption in the feline intestine: an electron-microprobe study on freeze-dried tissue. *Acta Physiologica Scandinavica*, **136**, 271–9.

Small, J.A., Heinrich, K.F.J., Fiori, C.E., Myklebust, R.L., Newbury, D.E. & Dilmore, M.F. (1978). The production and characterization of glass fibers and spheres for microanalysis. *Scanning Electron Microscopy*, **1978/I**, 445–54.

Smith, N.K.R., Stabler, S.B., Cameron, I.L. & Median, D. (1981). X-ray microanalysis of electrolyte content of normal, preneoplastic, and neoplastic mouse mammary tissue. *Cancer Research*, **41**, 1153–62.

Sparks, R.L., Pool, T.B., Smith, N.K.R. & Cameron, I.L. (1983). Effects of amiloride on tumor growth and intracellular element content of tumor cells *in vivo*. *Cancer Research*, **43**, 73–7.

Spencer, A.J., Osborne, M.P., Haddon, S.J., Collins, J., Starkey, W.G., Candy, D.C.A. & Stephen, J. (1990). X-ray microanalysis of rotavirus-infected mouse intestine: a new concept of diarrhoeal secretion. *Journal of Pediatric Gastroenterology and Nutrition*, **10**, 516–29.

Spencer, A.J. & Roomans, G.M. (1989). X-ray microanalysis of hamster tracheal epithelium. *Scanning Microscopy*, **3**, 505–10.

Statham, P.J. & Pawley, J.B. (1978). A new method for particle X-ray microanalysis based on peak-to-background measurements. *Scanning Electron Microscopy*, **1978/I**, 469–78.

Steinbrecht, R.A. & Zierold, K. (1989). Electron probe X-ray microanalysis in the silkmoth antenna – problems with quantification in ultrathin cryosections. In *Electron probe microanalysis: applications in biology and medicine*, ed. K. Zierold & H.K. Hagler, pp. 87–97. Berlin: Springer.

Stephen, J., Osborne, M.P., Spencer, A.J. & Warley, A. (1990). From HeLa cell division to infectious diarrhoea. *Scanning Microscopy*, **4**, 781–6.

Szállási, Z., Szállási, A., Boján, F. & Zs.-Nagy, I. (1988). Effect of enzymic (collagenase) harvesting on the intracellular Na^+/K^+ ratio of Swiss/3T3 cells as revealed by X-ray microanalysis. *Journal of Cell Science*, **90**, 99–104.

Szolgay-Daniel, E., Carlsson, J., Zierold, K., Holtermann, G., Dufau, E. & Acker, H. (1991). Effects of amiloride treatment on U-118 MG and U-251 MG Human Glioma and HT-29 Human Colon Carcinoma Cells. *Cancer Research*, **51**, 1039–44.

Tvedt, K.E., Kopstad, G., Haugen, O.A. & Halgunset, J. (1987). Subcellular concentrations of calcium, zinc, and magnesium in benign modular hyperplasia of the human prostate: X-ray microanalysis of freeze-dried cryosections. *Cancer Research*, **47**, 323–8.

Tvedt, K.E., Kopstad, G., Halgunset, J. & Haugen, O.A. (1989). Rapid freezing of small biopsies and standard for cryosectioning and X-ray microanalysis. *American Journal of Clinical Pathology*, **92**, 51–6.

von Euler, A. & Roomans, G.M. (1991). X-ray microanalysis of cAMP-induced ion transport in cystic fibrosis fibroblasts. *Cell Biology International Reports*, **15**, 891–8.

von Euler, A., Mörk, A.C. & Roomans, G.M. (1992). Effects of chronic and *in utero* treatment with diuretics on mouse exocrine cells studied by X-ray microanalysis. *Journal of Submicroscopic Cytology and Pathology*, **24**, 225–30.

von Zglinicki, T., Bimmler, M. & Purz, H.J. (1986). Fast cryofixation technique for X-ray microanalysis. *Journal of Microscopy*, **141**, 79–90.

von Zglinicki, T. & Roomans, G.M. (1989a). Element concentrations in the intestinal mucosa of the mouse as measured by X-ray microanalysis. *Scanning Microscopy*, **3**, 483–93.

(1989b). Sodium, potassium, and chlorine in electrolyte secreting cells in the intestine. In *Microbeam Analysis 1989*, ed. P.E. Russell, pp. 1205–6. San Francisco: San Francisco Press.

(1990). Effects of VIP secretory stimulation on cytoplasmic ion concentrations in mouse intestinal crypt cells. In *Microbeam Analysis 1990*, ed. J.R. Michael & P. Ingram, pp. 421–2. San Francisco: San Francisco Press.

von Zglinicki, T. & Uhrik, B. (1988). X-ray microanalysis with continuous specimen cooling. Is it necessary? *Journal of Microscopy*, **151**, 43–7.

von Zglinicki, T., Ziervogel, H. & Bimmler, M. (1989). Binding of ions to nuclear chromatin. *Scanning Microscopy*, **3**, 1231–40.

Warley, A. (1987). X-ray microanalysis of cells in suspension and the application of this technique to the study of the thymus gland. *Scanning Microscopy*, **1**, 1759–70.

(1990). Standards for the application of X-ray microanalysis to biological specimens. *Journal of Microscopy*, **157**, 135–47.

Wilson, S.M., Elder, H.Y., Sutton, A.M., Jenkinson, D. McEwan, Cockburn, F., Montgomery, I., McWilliams, S.A. & Bovell, D.L. (1988). The effects of thermally-induced activity *in vivo* upon the ultrastructure and Na, K and Cl composition of the epithelial cells of sweat glands from patients with cystic fibrosis. *Tissue & Cell*, **20**, 691–700.

Wroblewski, J., Anniko, M. & Wroblewski, R. (1983). Healthy and diseased epithelium of the oral and maxillary sinus mucosa studied by X-ray microanalysis. *Journal of Submicroscopic Cytology*, **15**, 593–601.

Wroblewski, J., Müller, R.M., Wroblewski, R. & Roomans, G.M. (1983). Quantitative X-ray microanalysis of semi-thick cryosections. *Histochemistry*, **77**, 447–63.
Wroblewski, J. & Roomans, G.M. (1984). X-ray microanalysis of single and cultured cells. *Scanning Electron Microscopy*, **1984/IV**, 1875–82.
Wroblewski, J., Wroblewski, R. & Roomans, G.M. (1988). Low temperature techniques for X-ray microanalysis in pathology: alternatives to cryoultramicrotromy. *Journal of Electron Microscopy Technique*, **9**, 83–98.
Zierold, K. (1988). X-ray microanalysis of freeze-dried and frozen–hydrated cryosections. *Journal of Electron Microscopy Technique*, **9**, 65–82.
Zierold, K., Tobler, M. & Müller, M. (1991). X-ray microanalysis of high-pressure and impact-frozen erythrocytes. *Journal of Microscopy*, **161**, RP1–2.
Zs.-Nagy, I. (1989). A review on the use of bulk specimen X-ray microanalysis in cancer research. *Scanning Microscopy*, **3**, 473–82.
Zs-Nagy, I. & Casoli, T. (1990). A review on the extension of Hall's method of quantification to bulk specimen X-ray microanalysis. *Scanning Microscopy*, **4**, 119–28.
Zs.-Nagy, I., Lustyik, G. & Bertoni-Freddari, C. (1982). Intracellular water and dry mass content as measured in bulk specimens by energy-dispersive X-ray microanalysis. *Tissue & Cell*, **14**, 47–60.
Zs.-Nagy, I., Lustyik, G., Lukácks, G., Zs.-Nagy, V. & Balázs, G. (1983). Correlation of malignancy with the intracellular $Na^+:K^+$ ratio in human thyroid tumors. *Cancer Research*, **43**, 5395–402.
Zs.-Nagy, I., Lustyik, G., Zs.-Nagy, V., Zarándi, B. & Bertoni-Freddari, C. (1981). Intracellular $Na^+:K^+$ ratios in human cancer cells as revealed by energy-dispersive X-ray microanalysis. *Journal of Cell Biology*, **90**, 769–77.
Zs.-Nagy, I., Pieri, C., Giuli, C., Bertoni-Freddari, C. & Zs.-Nagy, V. (1977). Energy-dispersive X-ray microanalysis of electrolytes in biological bulk specimens. I. Specimen preparation, beam penetration and quantitation. *Journal of Ultrastructure Research*, **58**, 22–33.
Zs.-Nagy, I., Tóth, L., Szállási, Z. & Lampé, I. (1987). Energy-dispersive, bulk specimen X-ray microanalytical measurement of the intracellular Na^+/K^+ ratio in human laryngeal tumors. *Journal of Cancer Research and Clinical Oncology*, **113**, 197–202.

18 X-ray microanalysis of cultured mammalian cells

Joanna Wroblewski and Romuald Wroblewski

A Introduction

Cell cultures are frequently used in biological studies because it is possible to control closely the experimental conditions. Several different *in vitro* systems are currently used in biochemical and morphological studies of the effects of growth factors, teratogens, toxic substances and heavy metals on cell physiology and morphology. Changes in the elemental composition of the cells can be measured by X-ray microanalysis (XRMA) which, combined with biochemical or morphological techniques, contributes to an understanding of the role of ions in physiological and pathological processes. XRMA has several important advantages over more widely used analytical methods, e.g. atomic absorption or flame spectrophotometry (Wroblewski *et al.*, 1989). It allows analysis of individual cells in a heterogeneous cell population, which cannot be achieved by the above mentioned methods. Cell cultures can be relatively easily prepared for XRMA without the use of costly equipment such as a cryostat or cryoultramicrotome. There is no need for dissection (compared to tissue samples) and the thickness of cultured monolayers, which is in the range of 5 μm, rarely exceeding 10 μm, improves conditions for good fixation by quench freezing. Due to the ease of specimen preparation, several XRMA investigations have been carried out on cultured cells (Lechene, 1989; Saubermann & Stockton, 1988; Wroblewski *et al.*, 1983b; Wroblewski & Roomans, 1984, Wroblewski *et al.*, 1987; Zierold, Gerke & Schmitz, 1989). Intracellular element localisation, epithelial transport, dynamic processes and pathological accumulation of extraneous and endogenous material have been studied. It has been well documented that various cell functions are regulated by changes in the concentrations of inorganic cations. For example, signal transduction in mitogenic stimulation is often correlated with ionic changes, and several groups have shown by use of fluorimetric methods and ion selective electrodes that changes in the intracellular concentration of free ions (H^+, Ca^{2+} and Na^+) occur as a result of interaction of the ligand (growth factor, hormone, drug) with its receptor. However, in most of such studies only one ion or one element can be analysed at a time. Thus, XRMA in the electron microscope may prove to be a more powerful method, because it allows simultaneous detection and quantitative analysis of several elements, even if it does not provide information on whether the element occurs as a free ion or in a bound state (Hall, 1979; Roomans, Wroblewski & Wroblewski, 1988).

In this chapter we will briefly review part of our own and other work carried out on cultured mammalian cells. We will concentrate on the discussion of the methodological aspects which are common to different applications of XRMA to cultured cells. Finally, we will present some data on the effects of heavy metals on mammalian cells.

B   Preparation methods

1   Cell culture

Cells for XRMA can be grown directly on a support (thick or thin) that is compatible with cell culture conditions and with XRMA, and is inert to the cells. The growth support for XRMA should not contain elements that contribute to the elemental spectra and/or interfere with analysis of elements of interest. For analysis in scanning mode (SEM) at the cellular level, a solid, bulk support such as silica chips, pure graphite plates (Fig. 18.1) or plastic coverslips (Thermanox) may be used. Titanium or gold stubs can also be used, but are rather expensive and difficult to polish. Fig. 18.2 is a schematic representation of the excitation volumes in cells grown on solid support analysed at different accelerating voltages. A disadvantage of using a non-transparent, solid support is that the cells cannot be easily observed in the light microscope during culture or experiment. A thin or transparent supporting film allows observation in all modes of light microscopy and is therefore preferable if one needs to observe or photograph cells during culture or experiment. A thin support must be used when the image is formed by transmitted electrons in scanning transmission (STEM) or transmission electron microscopy (TEM) where, depending on the cell type, analysis can be carried out separately on nucleus and cytoplasm (see Fig. 18.3). The cells can be seeded on a thin film of Formvar or Pioloform stretched over an electron microscopical grid or a hole in a carbon plate (Fig. 18.4). Titanium, carbon or nylon grids are most commonly used. Cells seeded on electron microscopical grids grow preferentially in the proximity of the grid bars, which renders quantitative analysis more difficult due to the higher background caused by the grid material, although the film placed on a grid is stronger and withstands freezing and freeze-drying better than a film stretched over a large hole in the carbon plate. The supporting film should be carbon coated, if possible glow-discharged (to make it hydrophilic) and sterilised before culture. The cells can be also cultured in plastic dishes that have been previously checked for the possible presence of elements that may interfere with analysis. This type of preparation has, however, a serious drawback. The plastic has insulating properties which result in electrostatic charging during observation and

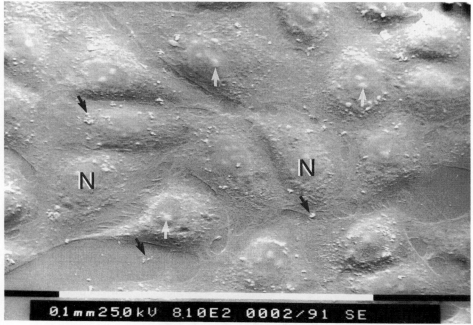

Fig. 18.1. Low power scanning electron micrograph of intestinal epithelial cells
cultured on a Formvar-coated carbon plate (solid support). The shape
of the nucleus (N) with a nucleolus (white arrows) can be seen. Some
remnants of the culture medium are visible on top of and around the
cells (black arrows). Bar = 0.1 mm.

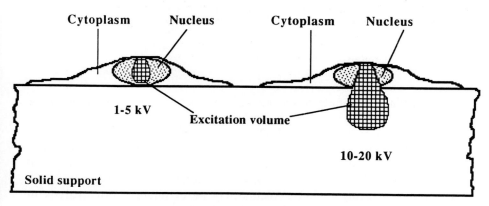

Fig. 18.2. Spatial (lateral and depth) resolution of analysis in cells cultured on a
solid support. Using low accelerating voltage (1–5 kV) the excitation
volume (checked area) is within the cell. Using a higher accelerating
voltage (10–30 kV) the electron beam passes through the entire cell and
the excitation volume includes the solid growth support.

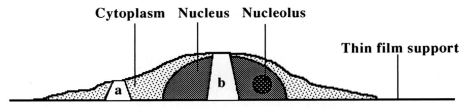

Fig. 18.3.   Approximate resolution of analysis that can be obtained in cells
cultured on thin film support. Using a high accelerating voltage (around
100–120 kV) the electron beam passes through the entire thickness of
the cell without significant lateral spread. Thus, analytical data can be
derived separately from both cytoplasm (a) and nucleus (b).

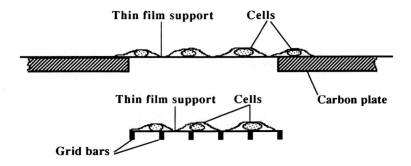

Fig. 18.4.   Cells grown on thin film support on a carbon specimen holder and on a
grid. The Formvar film stretched over a relatively large hole (2.5 mm in
diameter) in a carbon plate (top) is more fragile than the film supported
by the grid bars. However, the carbon plate is not contributing any
detectable peaks to the analytical spectra, rendering the quantitation
more straightforward. Cells cultured on grids tend to grow on the film
in the areas supported by the grid bars. In these cells a strong signal
from the grid material contributes to the background, which makes
quantitative analysis more difficult.

analysis, and contribute to the background in the low-energy part of the
spectrum.

The way in which the cultured cells should be prepared for XRMA depends
on whether the elements of interest are firmly bound or diffusible. For
example, in physiological studies the analysis is focused on the detection and
quantitation of mobile ions (elements), while in toxicological studies
accumulated, firmly bound elements may be of interest. For analysis of
diffusible elements only, anhydrous methods of specimen preparation can be
employed (Hagler, 1988; Roomans *et al.*, 1988; Warley, Kendall & Morris,
1986). Therefore, for XRMA of cells in functional states, cryotechniques
including cryofixation and freeze-drying or cryosectioning and freeze-drying

are most commonly used. For low resolution analysis (at cellular level) it is preferable to culture the cells directly on specimen holders for XRMA or on film-coated electron microscope grids. In the analysis of whole cells grown directly on the specimen holders (carbon plates or grids), the culture medium surrounding the cells has to be removed prior to freezing. Remnants of the culture medium or experimental buffer will otherwise in an uncontrolled way contribute, after freeze-drying on top of and around the cells, to the elemental spectrum.

High resolution XRMA (at organelle level) can only be carried out on ultrathin cryosections. Cultured cells can be either frozen on their support (Tvedt *et al.*, 1988; Zierold *et al.*, 1989; Zierold, 1991) or grown or prepared in suspension and pelleted before quench freezing and sectioning in a cryoultramicrotome (Warley *et al.*, 1986; Warley, 1989).

2 Removal of medium

The removal of the medium surrounding the cells (culture medium, experimental medium or buffer) has to be done very carefully in order to prevent redistribution and gain or loss of intracellular elements. This can be done by quickly rinsing the cultures with appropriate solutions. Volatile buffers, such as ammonium acetate, or sucrose solutions of the same osmolarity as the culture medium or the experimental buffer can be used (Wroblewski & Roomans, 1984). Some cell types withstand even a brief rinse with ice-cold distilled water (Lechene, 1989). However, handling of the cultures before cryofixation should be kept to a minimum and should always be tested and adapted to the cell type studied. Excess rinsing fluid can be absorbed with a filter paper. If too much of the rinsing solution is left on top of the cultures it will cause slower freezing with resulting ice-crystal damage or, after freeze-drying, form a thick film (in the case of sucrose) that may deteriorate the image and contribute to the background, lowering the sensitivity of analysis. The best way to determine the effects of the rinsing solution on the elemental content of the cells is to compare the spectra from rinsed whole cells with spectra obtained from cryosectioned non-rinsed cells. If no major differences are detected between the two types of cell preparations, one can be confident that the rinsing solution does not cause any leakage of ions from or to the cells. One may also compare with the results obtained from tissue sections (Warley *et al.*, 1986).

3 Fixation

Different fixation procedures may be recommended depending on (1) the type of planned analysis (firmly bound versus mobile ions) and on (2) the various ways of preparing the cultures. If only a firmly bound element or an insoluble inclusion is of interest, even conventional chemical fixation and dehydration

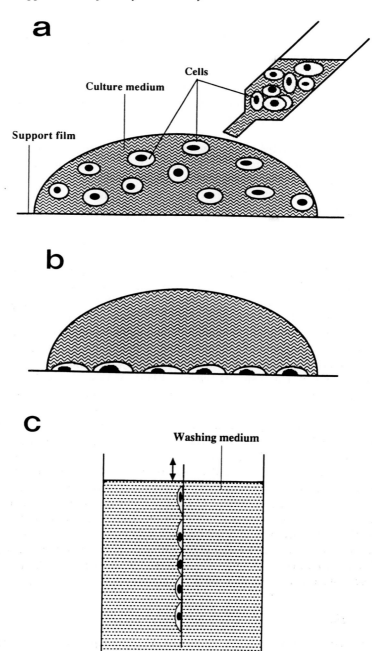

Fig. 18.5a–c.   For legend see facing page.

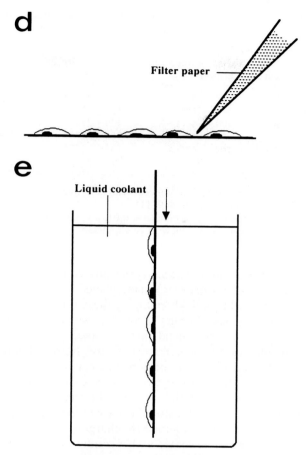

Fig. 18.5.   Major steps in the preparation of cultured cells for energy dispersive
X-ray microanalysis in S(T)EM and TEM. (a) Seeding on support
compatible with XRMA, such as Formvar film covering an electron
microscopical grid or a hole in the carbon plate; (b) cells attach and
grow on the supporting film; (c) removal of the culture medium by a
quick rinse after the experiment is completed; (d) removal of excess
rinsing solution with a filter paper; (e) cryofixation in liquid coolant.
After freeze-drying and coating with carbon, cells are ready for
analysis.

in an organic solvent may be acceptable. However, in most cases rapid
freezing is recommended, as it is the quickest and most reliable method of
fixation, which momentarily arrests all the ions in their *in vivo* state and does
not add any foreign element to the cell preparations (Hagler, 1988; Zierold,
1991; Wendt-Gallitelli & Isenberg, 1989). There are different freezing

Table 18.1. *Effects of rinsing solutions on the elemental content of*
chondrocytes

Element	0.3 M sucrose	0.15 M ammonium acetate	Culture medium	Distilled water
Na	109 ± 13	97 ± 34	946 ± 203***	162 ± 34
P	259 ± 36	461 ± 31***	157 ± 13*	335 ± 18
S	391 ± 25	409 ± 34	429 ± 27	410 ± 9
Cl	135 ± 9	19 ± 5***	356 ± 25***	1 ± 1***
K	419 ± 33	62 ± 5***	135 ± 12***	231 ± 10***
Ca	8 ± 1	19 ± 3**	11 ± 2	15 ± 1***

The data represent absolute elemental concentrations in mmol/kg dry weight;
mean and standard error are given ($n = 9$–11).
*, **, *** significantly different ($p < 0.05$, $p < 0.01$, $p < 0.001$ respectively) from
sucrose-rinsed cells. Data from Wroblewski *et al.*, 1983b.

techniques that may be employed. Among the most commonly used are liquid
coolants such as freons, propane or ethane cooled by liquid nitrogen. The
cultures can be cryofixed by quickly plunging the carbon plates or grids with
the cells into a liquid coolant (Fig. 18.5). The cells can also be frozen in
melting nitrogen or against a metal mirror cooled by liquid nitrogen or
helium. Subsequently, the cells can be stored or transported in liquid nitrogen
to a freeze-drier. Freeze-drying can be done in a conventional freeze-drier at
$< -85\,°C$ for 24 h or for more extended periods of time depending on
vacuum conditions and size of the sample. The freeze-dried cells may be
brought to room temperature and stored dry (over molecular sieves or silica
gel) before analysis. In order to minimise the charging phenomenon under the
electron beam, the cells can be coated with a conductive, thin carbon film. If
the electron microscope is equipped with a cryostage or a cryotransfer system
the cells can be freeze-dried in the column of the microscope or analysed in the
frozen–hydrated state.

For high resolution XRMA, sections of cultured cells can be obtained by
means of cryoultramicrotomy at temperatures preferably below $-100\,°C$
(Zierold *et al.*, 1989; Zierold, 1991; Wendt-Gallitelli & Isenberg, 1989). As the
methods of preparation are in general the same as for tissue samples, the
technical details will not be discussed in this chapter, although we want to
point out that removal of the culture medium is unnecessary, and that the
analytical spectra obtained on cryosectioned cells are not influenced by the
extracellular matrix or supporting material.

4   Analysis

Energy-dispersive and wavelength-dispersive X-ray microanalysis have been applied to cultured cells. Both methods of analysis have some advantages and disadvantages. For analysis of light elements, wavelength spectrometers provide higher sensitivity and therefore better possibilities of quantitation. However, the spatial resolution of analysis is inferior compared with energy-dispersive spectrometry. If possible (for instrumental reasons) a combination of both techniques is preferred.

In STEM, separate analysis of the cytoplasm and nuclei can be performed (Wroblewski *et al.*, 1983b). Quantitative analysis can be carried out according to Wroblewski *et al.* (1983a). The elemental concentrations are routinely expressed in mmol/kg dry weight.

For high resolution XRMA thin sections obtained by means of cryoultra-microtomy can be analysed in the frozen–hydrated state (the problems of mass loss not withstanding; see von Zglinicki, this volume) or after freeze-drying outside or in the column of the electron microscope. In such specimens XRMA can be carried out at the ultrastructural level and the elemental concentrations are calculated by the method of Hall (1979).

C   Applications

We have previously studied the technical aspects of using whole cultured cells for XRMA (Wroblewski *et al.*, 1983b). Our results indicate that each cell type and each new type of application requires adaptation of the cell culture and preparative steps. We have shown that cultures of chondrocytes from elastic cartilage do not withstand rinsing in distilled water or ammonium acetate and therefore only sucrose could be used to remove culture medium before cryofixation (Table 18.1). Rinsing with ammonium acetate caused loss of chlorine and potassium from the cells, and chlorine was extracted from the cells when water was used as a rinsing solution.

We have also used XRMA to characterise cell populations in the epiphyseal cartilage (Wroblewski *et al.*, 1987). This type of cartilage can be roughly divided into three major growth zones: resting zone composed of mainly immature cells, the proliferative zone enriched in dividing cells and the hypertrophic zone rich in degenerating cells. Chondrocytes from respective zones can be obtained by centrifugation of cells in a discontinuous density gradient. By this method we could recover three fractions of cells: fraction I, hypertrophic cells; fraction II, proliferative cells; and fraction III, immature to early proliferative cells. The cells were cultured on a thin film, which allows separate analysis of the nucleus and cytoplasm (Fig. 18.6). Chondrocytes spread well during culture and the portion of cytoplasm that is above and

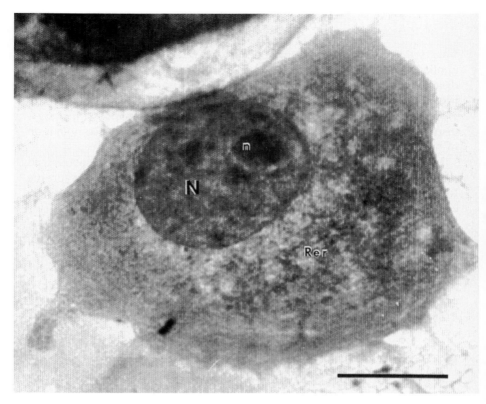

Fig. 18.6.   Scanning transmission electron micrograph of a chondrocyte cultured
on a Formvar film stretched over a relatively large hole in the carbon
plate. In the upper part, the nucleus (N) with a prominent nucleolus (n)
can be discerned. In the cytoplasm, the contours of the rough
endoplasmic reticulum (Rer) are visible. The cytoplasm around the
nucleus has lower density. The Formvar film outside the cells is almost
free of the extracellular matrix. Bar = 10 μm.

below the nucleus is very small and almost devoid of organelles, thus not
contributing significantly to the analysis of the nucleus. The degree of
differentiation of the cells could be defined by their elemental composition.
The hypertrophic cells had a significantly lower concentration of potassium
than the immature or the proliferative chondrocytes. These cells, attached
poorly to the culture support, did not divide or maintain membrane potential,
which may explain the low potassium level.

In our on-going investigation on the activation of cellular oncogenes by
heavy metals, the effects of $CdCl_2$ were studied in a skeletal muscle cell line
and in primary cultures of chondrocytes. In muscle cells, cadmium caused an

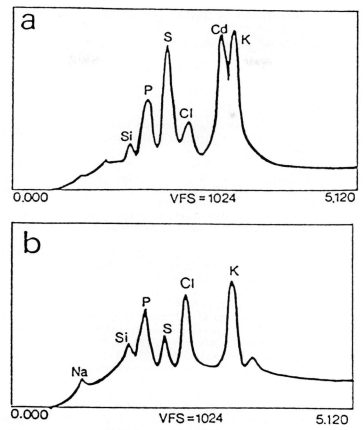

Fig. 18.7. X-ray spectra from the cytoplasm (a) and the nucleus (b) of a cultured skeletal muscle cell treated for 1 h with 5 μm CdCl$_2$. The spectrum acquisition time was set to 120 s and the accelerating voltage to 100 kV. The cell, grown on a Formvar film covering a hole in a carbon plate, was quickly rinsed in cold water and frozen in liquid nitrogen.

induction of oncogenes *c-jun* and *c-myc* (Pei & Ringertz, 1990). After 1 h incubation with 5 μm CdCl$_2$, cadmium could be detected by XRMA in the cytoplasmic dense granules of a sub-population of cells (Fig. 18.7a,b) together with a high concentration of sulphur. The increase in sulphur concentration may be a result of activation of metallothionein in response to cadmium. Metallothionein is a sulphur-rich protein that specifically binds cadmium and zinc, and is therefore assumed to prevent the toxic effects of these metals. The mechanism by which cadmium activates some proto-oncogenes and subsequent mitogenic activation is at present not known. However, we hope that by XRMA we will be able to analyse the distribution of cadmium in the cultured cells at different time points, and correlate our data with the biochemical data on expression of genes involved in the control of cell proliferation and phenotypic modulation.

References

Hagler, H.K. (1988). Artifacts in cryoelectron microscopy. In *Artifacts in biological electron microscopy*, ed. R.F.E. Crang & K.L. Klomparens, pp. 205–17. Plenum Publishing Corporation.

Hall, T.A. (1979). Biological X-ray microanalysis. *Journal of Microscopy*, 117, 145–63.

Lechene, C. (1989). Electron probe analysis of transport properties of cultured cells. In *Electron probe microanalysis*, Vol. 4, ed. K. Zierold & H.K. Hagler, pp. 237–49. Berlin, Heidelberg, New York: Springer-Verlag.

Pei, J. & Ringertz, N.R. (1990). Cadmium induces transcription of proto-oncogenes *c-jun* and *c-myc* in rat L6 myoblasts. *The Journal of Biological Chemistry*, 265, 14061–4.

Roomans, G.M., Wroblewski, J. & Wroblewski, R. (1988). Elemental microanalysis of biological specimens. *Scanning Microscopy*, 2, 937–46.

Saubermann, A.J. & Stockton, J.D. (1988). Effects of increased extracellular K on the elemental composition and water content of neuron and glial cells in leech CNS. *Journal of Neurochemistry*, 51, 1797–807.

Tvedt, K.E., Halgunset, J., Kopstad, G. & Haugen, O.A. (1988). Quick sampling and perpendicular cryosectioning of cell monolayers for the X-ray microanalysis of diffusible elements. *Journal of Microscopy*, 151, 49–59.

Warley, A., Kendall, M.D. & Morris, I.W. (1986). Problems associated with the preparation of cell suspensions for X-ray microanalysis highlighted by the comparison of results with those obtained from tissue sections. In *Science of biological specimen preparation 1985*, ed. M. Müller, P. Becker, A. Boyde & J.J. Wolosewick, pp. 141–5. SEM Inc., A.M.F. O'Hare, USA.

Warley, A. (1989). X-ray microanalysis of freshly isolated cells in suspension. In *Electron probe microanalysis*, Vol. 4, ed. K. Zierold & H.K. Hagler, pp. 169–79. Berlin, Heidelberg, New York: Springer-Verlag.

Wendt-Gallitelli, M.F. & Isenberg, G. (1989). Single isolated cardiac myocytes frozen during voltage-clamp pulses: A technique for correlating X-ray microanalysis data on calcium distribution with calcium inward current in the same cell. In *Electron probe microanalysis*, Vol. 4, ed. K. Zierold & H.K. Hagler, pp. 265–79. Berlin, Heidelberg, New York: Springer-Verlag.

Wroblewski, J., Müller, R.M., Wroblewski, R. & Roomans, G.M. (1983a). Quantitative X-ray microanalysis of semi-thick cryosections. *Histochemistry*, 77, 447–63.

Wroblewski, J., Roomans, G.M., Madsen, K. & Friberg, U. (1983b). X-ray microanalysis of cultured chondrocytes. *Scanning Electron Microscopy*, 2, 777–84.

Wroblewski, J. & Roomans, G.M. (1984). X-ray microanalysis of single and cultured cells. *Scanning Electron Microscopy*, 4, 1875–82.

Wroblewski, J., Makower, A.-M., Pawlowski, A., Madsen, K. & Friberg, U. (1987). Cell fractions from rat rib growth cartilage. Morphological and X-ray microanalytical investigation. *Journal of Submicroscopical Cytology*, 19, 269–74.

Wroblewski, R., Wroblewski, J., Lundström, H., Edström, L. & Jansson, E. (1989). Energy dispersive X-ray microanalysis, neutron activation analysis and atomic absorption spectrometry: comparison using biological specimens. *Scanning Microscopy*, **3**, 861–4.

Zierold, K. (1991). Cryofixation methods for ion localization in cells by electron probe microanalysis: a review. *Journal of Microscopy*, **161**, 357–66.

Zierold, K., Gerke, I. & Schmitz, M. (1989). X-ray microanalysis of fast exocytotic processes. In *Electron probe microanalysis*, Vol. 4, ed. K. Zierold & H.K. Hagler, pp. 281–92. Berlin, Heidelberg, New York: Springer-Verlag.

Index